A Contagious Cause

A Contagious Cause

*The American Hunt for Cancer Viruses
and the Rise of Molecular Medicine*

ROBIN WOLFE SCHEFFLER

THE UNIVERSITY OF CHICAGO PRESS   CHICAGO AND LONDON

The University of Chicago Press, Chicago 60637
The University of Chicago Press, Ltd., London
© 2019 by The University of Chicago

Published 2019

28 27 26 25 24 23 22 21 20 19   1 2 3 4 5

ISBN-13: 978-0-226-45889-2 (cloth)
ISBN-13: 978-0-226-62837-0 (paper)
ISBN-13: 978-0-226-62840-0 (e-book)
DOI: https://doi.org/10.7208/chicago/9780226628400.001.0001

Library of Congress Cataloging-in-Publication Data

Names: Scheffler, Robin Wolfe, author.
Title: A contagious cause : the American hunt for cancer viruses and the rise
   of molecular medicine / Robin Wolfe Scheffler.
Description: Chicago : The University of Chicago Press, 2019. |
   Includes bibliographical references and index.
Identifiers: LCCN 2018045367 | ISBN 9780226458892 (cloth : alk. paper) |
   ISBN 9780226628370 (pbk. : alk. paper) | ISBN 9780226628400 (e-book)
Subjects: LCSH: Oncogenic viruses—Research—United States—History. | Cancer—
   Etiology—Research—United States—History. | Virology—Research—
   United States—History. | Molecular biology—United States—History.
Classification: LCC QR372.06 S34 2019 | DDC 616.99/4019—dc23
LC record available at https://lccn.loc.gov/2018045367

FOR CAITLIN

The virologist is among the luckiest of biologists because he can see into his chosen pet down to the details of all its molecules. —David Baltimore, Nobel Prize Address, 1975

Contents

Acronyms

ACS	American Cancer Society
ASCC	American Society for the Control of Cancer
CCNSC	Cancer Chemotherapy National Service Center
CCRF	Children's Cancer Research Foundation
DHEW	Department of Health, Education, and Welfare
EBV	Epstein-Barr Virus
IARC	International Agency for Research on Cancer
NASA	National Aeronautics and Space Administration
NCI	National Cancer Institute
NFIP	National Foundation for Infantile Paralysis
NIH	National Institutes of Health
PERT	Program Evaluation Review Technique
ras	transforming gene of Murine Sarcoma Virus, first human oncogene identified
RSV	Rous Sarcoma Virus
src	The transforming gene of RSV
SV-40	Simian Virus 40
SVCP	Special Virus Cancer Program
SVLP	Special Virus Leukemia Program
UCSF	University of California, San Francisco
VCP	Virus Cancer Program

"An Infectious Disease—A Virus"

In the winter of 1961, a concerned mother contacted the American Cancer Society to report a "cancer epidemic" in the Chicago suburb of Niles, Illinois. The population of Niles had swelled after the Second World War thanks to an influx of new families seeking a safe and healthful environment in which to raise their children. Now, an outbreak of leukemia menaced their sense of tranquility. In the previous year, eight children at the St. John Brebeuf Parish school had died from leukemia and another five had been diagnosed with the illness, a combined rate of mortality and morbidity five times as high as the national average. An epidemiologist dispatched by the US Public Health Service attributed the cluster of deaths to "an unidentified infectious agent." On an August evening in 1963, two hundred people attended a packed meeting about this possible epidemic of leukemia. With memories of summertime polio outbreaks fresh on their minds, the residents of Niles found the possibility of a leukemia virus to be chillingly plausible. While urging calm, the Niles Board of Trustees moved to require reports of all leukemia cases within the town, a measure previously invoked only for infectious diseases.[1] When a photographer trailed a doctor collecting blood samples from survivors of the "epidemic," a young woman, fearing the stigma of becoming known as a contagious cancer carrier, would only agree to be photographed with her back to the camera.[2]

Members of the medical community did not assuage the residents' concerns. In the *Journal of the American Medical Association*, the director of Cook County Hospital's Hektoen Medical Research Institute, Steven Schwartz, announced that he had found unknown antibodies in the blood of some of the children's relatives and even in that of laboratory technicians who handled the samples. In restrained clinical prose, Schwartz

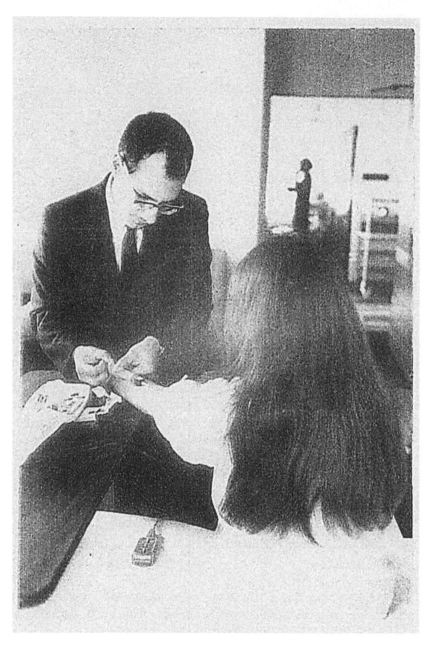

FIGURE 0.1. Dr. Paul Levine of the NCI draws blood from a young woman designated a survivor of the Niles, Illinois, leukemia cluster in 1968. Levine was joined by an unknown photographer from *Life* magazine, who took this photo for a story that was preempted by coverage of the assassination of Robert Kennedy in June 1968. Image courtesy of Paul Levine.

concluded that the Niles cases lent "further credence to the viral etiological theory" of human leukemia. He was more direct with a reporter for the *Saturday Evening Post*: "You can't see patients for twenty years without being convinced that certain things are so. . . . Leukemia looks to me like an infectious disease—a virus."[3] Alarmingly, as events in Niles unfolded, reports of similar leukemia "clusters" emerged in Buffalo, New York; Bergen County, New Jersey; Cheyenne, Wyoming; Louisville, Kentucky; Mount Prospect, Illinois; Seattle, Washington; and Orange, Texas— raising the chilling possibility of infectious leukemia outbreaks across the country.[4] Reflecting the sense of urgency fostered by these mounting reports, the American Cancer Society named leukemia its "top public enemy" in 1963.[5]

Yet the grim threat of a leukemia virus also contained a kernel of hope. If leukemia did have an infectious cause, then it could be brought under control, and possibly eradicated, by vaccination, as had happened in the case of polio, another feared childhood menace. Reflecting that optimism, the National Cancer Institute announced the creation of the Special Virus Leukemia Program in 1964. To capitalize on the possible discovery of a leukemia virus, the administrators of the program formulated a unique "superplan" to direct the process of vaccine development. Drawing on the experience of the Department of Defense, these administrators planned to divide up the entire process of vaccine development—from virus discovery to production—into contracts delegated to coordinated teams of doctors and researchers spread across university departments, government laboratories, and private industry. In a feature article praising the program, *Life* magazine explained that, unlike most medical research efforts, these administrators had a strategy "that would do more than give out research money and wait for results. It . . . would *plan* research and *make* results." In the ensuing fifteen years, the National Cancer Institute spent more than $6.5 billion (in 2017 USD) on cancer virus research, an effort larger than the Human Genome Project a generation later.[6] However, throughout this national effort to develop a cancer vaccine, not a single human cancer virus was known to exist.

* * *

I began the research that produced this book seeking to understand the contribution of a limited number of animal cancer viruses to molecular biology. These viruses interested me because they served as an essential bridge for the "migration" of molecular biologists from studies of bacteria to the study of complex organisms in the 1960s and 1970s.[7] Midway through this research, however, I learned of the outbreak in Niles and the bold actions of the National Cancer Institute. At first, I was confounded by the events. I thought of human cancer as a disease caused by genetic factors or environmental exposures. The website of the American Cancer Society assured me, as it still assures visitors, that "Cancer is NOT Contagious."[8] The many kinds of cancer in humans—more than two hundred by many counts—made the idea that the disease was caused by a singular "cancer virus" appear simplistic.[9]

Yet as I read further, I encountered the publicity and controversy surrounding the release of the human papillomavirus vaccine as a preventive against cervical cancer (and many other cancers) and the use of hepatitis B virus vaccines against liver cancer.[10] Overall, I was surprised to learn that contemporary estimates attribute nearly one in six cancers worldwide to infection by viruses.[11] In 2011, reflecting on the advances in the forty years since the declaration of the "War on Cancer" in 1971, *Science* commented that while progress against cancer as a whole had been frustrating, "we now know that viruses do in fact play a causal role in certain human cancers, and, thanks to decades of tumor virology research, vaccines against these viruses have been developed into successful cancer-preventive agents. That's something to celebrate."[12]

Delving into the associations between cancer and viruses in the present day, however, did little to ease my confusion as a historian. If anything, what I uncovered highlighted a central paradox that my research brought to light: American investment in the hunt for a cancer vaccine peaked in the decades after the Second World War, long before any human cancer viruses were identified in the laboratory. To understand this paradox, I turned from examining how viruses appeared in the laboratory to how they figured in American society's response to cancer.[13] Mounting investment in research into the biological roots of cancer was symptomatic of a new moment in midcentury American politics, a moment in which the promise of biomedical breakthroughs was seen as an appealing alternative to federal intervention in the medical marketplace. As I elaborate below, I came to see this development as a part of what I will call the *biomedical settlement*: the tacit promise that in lieu of providing health care to its citizens directly, the government could foster public welfare through biological investigations of disease. Identifying a viral cause of cancer and cutting it off with vaccination was one of the most compelling promises advocates of the biomedical settlement could offer in favor of this new approach to the nation's health.

Cancer viruses were in the vanguard of the midcentury campaign to provide biological solutions to the manifold problems of disease. The American hunt for these viruses brought together two different approaches to disease: an older tradition of public health vaccination, such as the campaigns waged against polio and smallpox, and a new tradition of molecular biology, whose advocates offered the hope that unlocking the mysteries of disease would usher in revolutionary new therapies. In this book I use the double life of cancer viruses to weave together political, medical, and biological changes that are often viewed in isolation. The promise of a cancer vaccine created the largest and most ambitious federal infrastructure for peacetime biological research seen in the twentieth century. The social and material resources provided by this infrastructure, culminating during the "War on Cancer," played a pivotal role in the migration of molecular biologists and their vision of life from simple bacteria to complex cells. However, as the study of cancer moved further into the laboratory, tension grew between molecular biology's pursuit of knowledge and the expectations fostered by the biomedical settlement. Defining the nature of the hunt for cancer viruses and evaluating its success provided a prominent arena in which American society grappled with the promise and frustration of molecular approaches to health and disease.

Defining the Cancer Problem

Cancer had long inspired a unique degree of fear. A cancer diagnosis promised a painful and solitary death, very much unlike the gradual demise from tuberculosis, the leading killer of the nineteenth century, romanticized by writers and artists. It was not unusual for flesh necrosis or secondary infections to take hold in the advanced stages of the disease; the pain and odor associated with terminal cancer denied sufferers the solace of friends and family and the clarity of mind to die with dignity.[14] At the start of the twentieth century, new concern about the "cancer problem" arose to join these long-standing fears.[15] Statistics revealed that the incidence of cancer was rising; by the middle of the twentieth century, it was the second leading cause of death in the United States and many other industrial nations. Optimistically, some physicians speculated that the rising rates were an artifact of improved diagnosis or longer lifespans resulting from the control of epidemic diseases, what later demographers called the "epidemiological transition" to chronic illnesses, but many others expressed concern that rising cancer rates were a symptom of modern society itself.[16]

A speaker at a meeting of the Brooklyn Surgical Society in 1909 captured this sense of foreboding: "The cancer problem . . . remains an enigma which thus far has baffled solution. By reason of the mystery which attends its origin, the apparent increase of its prevalence which seems to have followed civilization and luxury, and the steady course to a fatal termination that is its characteristic, it is a dark cloud which hangs on the horizon of every family."[17]

Despite international concern for rising rates of cancer, there was no single means of addressing the disease. Initially, nations pursued cancer treatment, education, and prevention in very different measures. The different approaches that these nations followed highlight the ways in which the solutions that the United States sought for the cancer problem were the product of its particular social and political concerns rather than responses dictated by the nature of the disease. In the 1920s and 1930s, the French government remained more concerned with diseases, particularly tuberculosis, that appeared to threaten the generation of young men that had survived the decimation of the First World War. When France did turn to cancer, it focused on making cutting-edge treatments, especially radium, available through a network of hospitals. Germany, from experience with illnesses caused by its burgeoning chemicals industry, embraced the theory of environmental carcinogenesis and emphasized prevention over treatment, an approach that continued under the Nazi regime. Britain feared that active public education campaigns would spark "cancerphobia," and so limited its outreach efforts. Later, it focused on providing cancer care through the National Health Service.[18]

In the United States, it was not the federal government, but rather the American Society for the Control of Cancer, founded in 1913, that first claimed responsibility for the disease. Its founders deliberately emphasized "control" to avoid the suggestion of curing or eliminating cancer. In the 1920s and 1930s the society launched wide-ranging public education campaigns aimed at encouraging early detection and early treatment, particularly through the advancing field of surgery.[19] Public concern for the disease prompted the creation of the federal National Cancer Institute (NCI) in 1937, but the pessimistic view that surgeons held regarding the medical returns of laboratory research and the resistance that doctors maintained to state intervention limited the scope of its actions.[20] Alone among the nations facing the cancer problem, the United States adopted a fourth solution in the middle of the century: biological research into the mechanisms of the disease. From its modest origins in 1937, the scope of biological research supported by the NCI expanded dramatically starting in the early 1950s. In 1971 Congress adopted calls for a "War on Cancer," quadrupling the budget of the NCI in only a few years and launching the widest-ranging program of biological research to serve medical ends seen before or since.[21]

While it might now seem to be an obvious approach to the cancer problem, the ardor with which midcentury Americans embraced biomedical research was truly remarkable. Although vaccination and antibiotics provided medical research with an aura of potential during its "golden age" after the Second World War, the rapid rise of support for biomedical research against cancer far outpaced demonstrable evidence of its medical uses.[22] Well into the twentieth century, research into the biological nature of cancer did not promise therapeutic insights; exploring cancer's hereditary roots did not provide a means of arresting the disease, and pursuing prevention by divining its environmental or chemical causes seemed to require daunting adjustments to society. Microbiology was the one branch of laboratory science that promised to unite biological understanding of the disease with the development of new cures, but the existence of cancer viruses was hotly contested.[23] Moreover, many prominent doctors and scientists remained vocally pessimistic about the possibility of curing cancer. Because of this dissonance, tracking political and social changes is essential to understand how and why Americans placed their faith in the biomedical approach to disease.

The American Biomedical Settlement

The rapid expansion of biomedical research in the United States provides a vantage point for considering the long arc of debates and discussions regarding the role of government in American society. Historians typically regard federal spending on biomedical research as a by-product of America's immense prosperity after the Second World War and its national faith in science. Rather than striking a tune of American exceptionalism in a medical key, I regard this spending as one facet of a discussion spanning the twentieth century concerning what avenues the state could follow to

protect and promote public health. In the late nineteenth century, indus-
trializing nations faced calls to assume a greater and greater role in protecting the welfare of their citizens through workplace regulations, education, unemployment insurance, old age assistance, or individual medical
care. The United States adopted some of these measures and not others.
Befitting the fear that cancer inspired, the American response to it from
the First World War through the end of the Cold War was interwoven with
these debates.[24]

At the opening of the twentieth century, doctors and reformers urged
the government to guard public health against the threat of infectious
disease by exercising its police powers—steps such as quarantine, vaccination, and sanitation. However, state authorities found that these tools
provided an inadequate means of responding to chronic diseases, whose
sufferers required sustained medical care rather than containment. In the
1930s, advocates of social welfare began to argue that the government
should take on a greater role in protecting the "health security" of its citizens, just as it protected the economic security of those in old age.[25] After
the Second World War, however, the United States did not follow other
industrialized democracies in embracing health security as a right that it
owed to its citizens—notably, the American Medical Association mobilized to block the national health insurance plan proposed by President
Harry Truman.[26]

This moment provided the grounds for the emergence of the biomedical settlement. I use this phrase to capture a new ideology adopted by
social welfare activists, scientists, doctors, administrators, and legislators
as they grappled with the question of what role the federal government
could assume in addressing the cancer problem. While there was never a
formal articulation of the settlement, it captures the vision of the role that
government could play in the fight against disease that emerged out of these
negotiations. Rather than protecting public welfare through the provision
of individual health care or the control of epidemic disease, the settlement
promised that the government would foster health by supporting the study
of illness at a biological level on an unprecedented scale. The parties to
the settlement joined the still novel idea of approaching disease as a biological event with the creation of new pathways for federal intervention
in the nation's health that skirted medical opposition.

The negotiations and maneuvers that brought the federal government
into cancer research, the first site of the biomedical settlement, opened
channels through which a flood of funding for research on other diseases

followed. Federal spending on biomedical research, though modest in comparison to spending for military or social programs, assumed an outsized status in demonstrating that the government was caring for its citizens. As a midcentury report for the Department of Health, Education, and Welfare observed, the world was in the midst of a "scientific and technological revolution." The future direction of "medical science" was the exploration of "fundamental physical phenomena related to the nature of life and to the growth and control of living organisms." "The fruits" of this research would be "the prolongation of life and the relief of suffering."[27] Although the federal government also sought to indirectly promote individual health through hospital construction, medical education, and pharmaceutical regulation, the most powerful sign of its commitment to health was its support of biological research directed at curing and preventing illness.[28]

Although federal spending on biomedical research skyrocketed starting in the 1950s, this did not mean all of the parties to the settlement were in accord with one another. The biomedical settlement did not satisfy all of the activists who sought national health insurance, nor did it placate opposition to federal intervention by the medical community or convince advocates of small government. The terms of the settlement were a matter of constant debate between zealous social welfare activists, socialism-wary doctors, fiercely independent scientists, power-seeking administrators, concerned legislators, and taxpaying citizens. Their ongoing negotiations drew and redrew the boundaries of what government interventions were possible.[29]

In the midst of this process of contention and compromise, the policies implemented under the aegis of the settlement had a transformative effect on the development of the biological sciences in the United States. Previously, the largest federal investment in biological research had been made under the auspices of the Department of Agriculture.[30] While other federal patrons of biological research arose after the Second World War, such as the Atomic Energy Commission and the National Science Foundation, their efforts were soon surpassed by the rising budget of the National Institutes of Health (NIH), the agency favored by the advocates of the biomedical settlement. University departments in fields ranging from chemistry to microbiology underwent rapid growth thanks to support from the NIH.[31] To this day, the government of the United States, through the NIH, remains the largest single patron of biological and biomedical research in the world.[32]

The scale of the support unleashed by the biomedical settlement bound together the scientific and political dimensions of biological research.[33] Biologists who had rarely if ever found it necessary to engage with the federal government now found that their intellectual and professional futures were closely tied to the fate of the NIH. They were now both the expert agents who advised the federal government and the objects of policies formulated in Washington with or without their consent.[34] In this context, biologists' definitions of biomedical research and its appropriate aims were inseparable from political questions about which biological studies to support and what responsibility they assumed to promote the health of the nation's citizens when they sought federal funding for their research. Where was the boundary between "pure" versus "applied" research? How much could the process of scientific discovery be managed? What constituted a "fundamental" understanding of disease, and did it come before or after useful treatments?[35] Under the settlement, biologists were neither the sole nor the most powerful arbiters of these questions.

Viruses and the Making of Molecular Medicine

No field of biology was more marked by the biomedical settlement and its conflicts than molecular biology. In 1949 physical chemist Linus Pauling published a paper describing how a change in a few of the amino acids in the chain that formed the hemoglobin protein of red blood cells caused them to collapse, or sickle; he concluded that the associated condition, sickle-cell anemia, counted as the first "molecular disease."[36] In the ensuing decade, the discovery of the structure of DNA and the role it played in synthesizing enzymes, a set of discoveries based on the study of *E. coli* bacteria and the viruses that preyed upon them, added to this enthusiasm. Molecular biologists appeared to be in a position to extrapolate from their laboratory studies to the nature of life as a whole: "Anything found to be true of *E. coli* must also be true of Elephants," one notable set of French molecular biologists declared.[37] Enthusiasm for the medical dividends of these molecular studies continued to build. In 1976 the President's Commission on Biomedical Research declared that the "biological revolution" of the previous quarter-century, sustained by "entirely new disciplines" and powerful "research technology," would bring changes to medicine "unlike anything in the millennia of its existence."[38] This "revolution" appeared to have reached its fulfillment with the start of the Human Genome Project of the 1990s, which spent billions of dollars sequencing the human genome. The project's premise was that decoding the human genome would allow new cures for disease based on comprehending the molecular mechanisms of illness.

Although these events might suggest a smooth progression from fundamental scientific research to promising medical applications, the advance of molecular biology into medicine was a far more complex process.[39] It took place not in a single leap, but in three major steps: the first, between the 1920s and 1940s, was dominated by biochemistry; the second, in the 1950s and 1960s, by DNA and RNA; and the third, starting in the 1970s, by genetic engineering. Moreover, each of these steps occurred unevenly across different nations and institutions. Each transition required considerable resources and the intervention of new groups—philanthropy in the first case, government in the second, and venture capital in the third.[40] Nor were the "molecules" of molecular biology or medicine the same in every case—objects ranging from microbes to vitamins to DNA all counted as molecules. Given the long duration and shifting grounds of its emergence, it is more accurate to say that molecular medicine did not come into being at a particular moment or with a particular discovery but emerged out of an ongoing process of "molecularization" that moved in different directions and at different rates depending on its context.[41] While molecularization advanced in many ways and focused on many objects, the aspiration to resolve biological and medical problems through the study of their fundamental mechanisms united these diverse efforts.[42]

Viruses provide an ideal point of departure for exploring the shifting scientific ideas and political maneuvers that attended the molecularization of the cancer problem as a whole. In the late nineteenth century, germ theory, the umbrella under which cancer viruses emerged, demonstrated the power of the microscopic world to cause disease, bringing older notions of contagion into the laboratory.[43] In the early twentieth century, the behavior of viruses seemed to hover between that of living and nonliving things.[44] Even as technological advances such as electron microscopy made it possible to "see" viruses in a manner similar to the way bacteria could be seen, classifying viruses challenged biology. As the French microbiologist André Lwoff commented, "Viruses are Viruses."[45]

Despite their ambiguity, viruses' simplicity made them well suited to serve the aims of early molecular biologists, who sought to produce general laws of biology by studying the simplest possible systems.[46] However, simple systems based on bacteria, it became clear, were more different from the complex eukaryotic cells that comprised animals and humans

than molecular biologists had initially suspected. Small animal viruses—
and cancer viruses in particular—played a critical role in the transfer of
the methods of molecular biology to these new cells. Before technologies
such as recombinant DNA, gene cloning, and polymerase chain reactions made it possible to manipulate the large genomes of animal cells in
the 1980s, cancer viruses offered one of the few means available by which
to isolate and study the genes that controlled cellular growth and development.[47] Through viruses, the process of molecularization became deeply
entangled with the growth of state-sponsored biological research against
cancer.[48]

The Infrastructure of Molecular Biology

Speaking at the start of the War on Cancer, James Watson, the codiscoverer of DNA, voiced the sense that molecular biology was in the midst of
a profound change. "Biology is beginning to look like physics," he said.
While his colleagues had been accustomed to small-scale research, they
now had to "think in terms of multimillion dollar sums . . . to stay with
the times." The future work of molecular biology would be in "large collective teams," as it moved to "still larger and larger labs." Biology, Watson concluded, "now runs fast . . . because it offers to improve on our
lives. Our main support comes from federal funds."[49] Just as the national
security concerns of the Cold War provided fundamental infrastructure
for the physical sciences, the campaign against cancer is essential to understanding how new kinds of biological knowledge were produced. In
following cancer viruses and the materials assembled for their study, I
demonstrate how the social and material infrastructure for experimental
biological research created by the biomedical settlement shaped the generation of biological knowledge after the Second World War.[50] The study
of biological objects, such as cancer viruses, took place in terrain prepared
by the ideological tensions of the American biomedical settlement.[51]

Scientific infrastructure ranges from buildings to standard materials
to annual meetings, but in all its forms it draws attention to the reality
that the practice of science, even when it concerns itself with the smallest objects, remains entwined with broader social and material processes.
Just as physics relied on reactors, accelerators, telescopes, and gravity-
wave detectors, biomedical research also required considerable material resources and the coordination of labor across multiple locations,

challenges that made the broader political environment an inescapable element of scientific practice.[52] With each step forward, the resource needs of molecular biology became more baroque, expanding from fruit flies and pea plants to radioactive tracers, purified enzymes, specially cultured cells, and specific mutant virus strains. The infrastructures that supplied these needs played an essential role in allowing individual laboratories to pursue their exploration of life at the molecular level.[53] For example, in the 1940s and 1950s, the US Atomic Energy Commission distributed radioactive isotopes to fields from biochemistry to ecology to highlight the peacetime uses of nuclear science at a moment when it appeared to be a science of death. The production and use of those isotopes illuminate relationships between scientists and Cold War politics that are hard to grasp from the perspective of the laboratory bench.[54]

Infrastructural relationships were not incidental to the practice of molecular biology, but integral to the kinds of knowledge about life and disease that the field was able to create. Starting in the early twentieth century, experimental biologists began to abandon the idea that they were studying phenomena in nature. Faced with the complexity and diversity of life, they responded by exploring living processes through the intensive study of a few chosen organisms and experimental systems, such as purebred mice, fruit flies, or select viruses. Biologists assumed that the findings from these models and systems were applicable to life as a whole.[55] However, the way that organisms—from microbes to mice—became the "right tools for the job" reflected factors such as the ease of access to particular species, protests from antivivisectionists, the transparency of cells under the microscope, and the tempo of reproductive cycles as much as a sense that they were representative of the phenomena biologists aimed to study.[56]

As a consequence, the development of our knowledge of life cannot be divorced from the broader social and political infrastructures that anchor this inquiry. The experimental systems of biologists were not natural objects but constructions combining practices, scientific theories, and materials. The events that those systems were designed to study were coherent only so long as the rest of the system stabilized them. Carrying out experimental work required both establishing and maintaining such systems.[57] As the twentieth century advanced, the capacity of researchers to explore life on the molecular level was closely tied to the broader material and social worlds in which they labored.[58] The infrastructure supporting this research allowed the study of disease to move from the clinic to animal models, from animal models to *in vitro* systems, and from *in vitro* systems to the molecular biology we are familiar with today.[59]

However, our current understanding of the ways that molecular biology functioned often stops at the threshold of the university—we do not yet have a strong sense of the broader social and material worlds that made the work possible.[60] The infrastructural dimensions of work in molecular biology have often been overlooked because of the individualistic rhetoric of scientists and because historians of biology have often selected topics of analysis—ideas, individuals, or institutions—that have obscured the entanglement of biology with larger-scale processes.[61] In the shadow of debates presenting the Human Genome Project as a historically unprecedented "big biology" enterprise, it seemed reasonable to assume that molecular biology had until then been a matter of "benchwork science conducted in small laboratories," unencumbered by the political or logistical concerns of the physical sciences.[62] In following the pursuit of viruses at sites across the nation, this book suggests the extent of the infrastructure that sustained molecular biology's migration to complex cells.[63] Tracing the shifting governance of biomedicine and health enriches our understanding of how our knowledge of life has been produced in the past three generations.

Governing the Future

In cancer research and for the biomedical settlement as a whole, the federal government spent vast sums on biological research with, at best, a distant expectation of therapeutic returns—its choices reflected future possibility rather than present knowledge. Cancer had long been associated with the idea of urgency, initially through appeals for "early detection" and rapid surgical intervention in the first part of the twentieth century. Yet the sense of urgency did not translate into optimism concerning biomedical research until decades later.[64] President Lyndon Johnson's Commission on Heart Disease, Cancer, and Stroke captured this new sense of promise in 1964. Although "biomedical science" could not promise that it would "lead inevitably to means of prevention and care," the commission's report urged that "*without* a major continuing research effort there is no hope of advance . . . no cure for those conditions beyond our grasp."[65] This pattern may confound our expectation that scientific agreement must precede state action. However, the new visions of the future developed to support biomedical research fit within the planning culture of the Cold War, an era when futurism and prediction were integral to the governing practices of the American state. In this light, cancer virus research is better understood in relationship to efforts that extend from economic development to space flight and nuclear strategy rather than to previous biological research.[66]

In this context, the parties to the biomedical settlement developed dramatically different ways of knowing the problem posed by cancer, which in turn shaped the legitimacy of possible solutions that the government might pursue.[67] These different ways of knowing the disease emphasized different aspects of the cancer problem toward different ends. For physicians and scientists early in the century, the complexity of cancer served to limit government intrusion and public demands for a cancer cure. As advocates of the biomedical settlement sought to mobilize the federal government against cancer, they wrestled with the fact that the medical experts they sought to enlist framed the disease as too mysterious to admit state intervention. Finding and highlighting ways to render cancer intelligible, either by evidence or by analogy with other successful projects, was pivotal to spurring state action against cancer. At the NCI, entities such as cancer viruses simultaneously served as objects of study in the laboratory and objects of bureaucratic control, as managers budgeted for anticipated discoveries and challenges.[68]

When new communities of scientists, such as molecular biologists, came into contact with federal anticancer efforts, they too sought to redefine the future of cancer research at both a biological and a political level. The molecular mysteries of cancer—its status as a "riddle"—served to deflate planners' visions of curing cancer even as it allowed molecular biologists to assert the relevance of molecular studies to solving the cancer problem in the distant future.[69] Although the process could be confounding, such as planning a vaccine for a virus that did not yet exist, it was also very powerful. Contention and negotiation over the future was generative: it fostered the growth of infrastructure and aided the emergence of new understandings of cancer.[70]

Nonetheless, during the decades examined in these pages, the promise of the biomedical settlement—that diseases were best addressed as a matter of fundamental biology—remained largely unfulfilled. The residents of Niles, Illinois, never found out what caused the mysterious cluster of leukemia deaths, nor did the NCI produce a leukemia vaccine or manage to reduce the incidence of cancer for most of the twentieth century. Critics of this approach charged that the time, money, and intellectual energy devoted to laboratory studies of cancer would be better spent addressing its social or environmental roots.[71] The difficulty of the cancer problem,

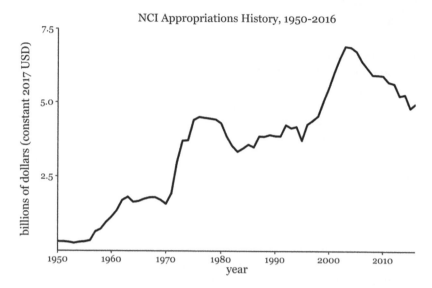

FIGURE 0.2. The history of federal spending on the National Cancer Institute illustrates the dynamic of "boom" and "bust" years that can prevail amid an overall increasing trend. Moments of slowing or declining spending caused acute concern among the biologists who came to depend on this funding.

its *recalcitrance*, offers a case different from those usually covered in the history of science and medicine, which focus on moments of success rather than frustration.[72] Advocates of biomedical approaches to cancer and other diseases worked ceaselessly to address this dissonance between their vision of future success and their present frustration.[73] This book therefore illustrates an important aspect of the development of modern biology and medicine: its growth was entangled with particular ways of managing the cycles of hope and frustration produced by the effort to spur government action by projecting optimistic visions of the future.[74] Following the path of cancer virus research suggests that these cycles of concern, hope, mobilization, frustration, and redefinition are not extraneous to the history of biology and medicine, but central drivers of its development.[75]

Hunting for Cancer Viruses

Throughout the twentieth century, cancer viruses provided an enduring target for biomedical solutions to the cancer problem.[76] Only a narrow segment of research resulted in the successful identification of human cancer viruses, but the activity associated with the American hunt for cancer viruses was much broader. Rather than focusing on the narrow "successes," therefore, I follow here the process of the search itself, the hopes for vaccination that inspired it, the means by which it was pursued, and the impact the search had on American approaches to the problems of health and disease. This book is organized into three major parts, each reflecting an important moment in the progression of the search for cancer viruses.

The first part examines how viruses figured in the definition of the problem posed by cancer. Overall, I aim to unsettle our idea that discoveries in the laboratory or the clinic should frame our understanding of how solutions to the cancer problem developed. In chapter 1, I offer a history of responses to the idea in the early twentieth century that cancer might be contagious. For the public as a whole, the idea of contagious cancer appeared to be a reasonable, if terrifying, extrapolation from theories of contagion and germ theory. For doctors and surgeons, however, this idea proved to be a dangerous framing of the cancer problem—their advocacy against the idea of cancer viruses reflected both clinical skepticism of germ theory and concern for their professional position. In chapter 2, I explore the development of the idea of cancer viruses in the first half of the twentieth century. Those viruses—first exciting extensions of germ theory, then ridiculed as illusions, then embraced as a target of vaccination—illuminate the contentious intellectual and social relationships between laboratory and clinic that accompanied this early effort to molecularize the cancer problem. Chapter 3 considers the cancer problem not from the perspective of scientists or doctors but from the perspective of medical philanthropists. This group was an influential arbiter of what actions the federal government later thought possible against cancer, actions that fit uneasily between its existing tools for the confrontation of epidemic disease and the new challenges of chronic illness. Philanthropic groups believed that there was more to be lost than to be gained by government intervention—a view that persisted through the expansion of the federal government during the New Deal and the Second World War.

The second part of the book examines how the biomedical settlement enabled the hunt for human cancer viruses. The settlement created spaces for charting the future path of cancer research unconstrained by the pessimism of many experts and, in turn, brought into being new forms of infrastructure for biomedical research. Chapter 4 explores how the mobilization of the federal government against cancer resulted from negotiations among the parties to the biomedical settlement, especially around the promise that cancer was a curable disease. A commitment to curing rather than controlling cancer set up new futures and offered new resources to biomedical research. However, if a cancer cure were possible, the federal government became accountable to its citizens for progress toward that cure. Chapter 5 follows the way concerns for the efficacy of federal action inspired the development of the Special Virus Leukemia Program, and how the effort to plan for a leukemia vaccine created a new sense of cancer viruses as "administrative objects." Chapter 6, through three case studies, illustrates the impact that the National Cancer Institute's planning effort had on the infrastructure of biomedical research. Most importantly, when the first attempts at a search for cancer viruses were frustrated, the administrators of the NCI chose to delve deeper into molecular biology and genetics.

The final part of the book demonstrates the importance of cancer viruses to the rise of molecular medicine as the expansion of molecular biology brought its members into contact with the cancer problem. This process worked both along and against the grain of the biomedical settlement, as molecular biologists sought to redefine the nature of biomedicine in accord with their own communal values. Chapter 7 uses the question of cancer viruses during the start of the War on Cancer as a means of illuminating the political and professional debates over the nature of biomedical research that were instigated by the financial crises of the 1960s and new skepticism regarding the slim returns of the biomedical settlement. Chapter 8 looks at how molecular biologists came to define themselves as a political community through their resistance to the efforts of the NCI to plan and direct cancer virus research on a large scale during the War on Cancer.

While chapters 7 and 8 emphasize the role that molecular biologists played in opposition to the government, the final two chapters explore how the infrastructure of virus studies aided the political and intellectual expansion of molecular medicine at a decisive moment in its development. Chapter 9 focuses on a single laboratory in San Francisco to show how the national infrastructure for virus studies created by the War on Cancer shaped a particular experimental system that produced a major advance in molecular biology: the discovery of cellular oncogenes. Chapter 10 returns to the national stage, considering how in the "bust" of the War on Cancer in the 1980s, its infrastructure provided an essential political resource for molecular biologists and other advocates of molecular medicine even as concern for the environmental and social roots of cancer reemerged. The ways that the communities participating in the biomedical settlement grappled with disappointment are important to understanding how molecular knowledge redefined what counted as a successful solution to the "riddle" of cancer—shifting our focus away from the expectation of therapy and toward a deeper understanding of the processes at work in the development and growth of life as a whole.

CHAPTER ONE

Cancer and Contagion

In 1911 Peyton Rous, a researcher at the Manhattan-based Rockefeller Institute for Medical Research, observed a "non-filterable" agent capable of transmitting tumors in chickens. Rous struggled to characterize the ambiguous nature of the agent, which appeared to be neither microbial nor chemical; but with the advantage of considerable hindsight the Nobel Foundation recognized his observations in 1966 as the discovery of the first tumor virus.[1] Two years after his observations, the Rockefeller Institute's front office passed Rous an inquiry from Richard Boardman, a lawyer across the Hudson River in Jersey City. Boardman had read in the newspaper about Rous's discovery of a "cancer parasite" and was moved to write in the hope that Rous might settle an ongoing dispute between him and his wife, Dorcas Boardman.

Both Richard and Dorcas were concerned that there were cancer "germs" in the mattress that the couple had inherited after caring for Dorcas's aunt, who had died a decade earlier after a "long illness"—a common euphemism for cancer. Richard sought an answer for what he presented as his wife's concern that there was "the danger of communication of the disease of cancer . . . lurking in the use of" a mattress that had been "stained by the drain from the cancer." Richard was inclined to believe that the mattress was safe, not because he denied that cancer was infectious, but because he doubted that cancer "germs" could survive for so long in the mattress. Dorcas maintained that they might have survived. Yet her concerns were not so great as to prevent the couple from offering the mattress to members of their household staff.[2] With Rous's guidance, the institute's business manager sent a reply, sympathizing with the "tyranny that germ theory may exercise over the imagination" and assuring Richard that while "it is usually impossible to prove a negative," in this instance there was no "danger of infection whatsoever."[3]

In approaching the history of biomedical objects such as viruses, we may feel a strong temptation, rooted in our present understanding of these objects, to identify with the views of physicians and scientists. Rous's contributions to the virus theory of cancer, discussed in chapter 2, are well known to historians of science and medicine, but the Boardmans' concerns are not. Our first instinct is to regard the Boardmans' fears of cancer germs lurking in a mattress as misplaced, because they fall so far from the attitudes we have been trained to cultivate toward cancer. As one early twentieth-century textbook on cancer noted, although the idea that cancer might be contagious was "the oldest hypothesis of the origin of cancer," it was a hypothesis that "few competent observers" credited in the wake of advances in microbiology.[4] From the perspective of later scientific observers, cancer viruses fit the mold of a classic "unpopular" theory later redeemed by new experimental methods.[5] As Rous accepted the Nobel Prize in 1966, he attributed the record-setting fifty-five-year delay between his findings and his award to the "downright disbelief" that other cancer researchers directed toward his theories of viral carcinogenesis. Only the diligent work of a small number of experimentalists exorcised the theory's unpopularity.[6]

However, taking the skepticism expressed by a small number of doctors and scientists as our guide to the associations between cancer and contagion throughout history can be misleading, no matter how much such views resonate with our own. Scientists, physicians, and laypeople addressed the larger problem of cancer through different "regimes of perceptibility," combinations of scientific and social practices that they used to make sense of the disease and its causes. Identifying, or failing to identify, a cancer virus using new experimental methods becomes meaningful only in reference to the importance ascribed to these results by others— the technology does not speak for itself. Moreover, different social and scientific factors can align to create moments of imperceptibility, where particular causes of disease are harder to study.[7]

As the Boardmans' debate over their mattress indicates, the theory that cancer was a viral disease drew upon a deep reserve of public belief in cancer as contagious disease. That reserve exerted a powerful influence on how the public received scientific research on cancer viruses and tinged how cancer specialists approached the question of cancer's potential infectious causes. Although the laboratory techniques that Rous borrowed from microbiology struggled with the nature of viruses, the vehemence of the skepticism expressed by members of the oncology community was less about Rous's theory itself than about broader questions of how medicine would relate to laboratory science and public concern about cancer.[8]

The association between cancer and contagion proved enduring and controversial precisely because it existed at the intersection of the different regimes of perceptibility created by the techniques of the laboratory, the practices of physicians, and the customs of the public. Preserving the tension between these perspectives is vital for understanding how cancer viruses traveled through early twentieth-century American society. We should follow the ways in which different individuals and institutions brought cancer viruses into being through their actions and habits rather than defaulting to one of those perspectives.[9] Cancer viruses became tangible to different groups through perceptual regimes instantiated in personal habits, architecture, fund-raising, legislation, and education. These regimes were shaped not only by clinical practice, hygiene, and laboratory analysis, but also by fears of death, concerns about professional authority, and hopes for a cure. Approaching the history of cancer viruses with the full range of these regimes in mind underscores that the development of our present understanding of cancer, contagion, and viruses was far from inevitable.

This chapter examines different communities as they approached the question of viral carcinogenesis—both before Rous's discoveries and afterward, as the hunt for cancer viruses continued—in the context of the expansion of laboratory-based microbial theories of disease associated with the so-called bacteriological revolution.[10] In addition, it highlights how popular ideas about cancer as a contagious disease shaped the reception of cancer virus research by both communities of professionals and the public. Centuries-old beliefs, habits, and practices coexisted with new efforts to identify the agents of disease in the laboratory. Nor were physicians and biologists unified in their approach to cancer and contagion. At different moments they contended with the interests and concerns of many other groups regarding the problem of cancer as a contagious disease. Cancer specialists struggled not only with concerns regarding standards of proof but also with the implications of their new theories for the social status of the medical profession. In fact, during the years of its eclipse as a credible scientific theory, cancer specialists spent considerable time and energy campaigning against the idea that cancer viruses existed. Their actions strongly suggest that technical debates concerning cancer viruses were haunted by the continuing resonance in the public mind between cancer and contagion.

Contagion and Cancer

In 1741 the inhabitants of the Saint-Denis neighborhood in Reims, France, gathered to defeat a great danger to their community: the first hospital in Europe dedicated exclusively to the treatment of cancer. The wealthy Maillefer family had planned their hospital on the model of institutions for the treatment of consumption, and the hospital's organizers had identified what they thought was an ideal site within Saint-Denis: a large building on a quiet street with extensive gardens that would console patients. For the hospital's prospective neighbors, however, the possibility of such a dense concentration of patients with cancer raised the terrifying concern that it could spread into the community—particularly that the odors associated with the rot of advanced tumors would carry the disease beyond the hospital walls. The residents of Saint-Denis strenuously protested, petitioning King Louis XIV to demand that the hospital be either closed or moved far outside the city walls. The hospital's final location, far from the center of Reims, bears vivid witness to the power of popular fears of contagious cancer.[11]

In eighteenth-century Europe, the association between cancer and contagion thrived because both ideas were much more loosely bounded than their twentieth-century counterparts. Cancer was capable of manifesting itself in numerous terrifying forms. Ulcerating tumors stank, and individuals stricken with cancer suffered from vomiting or convulsions in the late stages of the disease. Rot and corruption dominated descriptions of the illness. Cancer "ate" its way into surrounding healthy flesh and "dissolved" ligaments, bones, and tissue. Surgeons who attempted to amputate external tumors were confounded by the apparent ability of "seeds" of tumors to spread into other parts of the body. The grisly deaths caused by cancer and the limited treatments available fostered the sense of taboo and fear that followed the disease well into the twentieth century.[12]

Meanwhile, the doctrine of contagionism, which developed during the fifteenth century to explain the spread of plague and other diseases, readily encompassed cancer. Contagionism called attention to the transmission of disease through contact with a wide range of objects. The idea of tumor "seeds" within the body merged easily with the general idea of seeds of contagion.[13] Moreover, the classification of disease under humoral theory grouped cancer with other potentially contagious inflammatory diseases, such as syphilis and tuberculosis. Writings from Babylonian, Persian,

Indian, Greek, Arabic, Roman, and European sources all described instances of "tumors," which included lumps, cysts, inflamed masses, and
other kinds of swelling, and their occasional treatment.[14] Seventeenth-
and eighteenth-century observers reported numerous cases in which cancer passed from one person to another by different mechanisms varying
from sexual intercourse to sharing a pipe or a cup.[15]

Further study of cancer in the nineteenth century did nothing to dispel the fear of contagion, especially the chance of transmission through
sexual activity. The first deliberate efforts to collect statistics on the incidence of cancer strengthened its association with sexually transmitted infections. In 1842 Domenico Antonio Rigoni-Stern, the provincial surgeon
of Verona, Italy, published a paper reporting the results of his effort to
establish relative rates of cancer deaths. He claimed that cancer was eight
times as common in women as in men, a finding likely explained by the
relatively greater ease of diagnosing cancers of the breast and women's
reproductive organs. Rigoni-Stern became infamous for the claim that
married or widowed women, who had presumably been sexually active,
were far more likely to die of "uterine cancer" than nuns, who were presumably celibate.[16] This finding fit well into the views of female sexuality
and health common among male European medical authorities. Cancer,
like syphilis and other venereal diseases, was a disease best prevented by
avoiding promiscuous sexual behavior.[17] Associations between sex and
cancer infection endured into the twentieth century. The life insurance
actuary Frederick Hoffman, whose own compilations of statistics played
a prominent role in the discussion of cancer as a public health problem,
addressed the chances of the transmission of the disease via "marital infection" in 1915, albeit with the aim of assuring his readers it was not a
possibility.[18]

Other efforts to collect systematic epidemiological data about cancer
also deepened the sense that cancer was contagious. In the late nineteenth
century, the British physician Alfred Haviland collected detailed statistics
regarding deaths from cancer, heart disease, and other diseases throughout England. Haviland used these statistics to argue for the importance
of local geography to the incidence of cancer. He explained that higher
rates of cancer in valleys were due to a lack of ventilation by winds, creating conditions similar to those where the "malarial air of rheumatism
lurks."[19] This suggestion, which drew on the miasmic theory that disease
might spread through contact with unhealthy odors, recalled the potential
connections between cancer transmission and odor that had concerned

the denizens of Saint-Denis. A contemporary of Haviland speculated that the smell of advanced cancer could spread the disease by traveling throughout a house and down into the stomachs of healthy residents. Drinking brandy was considered an effective way to ward off those dangerous odors.[20] This recommendation spoke to the expansive understanding of contagion and cancer before the advent of germ theory: while the means of cancer's transmission might be particulate, as in the case of the passage of cancer seeds, it could also draw on broader associations of disease transmission with miasma, odor, and rot.[21]

Shifts in the classification of cancer associated with tissue and cell theory did not displace its associations with contagion. At the start of the nineteenth century, medical theorists focused on the outer symptoms of tumors, which they placed alongside other forms of inflammation. Since many cancers developed within the body and techniques of surgery were limited, the physiological structures associated with the disease remained obscure. At best, surgeons might examine tumors in the course of an autopsy. In the 1840s the development of the compound microscope and the expansion of physician training at sites such as the Paris clinics allowed the study of tumor cells by anatomists and pathologists. Cell theory provided a new approach to the study of disease, one that sought to determine the structural differences between "normal" and "pathological" tissues. However, even with the microscope, pathologists continued to engage in heated debate regarding the classification of cancer cells and the degree to which malignancy could be determined from cell structure alone.[22] Techniques attending the use of the microscope—such as tissue staining to highlight cell structures—remained contentious. These points of controversy and confusion left cell theory on the margins of medical approaches to cancer.[23]

Even after the microscope and cell theory gained greater acceptance, observers debated whether cancer was in fact different from other diseases of inflammation such as tuberculosis or syphilis. One British surgeon saw all three of these diseases as part of a pathological progression. He maintained, based on advances in the microscopic examination of diseased tissues, "it is now thoroughly well established that there exists every possible gradation between simple glandular enlargement and cancer of the glands. It is impossible to say where one begins and the other ends."[24] A generation later, another British physician emphasized the striking resemblance between the "chronic swelling" produced by tuberculosis or syphilis and the tumors of "malignant disease."[25]

Cancer after Germ Theory

The advent of germ theory, discussed further in chapter 2, dramatically redefined the effort to identify the causes of disease. Germ theorists promised that they could isolate and guard against agents of illness that could not be perceived through sight, touch, or smell, ushering in what historians have identified as a quintessentially "modern" view of disease grounded in the laboratory rather than in the clinic.[26] Public concern about intangible germs reshaped everyday practices ranging from toilet design to the taking of Communion.[27] Germ theory also implied relationships between the laboratory and clinic and between the laboratory and the natural world that were the topic of impassioned debate. As different communities became embroiled in these controversies, the range of appropriate applications of germ theory remained far broader than in its later incarnations.[28]

The protean nature of germ theory as it moved among these different communities permitted the search for a cancer microbe to coexist alongside the search for the microbial causes of numerous other diseases. Suggestively, cancer bore similarities to certain kinds of tuberculosis and syphilis. All three diseases killed slowly, produced strange growths, and seemed to be caused by a complex mixture of environmental, hereditary, and behavioral factors. To some medical practitioners, the behavior of tumors within the body, including their ability to distribute "seeds" in the blood, seemed remarkably like the behavior of microbial infections. "From a surgical point of view cancer is a spreading infective process, and the cancer cell contains elements of infection," one paper explained.[29] Anecdotal evidence suggested that transmission occurred. In several circumstances, surgeons operating on tumors appeared to subsequently develop cancer from cuts or needle pricks.[30] A Canadian physician warned that "a woman suffering from cancer of the uterus may, during a year or two before she dies, infect her friends and neighbors with cancer of the face, lips, throat, stomach and intestine."[31]

As with other diseases, efforts to identify a cancer germ repurposed older theories of its transmissibility grounded in contagion or miasma, but it did not displace them. In the 1880s the leadership of the New York City Women's Hospital rebuffed the offer of the wealthy Astor family to fund the construction of a cancer treatment pavilion, on the fear that it might contaminate the rest of the hospital. While the trustees of the hospital

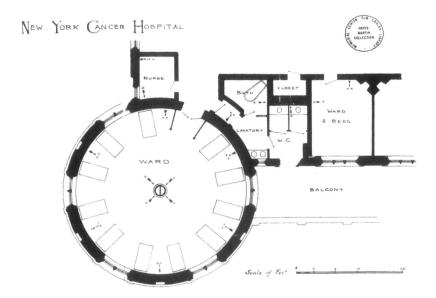

FIGURE 1.1. Hospital architecture embodied concerns for cancer's transmissibility. The open floor plan of this ward of the New York Cancer Hospital incorporated large, light-filled spaces that were intended to slow the spread of contagious diseases. Image courtesy of Memorial Sloan Kettering Cancer Center Archives.

were not convinced that cancer was contagious, they felt that caution was best. When the New York Cancer Hospital opened a few years afterward, its austere architecture adopted design features of contagious disease wards—its circular towers were designed to inhibit both the growth of germs and the spread of odors.[32]

Likewise, the miasmatic cancer regions that Haviland and other medical geographers had identified offered medical thinkers evidence of cancer infection clusters. In the early 1890s, after reading reports of cancer houses in Normandy, France, D'Arcy Power, a fellow of the Royal College of Surgeons, started to investigate regions of high cancer incidence in England and Wales. Power collected soil from the sites that he visited with the aim of cultivating a "hypothetical cancer organism."[33] Power's hypothesis blended contagion with other possible causal factors. "No one," Power conceded, "imagines that cancer is directly contagious. It is possible, however, in epidemic cases that there may be some condition of earth or water common to all the individuals attacked, in which the organism, if such there may be, may pass a part of its existence."[34] Indeed, during the

1880s and 1890s, numerous researchers in Europe and the United States attempted to identify a cancer germ or parasite.[35]

In 1894 Samuel Shattock, the recipient of the prestigious Morton lectureship of the Royal College of Surgeons, devoted his talk to the question of cancer as a "microparasitic disease." This question raised substantial conceptual as well as experimental issues. Was cancer one disease or many? If it was many, was each type caused by its own parasite? Shattock described his efforts to isolate and cultivate a cancer parasite from tumors as well as from the soil of cancer houses, similar to those in Normandy, and use the parasite to inoculate experimental animals. Fortunately for these animals, no cancer developed. However, the inconclusive nature of Shattock's experiments did not amount to evidence of the parasite's nonexistence. Rather, he concluded, "the conditions were not sufficiently natural" in laboratory cultivation.[36]

Buoyed by the therapeutic potential of discovering cancer microbes or parasites, several cancer research institutions established in Germany and the United States shortly after the turn of the twentieth century centered their efforts on finding such organisms.[37] Roswell Park, the famed director of an eponymous institute of cancer at the University of Buffalo in the state of New York, assured the American Medical Association that "the parasitic or infectious theory of cancer is the only one which satisfies the needs of both the pathologist and the clinician. . . . Are we now to go further and say that parasites have been discovered? For myself I do not hesitate to answer positively in the affirmative."[38]

Many critics of germ theory were unconvinced by these promises. William Roger Williams, author of the widely cited *Natural History of Cancer* (1908), provided a withering capsule history of the search for cancer microbes: "The original investigators of the so-called parasites of cancer entered on their quest . . . armed with novelties—with new microscopes of unprecedented magnifying power, with new stains of great complexity, and with new techniques of equal elaboration. Instead of first testing these inventions on normal tissues, on known morbid conditions, and on known microbic diseases, they rushed straight away to the study of the minute anatomy of cancer—a disease that had never before been investigated with this degree of elaboration."[39] Similarly, a magazine article entitled "More Cancer Germs" editorialized that the "cocksure" faith of new investigators in their "own particular germ[s]" provided "sufficient proof of the extreme uncertainty of the whole subject."[40]

The flexibility of the germ and parasitic theories often frustrated these critics. "It does not appear that any of the numerous observers in this field agree with others as to what a cancer parasite is," a vexed pathologist wrote. "Bacilli, cocci, protozoa, blastomycetes, and moulds have all had their day. Each new observer points out the mistakes made by his predecessors and brings forward a new parasite to be in turn demolished by his successors."[41] Of the many attempts to find an infective agent responsible for cancer, a pathologist wrote that "the laborious search for such an extremely hypothetical creature should no longer be suffered to divert the energies of investigators from those numerous domains of cancer pathology in which valid work remains to be done."[42]

Confronted by these criticisms, advocates of the idea that cancer was contagious could point to the difficulties of applying germ theory to other contagious diseases in order to maintain faith in their search. For these advocates, the absence of evidence for a cancer microbe in particular instances did not count as evidence of its absence in the disease. An editorial in the *Lancet* reminded readers, "That we have not yet found the organism and that we cannot cultivate it on artificial media cannot be advanced as arguments against the microbiotic theory" of cancer, since similar arguments had been raised and dismissed in the case of infectious diseases such as tuberculosis and tetanus. "There is sufficient probability of its truth to make it our duty to be very careful of spreading the disease," the editorialist continued, warning readers to treat "all discharges of cancerous growth" as well as the "stools of patients with cancer of the bowel" as potentially hazardous. Surgeons treating cancer should take it as their duty to be "very careful of spreading the disease."[43]

Shortly before Rous's observations, an informed layperson assessing the possibility of a cancer microbe had no reason to credit the theory any more or less than the microbial explanations advanced for a host of other diseases. *The Control of a Scourge* (1907), a book on cancer aimed at the public, deferred to this sentiment. While asserting that readers "may take it as a fact that the chance is infinitesimal, if indeed it exists at all, of their 'catching' the disease from anybody suffering from it," the book's chapter on prevention nevertheless contained "simple directions to follow by those who are in attendance on patients suffering from cancer" or "are about to occupy rooms previously inhabited by a victim of the disease."[44] In a satire on the "bacillus of love," a writer for the *Baltimore Sun* conceded that though bacteriological researchers had not yet identified such an amorous microbe, "the same is true of the cancer bacillus, of whose existence little doubt is entertained."[45]

Cancer, Contagion, and the Public

After a generation of interest in the possible applications of germ theory to cancer, such theories went into abeyance around 1915 and then revived in the 1950s. In this sense, Rous's observations and their reception marked the end of a historical moment rather than the beginning of a new research path. Chapter 2 addresses the scientific and medical challenges that attended Rous's and others' efforts to identify cancer viruses in the laboratory, but these were not the only barriers theories of viral carcinogenesis faced. The principal resistance to further inquiries into cancer and contagion came from the emerging discipline of oncology, from the Greek root *onkos*, or mass, a term selected to minimize the taboo associated with cancer in the public imagination.[46] The surgeons and clinicians of the oncology community viewed the theory of contagious cancer as both incorrect and a menace to the relationship they sought to build with the public. In fact, a key aim of the surgical profession's broad-ranging campaign to educate the public about cancer was the effort to break the long-running connection between cancer and contagion. This education campaign helped explain the apparent eclipse of viral theories of cancer, but it provides an ironic testament to the persistence of contagious cancer in public consciousness.

In the same decades when germ theorists were drawn to the idea of preventing disease through understanding its microbial causes, surgeons became convinced that new operational techniques could treat cancer. The introduction of antisepsis and anesthesia during the nineteenth century allowed patients to endure and survive far more brutal operations than ever before, including those reaching deep within the body to remove previously inaccessible tumors. The response of the clinical community to cancer in the late nineteenth century was powerfully shaped by these surgical advances. Famously, William Halsted, a surgeon at Johns Hopkins University Medical School, popularized the technique of radical surgery, aiming at the total removal of tumors. Based on his apparent success treating women with breast cancer, and the success of others who adopted similar methods, surgeons came to believe that the treatment of cancer in its early stages offered the best hope of survival, a conclusion with long-lasting implications for how they discussed the cancer problem with the public. The surgical approach to cancer emphasized early detection, a message that sought to minimize the general sense of stigma associated with cancer and bolster medical authority.[47]

Before educating the public, surgical oncologists would need to address the obstacle presented by general medical practitioners, whom they regarded as untrained in the diagnosis of cancer and unduly pessimistic about the chances of a cure. The ignorance and pessimism of the medical community, surgeons asserted, created a further problem: cancer served as a site of attacks on the authority of the medical profession as it attempted to contain the "quackery" of alternative health practitioners. Faced with the meager offerings of "orthodox" cancer care, patients favored treatments ranging from mineral waters to electrical stimulation. Physicians struggled to impose distinctions between the promising new treatments and fraud.[48] Nor did doctors possess many means to bring heterodox cures under control. It took almost three decades of legal action to bring about a ban on one notorious cure, a caustic chemical paste applied to tumors, known as the Hoxide treatment.[49]

New surgical treatments promised to resolve these challenges. An article in the *Ladies' Home Journal*, published with the endorsement of the Congress of Surgeons of North America, provides a template for the early detection message. Scientific research into cancer's causes, microbial or otherwise, appeared to be of little use: "The nature of cancer is as hidden as the methods to be pursued in cure of it are clear. Some believe it to be caused by a germ, others not. There is no proof. Some consider it due to diet, others deride this. There is no proof. Some hold by environment, others by hereditary tendency. Each may freely make his own guess, for nowhere is there proof." Confronted by the "inexplicable" origins of cancer, the best advice was to never "defer an advised operation even for a day . . . [or to] shrink from the merciful knife, when the alternative may be the merciless anguish of slow death." The principal danger was "not in surgery, but in *delayed* surgery."[50]

This educational mission was taken up by the American Society for the Control of Cancer (ASCC), the precursor organization to the American Cancer Society, when it was founded in 1913 by a combination of surgeons and philanthropists. Alongside its other activities, which are discussed further in chapter 3, public education was a critical part of the ASCC's work. As it took form, the ASCC prioritized shaping the public perception of cancer.[51] The organization's vice president noted that the ASCC's first task was to confront "skepticism both in the ranks of the profession and among the people."[52] An organizer foresaw that the control of cancer would arrive only after the "education of every single adult of the community."[53] The society pursued two major goals: attacking what it considered overly

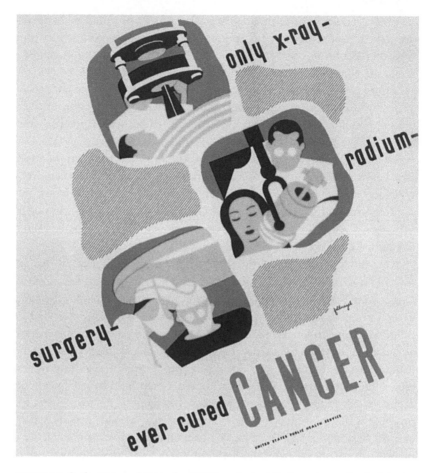

FIGURE 1.2. In the 1920s and 1930s, the ASCC emphasized the importance of early detection and early treatment by radiation or surgery. Image courtesy of US Public Health Service.

optimistic, or "quack," promises of cancer cures, and promoting the message that with early detection, surgery was the best means of treating cancer. In light of these aims, the creators of the society's program of public education were chiefly concerned with avoiding or minimizing discussions of the causes of cancer that undermined its emphasis on early detection. During the interwar years, the ASCC took particular pains to minimize the idea that cancer might be hereditary or contagious. Indeed, rebutting these notions alongside an attack on heterodox treatments were major points of the ASCC's first educational film, *The Reward of Courage*, in 1921.[54]

In the eyes of the ASCC's representatives, the suggestion that cancer might be contagious committed the double sin of stigmatizing those who had cancer and promising the possibility of a nonsurgical "cure"—the kind of promise that blurred the lines between medical authority and "quackery."[55] The doctors whom the society arranged to speak to local groups, from Manhattan to Huron, South Dakota, affirmed these points during the society's annual "Cancer Week" education campaigns in the 1920s.[56] At an international meeting on cancer, oncologist James Ewing, speaking for the ASCC, was especially quick to dismiss any claims that cancer might be caused by a "universal parasite."[57] The ASCC reminded readers of its bulletins that although doctors and nurses were in daily contact with patients with cancer, there was "no recorded instance of one case of cancer giving rise to another." Cancer patients needed sympathy and comfort rather than the "unnecessary and uncharitable" treatment that belief in their contagiousness fostered.[58] At the conclusion of an international meeting on cancer convened by the ASCC at Lake Mohonk, New York, in 1926, the assembled experts issued a statement intended

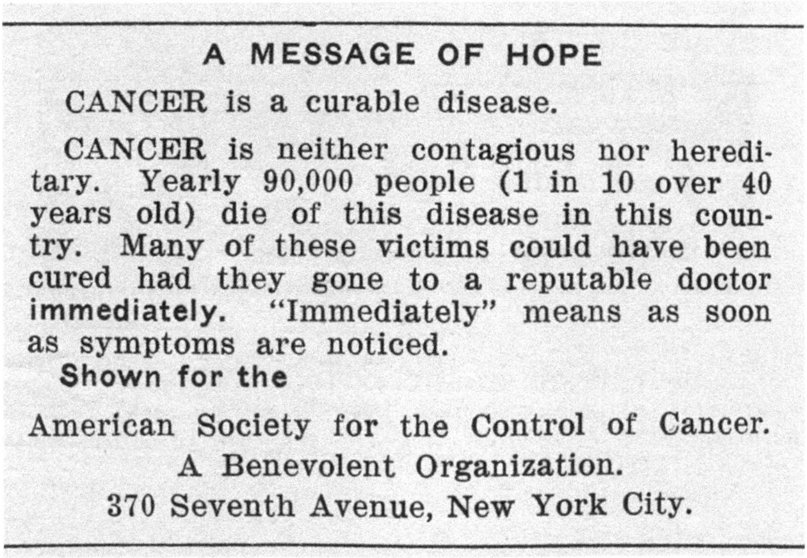

A MESSAGE OF HOPE

CANCER is a curable disease.

CANCER is neither contagious nor hereditary. Yearly 90,000 people (1 in 10 over 40 years old) die of this disease in this country. Many of these victims could have been cured had they gone to a reputable doctor immediately. "Immediately" means as soon as symptoms are noticed.

Shown for the

American Society for the Control of Cancer.
A Benevolent Organization.
370 Seventh Avenue, New York City.

FIGURE I.3. Early educational materials by the ASCC. The need to deny the contagious or hereditary nature of the disease supplies proof of how widespread the ASCC thought these beliefs were in the 1920s. Courtesy of the US National Library of Medicine.

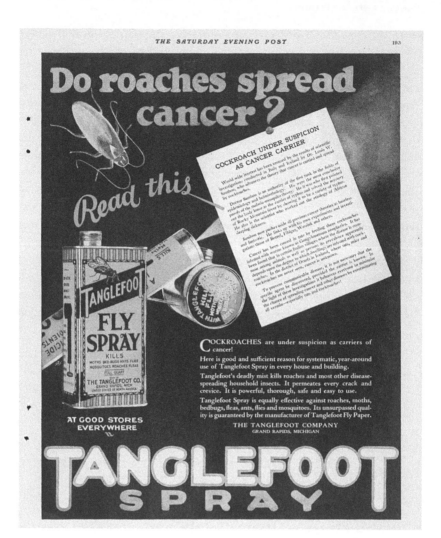

FIGURE 1.4. This DDT advertisement, from the *Saturday Evening Post* in 1926, references the Nobel Prize awarded in that year for proof that a parasite in cockroaches could transmit cancer from one rat to another, which readily fit cancer alongside other vector-borne illnesses that pesticides could combat. Courtesy of Science History Institute, Phil Allegretti Pesticide Collection.

to convey the most important facts about cancer to the public. The first two points denied the possibility that cancer was contagious or inherited. Rather than vaccination, the statement suggested that control of cancer hinged on "personal hygiene . . . and the intelligent cooperation of patient and physician."[59]

Even as the ASCC's program of education unfolded, cancer continued to resonate with contagion. For every article encouraging early cooperation with doctors, many others on cancer germs or cancer houses continued to appear.[60] Associations between cancer and food remained fraught. State and municipal sanitary codes barred individuals with cancer—like those with infectious or contagious disease—from handling, preparing, or serving food.[61] When Rous wrote to Campbell's Soup of Camden, New Jersey—a major processor of chickens—to seek further examples of tumors to study, the manager of the plant reluctantly agreed to provide space for one of Rous's assistants to examine birds. Fearing that customers might associate their soup with cancer, Campbell's staff insisted that the company's name should not be "mentioned directly or indirectly in connection with the results, or that such experiments were carried out in our establishment."[62]

A similar tension emerged around the 1926 Nobel Prize awarded to Danish doctor Johannes Fibiger, who claimed to have identified a cancer-causing parasite in cockroaches.[63] While the dramatic collapse of this theory in the wake of subsequent replication attempts further diminished the scientific standing of the hunt for cancer microbes, the association of these findings with advertisement campaigns for pesticide showed the rapidity with which such theories could be adapted to the interests of the broader public, and thus the wariness that researchers felt in discussing theories associating cancer and contagion.

Viruses as Heterodoxy

The ASCC's vigorous efforts to decouple cancer from contagion betrayed its anxiety about the fact that a substantial portion of the public still maintained these associations. Cancer viruses provided a way of talking about treatment that did not emphasize surgery, amplifying their association with other, heterodox therapies that the ASCC sought to discourage. As chapter 2 discusses, between the two world wars viruses existed at the limits of laboratory regimes of perception. Given that viral theories of cancer still resonated with widespread beliefs in contagion, the reception of claims regarding their existence was as much a matter of the observer's status relative to the growing community of cancer specialists as of any set of laboratory findings.[64] The surgeons and physicians who formed the core of that community were more concerned with the implications of microbial theories of cancer and how they were communicated to the public than with their validation within the laboratory.

This was particularly true of attempts to visualize cancer viruses. Efforts to provide visual proof of the association of viruses with cancer seemed to be the most direct analog with the laboratory conventions of microbiology. Experimentalists often announced the identification of cancer "germs" and provided images of those agents.[65] However, the size of the most likely candidates for an infectious agent for cancer were beyond the resolution of the optical microscope, requiring new kinds of visualization technology. In 1925, for example, William Gye, a British pathologist working at Britain's National Institute for Medical Research, announced that he had confirmed Rous's results on the infective nature of "filterable agents" obtained from chicken tumors. Furthermore, Gye's colleague, using an ultraviolet microscope, claimed to have seen the virus itself in tumor samples, promising that a causal link might be established.[66] These results were met with immediate excitement, and a crowd of thousands greeted the announcement of the results in London. Gye was reported to be hard at work developing a vaccine or serum against cancer.[67]

Among cancer specialists in the United States, the skeptical reception accorded to Gye's experimental findings was closely connected to disapproval of the manner in which he had made them known. In the eyes of this community, Gye had committed two cardinal sins: announcing his findings to the press before they were officially published, and actively discussing a nonsurgical cure for cancer. "It was unfortunate," one pathologist wrote, "that the normal process" of preparing for publication had been "totally reversed."[68] Another writer noted tartly that Gye should have "shut up and cured one hen first," before "talking about a cure in human beings."[69] One of Rous's colleagues at the Rockefeller Institute, James Murphy, appealed to the *London Times* that "speculation and anticipation . . . should be strongly deprecated."[70] An editorial in *The New Republic* castigated newspapers whose coverage had raised Gye's results to the level of sensation: "Who cares for scientific caution when cancer is concerned. . . ? Who of the millions that are devouring the stories of 'the great new scientific discovery' have paid any attention to the actual details

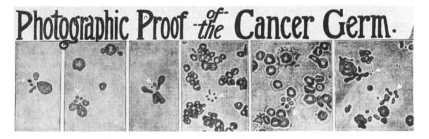

FIGURE 1.5. After the rise of the optical microscope, numerous claims emerged that cancer microbes had been detected. These efforts speak to the resonance of "seeing" viruses, even if they proved elusive under the microscope. This headline is from the *Salt Lake Tribune* (1912).

set forth in the news? . . . The cause of excitement is the dreadfulness of cancer, not the magnitude of scientific achievement."[71] The following year, 1926, marked the ASCC's most strenuous denials of the idea that cancer was contagious.

Claims to have seen cancer viruses emerged from less orthodox quarters as well. In 1931 the *Science News-Letter* reported the invention of a "super-microscope" by First World War navy veteran Royal Rife in Southern California. Incorporating polarized light and quartz prisms into his design, Rife claimed, allowed him to view viruses where previous microscopes had failed.[72] Soon thereafter, in *Science*, a doctor using the microscope claimed to have seen images of minute "filterable bacteria," a synonym for viruses.[73] While the "Rife Microscope" now has a reputation as a heterodox cancer treatment, given that Rife claimed to be able to destroy cancer viruses as well as visualize them, many of its original advocates were also involved in mainstream debates over the nature and detectability of viruses in the 1930s.[74]

Boundaries that appear clear in retrospect were more difficult to draw at the time of these observations. While many cancer researchers had no trouble dismissing Rife himself, claims that the Rife Microscope enabled researchers to "see" viruses substantially resembled the claims of advocates of another new laboratory tool: the electron microscope.[75] One set of authors wrote of both the Rife and electron microscopes that "with the aid of its new eyes . . . Science has penetrated at last beyond the boundary of accepted theory and into the world of viruses with the result that we can look forward to discovering new treatments and methods of combating the deadly organisms."[76]

Nonetheless, the broader efforts by the ASCC and medical authorities to decouple cancer from its long-standing associations with contagion succeeded in chilling the reception of viral theories of cancer during the 1920s and 1930s. For instance, in the course of foster nursing experiments with mice at the Jackson Laboratory, a center of experimental genetics, a researcher identified a "milk factor" that caused a higher incidence of mammary tumors. He feared, however, that his fellow researchers would receive a viral explanation as an "unrespectable proposal," so he demurred from offering it.[77] Concern regarding the propriety of discussing cancer viruses shadowed researchers long after scientific opinion on cancer viruses started to shift. Speaking with a representative of the US Public Health Service arranging a lecture on his tumor virus work in 1957, Richard Shope, one of Rous's colleagues in the 1930s, cautioned, "The address is written for a professional audience of knowledgeable scientists." He resisted passing a copy to the *New York Times* on the grounds that it might cause "alarmist headlines if the material is lifted out of context."[78]

The Persistence of Contagionism

Many of the points of contention and ambiguity that shadowed cancer viruses in the early twentieth century were resolved by the revision and extension of laboratory regimes of perceptibility. Yet public actions and attitudes continued to demonstrate the resonance of thinking about cancer as a contagious disease. In 1962, *Life* magazine ran a feature article, subsequently republished in *Reader's Digest*, on cancer viruses. It prompted an outpouring of correspondence to Rous, then recognized as a leading cancer virologist. Rous complained that "a host of letters has come to me, fearing or imploring what to do."[79] The questions reflected a remarkable continuity in the concerns that the general public brought to cancer germs, despite years of education efforts and major changes in expert understandings of the nature of viruses and cancer. One writer speculated that the cancerous growths he observed in cattle contained viruses that were passed to people who ate beef.[80] A worried husband wrote to Rous asking whether cancer could "be transmitted from husband to wife in normal sexual intercourse."[81] Another correspondent wrote that she was so troubled that she "could not sleep" after reading the article and learning that the previous inhabitants of her apartment had all died of cancer. "Is there anything I can do to kill the germ or whatever it is that cancer leaves behind it[?]," she implored Rous.[82]

FIGURE 1.6. The association of cancer with contagion has persisted for more than a century in the face of medical denials. This t-shirt appeared for sale in 2017. Image courtesy of Zinga-WorldWide LLC.

Rous responded to all of these with assurances that even if cancer viruses existed, cancer was not contagious. However, such exchanges show the degree to which contagion and germ theory persisted as regimes for thinking about cancer viruses in society as a whole. A writer diagnosed with cancer in the early 1970s reflected on her fear that a cancer virus had been passed to her by her pet dog, despite assurances "over and over again that this was impossible."[83] More tragically, in 1973 a sufferer from Hodgkin's lymphoma committed suicide shortly after reading a report suggesting that there was a virus responsible for the disease. He left a note stating that "he couldn't bear the idea of infecting his family."[84]

For many, approaching cancer as a viral disease did nothing to lift the fear and pessimism that the illness inspired. However, in the middle of the twentieth century, these two responses to the idea of cancer viruses, one concerned with medical authority, the other with long-standing beliefs in contagion, were joined by the biomedical view of cancer viruses formed in the laboratory. Unlike previous moments of fear or anxiety, this vision drew strength from the promise that identifying the agents of cancer marked an important step toward protection from the dread disease.

Cancer as a Viral Disease

Speaking at a women's club in Columbia, Missouri, on a winter evening in 1922, M. P. Neal, a local surgeon, endeavored to refute the belief that cancer was contagious. He offered a grisly but dramatic report as proof. One of "America's foremost surgeons" had inserted fragments of tumor tissue into cuts in his arm but had not developed cancer.[1] Neal's attempt, if not this dramatic report, was typical of the efforts of clinicians and surgeons to disassociate cancer from contagion in the 1920s and 1930s, as traced in chapter 1. While the link between cancer and contagion might have been firmly fixed in the public mind, the use of the biomedical sciences to study cancer faced considerable skepticism from doctors. Since the advent of germ theory, experimentalists' claims about the microbial causes of disease had faced recurrent skepticism from clinicians, who argued that germs were either unable to cause disease on their own or irrelevant to its treatment. As Neal no doubt knew, the dramatic self-experiment he cited recalled a notorious demonstration by Max Joseph von Pettenkofer, a nineteenth-century Bavarian anticontagionist, who had ingested samples of the bacteria *Vibro cholerae* to disprove its association with the disease cholera. Tensions between clinicians and experimentalists concerning the application of germ theory to cancer were especially strong. "Cancer," wrote the surgeon-author of *The Cancer Problem* (1914), "fails to come within the category of the revolutionary change in the conception of etiology and treatment which has marked the discovery of the infective nature of many diseases."[2]

The objections encountered by advocates of cancer virology as they sought to establish the experimental and clinical credibility of their work provide a case for thinking about the challenging process of molecularizing a disease.[3] As several generations of scholarship have now demonstrated, there was no singular "bacteriological revolution" in the practice of medicine after the advent of germ theory. Germ theory was both a cultural phenomenon drawing on long-standing views about how disease spread and the bellwether of a new way of pursuing medicine in the laboratory. Belief in microscopic agents of disease encountered many challenges from physicians, even those sympathetically inclined toward "scientific medicine."[4] The challenges were even greater as studies of cancer moved to virology, which grappled with the ambiguous nature of viruses for far longer than bacteriology did with bacteria.[5] These first studies opened inauspiciously with the difficulties that Peyton Rous encountered in his study of the viral origins of chicken tumors, and they floundered further on the ongoing doubts that clinicians expressed toward microbiology as a whole.

The revival of theories of viral carcinogenesis among physicians and biologists in the 1950s rested on two major developments in the regimes of perceptibility that these communities brought to bear on the microbial world. First, new technologies and techniques created a strong set of analogies between bacteriology and virology and suggested the abundance of cancer viruses in nature. However, those findings would have had a much more limited impact without a second development: the acceptance of the relevance of biological research to the practice of clinical medicine as a whole. The growth of bacteriology created a new set of spaces for biomedical research at universities and philanthropies, in which the pursuit of cancer viruses as molecular entities could continue independent of medical skepticism. Later, vaccination provided a framework in which to understand the therapeutic potential of virus research. The techniques of the laboratory, coupled with the growing credibility of vaccination as a strategy against viral diseases, combined to elevate cancer viruses as targets of biomedical study at midcentury.

Germ Theory and the Clinic

At the outset, the relationship between microbiology and practicing doctors was much more contentious than its later widespread use might suggest. Unlike other efforts to incorporate microscopy or chemistry into medical practice, germ theory proposed a new way of thinking about disease that was as controversial as it was potentially powerful. Nor were the objections raised by its critics tangential: they highlighted important questions about infection, causality, and the relationship of biology to medical practice that bedeviled cancer virus studies well into the twentieth century.

Robert Koch's identification of a tuberculosis microbe in 1882 provided an exemplary case for the use of laboratory methods to identify the causal agents of disease. Koch's success with tuberculosis rested on two innovations, one conceptual and one experimental. Experimentally, from his previous studies of pathology, Koch devised a means of growing, or culturing, tuberculosis microbes in petri dishes and then revealing their existence under the microscope by means of new methods for dyeing, or staining, microbes in the cultures. Conceptually, he devised a set of steps for establishing a causal relationship between the presence of microbes and the manifestations of disease; later these became known as Koch's postulates. Koch prepared pure cultures of microbes from a case of tuberculosis and used the pure cultures to infect healthy laboratory animals (guinea pigs were the unfortunate subjects in his tuberculosis experiments). After the animals sickened, Koch used his culture and staining techniques to show that their tissues now contained tuberculosis microbes. These steps—associating an agent with a disease, making a pure culture of the agent, infecting a healthy organism, and re-isolating the agent—created a powerful technique for revealing the causal power of the unseen world. Koch, however, took pains to claim only that he had shown that bacteria could be a cause of tuberculosis; he did not claim they were the sole cause. The extension of this potent framework to human diseases faced obstacles. Experimental steps such as infecting healthy organisms presented ethical challenges if using human subjects directly and practical challenges if substituting animal proxies for human bodies.[6]

While Koch might have focused on how to infer causality, his French counterpart, Louis Pasteur, promoted the therapeutic potential of using laboratory methods to isolate the agents of disease. In 1880 Pasteur informed the Paris Academy of Sciences that he had managed to produce a "vaccin," or a weakened batch of microorganisms, that protected chickens against fowl cholera. Based on his accomplishment, Pasteur promised a future in which vaccines could be used to prevent any disease whose cause could be isolated.[7] Fowl cholera vaccine was the first of many advances that Pasteur and Koch claimed for the new science of bacteriology during the 1880s and 1890s. Their successes stoked national pride and exemplified the promise that "germ theory" would harness laboratory research for human health. The promise of vaccines featured by Pasteur's work dramatically expanded popular appreciation of the possibilities of scientific medicine.[8]

Pasteur's work on vaccines continued to generate striking icons of microbiology's therapeutic potential. While other diseases in humans, notably smallpox, had earlier been the target of prevention through inoculation, the process of inoculation remained risky. Germ theory promised to change that situation, even for feared diseases. In the fall of 1885, Pasteur announced that he had developed a vaccine against rabies and had used it to cure a young boy who had been bitten by a rabid dog. Although this pronouncement concealed considerable uncertainty as to the efficacy or safety of the vaccine itself, public reception of his claim was enthusiastic.[9] His method promised that almost any disease whose microbial cause could be identified might be similarly prevented. Audiences on both sides of the Atlantic were enchanted by this possibility. In December 1885, readers flooded the *Newark Daily Journal* with donations for the treatment of six children bitten by a rabid dog. Reports of the children's journey to Paris appeared as far west as St. Louis. In the American case, this episode established enduring media interest in the medical promise of laboratory-based medicine.[10]

The attention drawn by Pasteur's and Koch's work inspired considerable enthusiasm for the benefits of biological studies of human illness writ large. Yet as a means of proving the cause of disease, germ theory continued to encounter conceptual, cultural, and practical challenges. The criteria for disease causation proposed by bacteriology also entailed a dramatic change in disease classification. Diseases were not to be defined by their symptoms as observed at the bedside but by their invisible microbial causes. A clinician's version of a disease could differ widely from a bacteriologist's definition.[11] After Koch's announcement that he had identified a tuberculosis microbe, many British physicians claimed that the infectious theory of the disease flew in the face of their experience with the broader category of consumption, as many cases of tuberculosis were known. In one survey, nearly two-thirds of the physicians denied ever observing an instance of tuberculosis being "communicated" from one individual to another.[12]

Indeed, the casual links proposed by germ theory in its unqualified terms were difficult to demonstrate. Many variants of germ theory assumed that exposure to bacteria and the occurrence of illness coincided. In this view, for example, the disease of tuberculosis was synonymous with infection by tuberculosis bacteria. That form of causation awarded no role for bodily resistance; immune responses severely complicated the causal link between infection and the incidence of disease. Critics of germ theory staged striking demonstrations—the chemist Max Joseph von Pettenkofer notoriously ingested samples of cholera bacteria in front of an audience—to show that exposure to a microbe did not entail the onset of disease. Infection without apparent illness happened frequently. Moreover, the passage of certain diseases between generations, such as syphilis, could be explained equally well by the idea of inherited constitutional weakness as by the transmission of an infection from parent to child. During the twenty years following the introduction of germ theory, doctors, pathologists, bacteriologists, and others interested in disease arrived at an understanding of disease causation that allowed for the interaction of hereditary, environmental, and microbial factors rather than assuming that microbes were the sole causes of illness.[13]

The multifactorial model of causation allowed accommodation between emerging biomedical disciplines such as physiology, bacteriology, and biochemistry and the clinical communities and medical schools that still provided the majority of scientists' support. These communities were unwilling to cede their control over the diagnosis and treatment of disease. "Nothing has so injured the status of the practitioner and of the medical profession," an influential German doctor wrote, "as the eagerness of bacteriologists to transfer decisions from the bedside to the laboratory . . . according to an artificial scheme . . . that diagnosis becomes subject to the infallible arbitriment [judgment] of the microbe cultivator remote from the bedside."[14]

Such skepticism mattered for cancer virology in particular because of the close relationships between the biomedical sciences and the institutional growth of academic medical schools. Researchers working in these settings were expected to pursue avenues of inquiry with clear relevance to the clinical problems faced by physicians. Within this environment, the exploration of disease theories that did not appear to reflect clinical concerns or clinical experience was discouraged.[15] Therefore, most of the sites that pursued cancer virus research in the early twentieth century were found not in university departments but in biomedical research communities sustained by philanthropic donors seeking to "reform" medical practice rather than to serve it. Several cancer research institutions established in Germany and the United States around 1900 took cancer microbes or parasites as the central aim of their efforts.[16] Scientists in the emerging fields of genetics, biochemistry, and bacteriology all grappled with establishing and defending how their approaches might help them comprehend cancer.[17]

Peyton Rous and the Challenges
of Viruses in the Laboratory

The Rockefeller Institute for Medical Research, where Rous carried out
his observations, was among those new institutions. With backing from
John D. Rockefeller's oil fortune, the Rockefeller Institute opened in
1901 to give American medical science the same kind of environment for
independent research as that enjoyed by academic scientists in Germany.
Its leaders were exponents of the idea that laboratory biology could advance medicine.[18] Rous arrived at the institute after training as a doctor
at Johns Hopkins University Medical School, another supporter of the effort to apply laboratory methods to medicine, and serving in a pathology
laboratory at the University of Michigan. He started working at the institute in 1908 as a laboratory assistant. Before gaining his own laboratory
in 1910, Rous was an assistant in both pathology and bacteriology—two
experimental subfields of biology whose medical applications appeared
most immediate. Unlike medical schools, the institute provided an environment where both of these approaches could be brought to bear on
cancer.[19]

Above all else, experimentalists approaching cancer struggled to find
a form of the disease that could be reliably reproduced in the laboratory.
It would be much easier to study tumors of this type than to wait for the
occurrence of unpredictable "spontaneous" cancers in nature. When Rous
began to study cancer, no means existed of causing tumors under experimental conditions or of cultivating tumor tissues outside of living bodies.
The first instance of chemically induced cancer in the laboratory occurred
in 1915. Other researchers had only embarked on a decades-long effort
to breed mice predisposed to cancer in 1909. The first stable system for
studying tumors in the laboratory, developed in 1901, entailed the transplantation of tumors from one rat to another. Those studies, especially
the immunological study of "resistance" to transplanted tumors, were one
of the primary pathways into experimental work with cancer in the early
twentieth century.[20] The pressing experimental question raised by transplantation was not in regard to how these tumors were caused; it pertained
to the identification and propagation of similar tumors in other species.[21]

Rous's first efforts followed that approach. He was excited when a
Long Island poultry breeder brought a hen with a "large, irregularly glob-
ular mass" to the institute in September 1909. No one had yet succeeded

in finding a transplantable tumor in birds, and Rous approached the tumor with this aim. When he first discussed his findings, Rous was pleased that he had managed to keep the tumor alive through four generations of transplantation into healthy chickens and that the tumor had remained "true to type" throughout.[22]

Rous's work on tumor transplantation turned to bacteriology when he sought to determine the smallest fragments capable of transplanting the tumor from one chicken to another. From other researchers at the Rockefeller Institute, Rous acquainted himself with different means of filtering tumor extracts. For microbiology, the primary technology for determining the size of infectious agents involved filters of varying fineness. Rous started with filter paper, which "held back all but a few red blood cells and lymphocytes," yet the filtered solutions still transmitted tumors when they were inoculated into healthy chickens. Next, Rous turned to a common feature of bacteriology laboratories, a British Berkefeld water filter. This was a case of scientists drawing on public consciousness of germs, rather than vice versa. Before the construction of municipal water filtration facilities became commonplace, individual households concerned about the presence of microbes in their water used filters like the Berkefeld. By forcing water through fine sand, those filters were widely believed to be capable of blocking disease-causing microbes. The filters also quickly found use in laboratories. Rous passed extracts from his tumor through Berkefeld filters, and the filtered extracts retained their ability to cause tumors. In somewhat counterintuitive terminology, Rous determined that the tumor agent was "filterable"—which denoted that it could not be removed by filtration. In light of his results, Rous wrote that the first conclusion would be "to regard this self-perpetuating agent active in this sarcoma of the fowl as a minute parasitic organism." However, Rous continued, "it is conceivable that a chemical stimulant, elaborated by the neoplastic cells, might cause the tumor."[23]

Rous's announcement met with widespread interest. He fielded requests for samples of his tumor from correspondents in Russia, Japan, the United Kingdom, and Germany.[24] A prisoner at New York's Sing Sing Prison wrote to the warden volunteering to be inoculated with a "cancer germ" to test whether cancer could be passed from one human to another in a similar fashion.[25] In an internal report on his research, Rous elaborated that he now thought the agent was a "virus" smaller than a bacteria.[26]

Rous's claim to have identified a tumor virus appeared to continue the institute's efforts to show the utility of laboratory research on viruses for

48

CHAPTER TWO

understanding disease. At the institute, microbiology research, of which bacteriology was a subfield, was a central part of the campaign to demonstrate the contributions that laboratory researchers could make to clinical medicine. Facing a skeptical, and in the case of animal welfare advocates who attacked the use of laboratory animals, a hostile, audience, the institute's workers sought to show that the laboratory could identify the microbial causes of human diseases. In 1909 Simon Flexner, Rous's predecessor in the institute's microbiology laboratory, was one of several scientists to demonstrate that polio was caused by a virus.[27]

However, the nature of this virus was still intangible. From the perspective of a medical observer in 1910, the distinctions between different types of filterable entities were confused and contentious. Clinicians and biologists were able neither to observe the filterable agents directly in the laboratory nor to cultivate them outside of their hosts, if indeed they were capable of reproduction. Writing for the *Journal of the American Medical Association*, Flexner explained that his conclusion that the agent involved in polio was a virus rather than a bacterium was based on the observation that it remained active even when exposed to glycerin (which bacteria did not) and that it could not be filtered. The agent, Flexner concluded, "belongs to the class of the minute and filterable viruses that have not thus far been demonstrated with certainty under the microscope."[28]

Unlike the processes of chemical staining and inspection with the optical microscope that were used to identify the presence of bacteria, the existence of viruses could only be inferred through the biological effects they produced in their animal or plant hosts. The tobacco mosaic virus killed tobacco leaves, the rinderpest virus killed cattle, and the rabies virus caused rabies. In the laboratory, viruses were defined by what they were not: they could not be filtered out of solutions, as bacteria could, they could not be grown in culture, and they could not be observed under the microscope. There were almost as many ways of defining viruses, therefore, as there were efforts to isolate examples of them.[29] It was uncertain whether these entities should be considered living parasites or enzymes, and debate over that question swelled in the decades following Rous's observations.[30] How could such a confusing and mysterious entity be the cause of any disease, let alone cancer?

Previous debates concerning the relevance of microbiological studies to cancer shadowed the reception of Rous's report in other quarters. The *New York Times* linked the new observations to previous studies of contagion and cautioned readers that "the germ theory of cancer is not new" and was "not generally accepted."[31] For fear of inciting public panic,

Flexner was quick to avow that "cancer is certainly not readily infectious" and that no "clinical evidence" existed that infection could pass cancer from one person to another.[32] Experimentalists working on tumor transplantation raised numerous concerns about Rous's findings: their applicability to other tumors, the potential that the filters had allowed larger particles of tumor to pass than Rous assumed, and even whether the growth Rous had identified was a tumor.[33] Even the stability of transplantable tumors suggested that they were laboratory artifacts rather than reliable proxies for natural cancer processes. As one member of the Pasteur Institute cautioned, the study of these grafts showed only the "second act" of cancer rather than revealing the origins of cancerous cells.[34]

James Ewing, a respected oncologist at New York's prestigious Memorial Hospital and a forceful skeptic of laboratory-based approaches to cancer, was an especially prominent and influential critic.[35] "One must hesitate to apply the standards of human pathology to the tissue reactions of the chicken," he cautioned.[36] In his widely read oncology textbook, Ewing enumerated the theoretical challenges that made demonstrating the infectious cause of cancer so difficult. Infectious diseases, Ewing maintained, were of a different character than cancer. Cancer appeared to have many causes, and cancer tumors themselves were so diverse as to baffle attempts to draw parallels between different cases. Clinical experience, Ewing continued, suggested that tumors typically had chemical or hereditary causes. "The temporary popularity of the search for a specific parasite must be attributed to the undue influence of the germ theory of disease which can be effectually combated only by further knowledge of the biology of the cell," he concluded.[37]

Trained in pathology rather than bacteriology, Rous was concerned that he could not master the techniques of microbiology necessary to provide definite proof of the link between the filterable agent and cancer. The leadership of the Rockefeller Institute for Medical Research, especially Flexner, was also keen to see his efforts focused on areas of cancer research that promised a more immediate therapeutic payoff. Rous therefore moved on to other research in 1915 rather than further exploring his observations of a cancer virus.[38]

Establishing Laboratory Approaches to Cancer

Rous's frustration was symptomatic of the slowing tempo of experimental cancer research between the First and Second World Wars. After the excitement of germ theory and the hope of surgery, the beginning years of the twentieth century were, in the words of one historian, the "Wilderness Years" of cancer research and treatment.[39] Beyond virus hunting, experimental inquiries into the causes of cancer did not suggest a therapeutic reward sufficient to justify the mobilization of the anticancer community as a whole.

In the United States, interest in Mendelian genetics and eugenics supported a drive to breed strains of mice predisposed to cancer. In many instances, the study of heredity was explicitly aimed at rebutting the idea that cancer had contagious origins.[40] In parallel with virologists, geneticists held their own fierce debates over the material identity and causal power of "genes." It was by no means clear what it meant to reveal a gene responsible for a disease, as there was no way to visualize or manipulate genes as material entities.[41] With support from the Ford and Hudson motor companies, advocates of the experimental study of cancer heredity opened the Jackson Laboratory at Bar Harbor, Maine, in 1929. Its operations involved a large-scale effort to maintain purebred colonies of mice for studies of the heritability of cancer, an early example of big science in experimental biology.[42]

However, even if cancer genes were identified in humans, this kind of knowledge offered little therapeutic hope. A Chicago physician attacked one early set of studies on the heritability of cancer in mice, remarking, "To destroy all hope in patients with cancer or supposed cancer, as well as to get their children morbid on the subject as to the likelihood of their children having cancer, is to engender an untold amount of mental torture in the supposed victims."[43] Many studies of hereditary cancer did nothing to dispel this opprobrium. Their only therapeutic application appeared to be selective breeding along eugenic lines. The author of one study on the inheritance of a rare cancer, retinoblastoma, recommended that parents with a family history of the cancer not have children. If, against this advice, the couple had children and one of them developed the cancer, the child should be sterilized as a condition of treatment.[44] Indeed, the director of the Jackson Laboratory, C. C. Little, was also an advocate of eugenics.

In a broader sense, research into the causes of cancer embraced many possibilities inside and outside the poorly understood biology of cancer. Psychological, dietary, and behavioral explanations for cancer—too much or too little sexual activity, depression, overindulgence, or psychological repression—all appeared in medical and popular discussions.[45]

The discovery of chemical carcinogenesis by coal tar compounds in 1915 opened not onto therapy but onto contentious debates over the safety of workers in the dye and chemical industries.[46] Professional groups such as the American Society for the Control of Cancer hesitated to promote the therapeutic potential of cancer research. For that reason, the founders of the ASCC intentionally omitted research as one of the activities in the organization's charter in 1913.[47]

Research into the causes of cancer did not provide a means of resolving the tension between hope and pessimism faced by the leadership of the ASCC in its mission to educate the public. Discussions of cancer causes and cures seemed to steer the organization into the ongoing contest between doctors and alternative healers regarding who could best address cancer.[48] In 1926 William L. Saunders, a wealthy businessman, wrote to the ASCC offering to sponsor a prize of $50,000 for anyone who could discover the nature of human cancer, to be followed by a further $50,000 to any person or organization who could offer a cure. The letter caused considerable debate within the ASCC. James Ewing doubted that the ASCC would have the capacity or expertise to study so many proposals. He feared that the wide call for submissions would be an "encouragement to quackery." Indeed, the only reason that the ASCC accepted Saunders's request to serve as a judge of possible cures, in the words of its director, was to "demonstrate the futility of seeking to solve the cancer problem through a monetary reward." The ASCC eventually fielded more than five thousand submissions—none of which it found meritorious.[49]

Virology was no exception to this general sense of frustration with cancer research. Virologists continued to contend with the difficulties of observing viruses in the laboratory and associating them with diseases in a manner that satisfied the criteria of Koch's postulates, even in the case of diseases whose contagious nature was widely accepted. In frustration, some prominent virologists suggested discarding Koch's postulates altogether.[50] Studies of cancer viruses continued alongside these debates in the 1920s and 1930s. Seeking to address the difficulty of isolating filterable agents from other tumors, advocates suggested that viruses were "masked" or "latent" within tumor cells, a stance whose assumptions demanded more, not less, credulity from skeptics of viral carcinogenesis.[51] In the face of such challenges, many experimental cancer researchers preferred to incorporate viruses into models of cancer causation that allowed for many possible factors, including chemical irritation and hereditary predisposition—that was the approach adopted by the landmark interwar virology text *Filterable Viruses* (1928).[52] Research into the viral causes of cancer continued at a few institutions, notably the Rockefeller Institute for Medical Research, where Rous's colleague Richard Shope investigated the viruses responsible for rabbit papillomas.[53]

However, this work appeared distant from human cancer. From the perspective of scientific and medical observers in the 1930s, discussions of cancer viruses were prone to swing between doubt concerning laboratory findings and dismissal of the therapeutic claims of treating cancer microbes. The nadir of the virus theory of cancer among cancer experts came in 1938 when the US surgeon general tasked a committee of leaders in the field of experimental medicine to suggest "fundamental aspects" of cancer research that the newly created National Cancer Institute should support. Advocates of cancer virology worried that a draft of the report minimized the importance of viruses, despite new and promising experimental results.[54] The committee's leader replied with an emphasis on probity and restraint: "Present knowledge in the field is too full of uncertainties to justify . . . drawing very definite conclusions" about cancer viruses.[55] The final report elided discussion of viruses and emphasized that any approach to cancer should involve "patience and the adoption of a long-term point of view."[56]

New Potential for Cancer Viruses

Biomedical and popular attitudes toward cancer virology both revived during the 1950s. The National Cancer Institute reversed its initial skepticism toward cancer viruses and launched the first of several large-scale investments in cancer virology at the end of the decade. The newfound credit given to viruses as agents of disease and their contribution to the molecularization of cancer was heavily indebted to new instrumentation and new analogies with other viral diseases. Unlike the tools available to Rous and his associates in the 1920s and 1930s, these instruments and techniques succeeded in strengthening the analogy between viral infection and bacterial infection. Beyond specific laboratory advances, this "virus age" aligned the idea of cancer viruses with the aims of biomedical research shaped by microbiology, recapturing the broader excitement of seeking a cancer cure. Significantly, most of the new excitement associated with cancer virus studies was not generated by participants in earlier controversies regarding contagious cancer but emerged from contact with communities that took the therapeutic potential of virology and vaccines as a given.

The encounter of biologists with instruments that had previously been expensive and rare during the Second World War constituted the first pillar of this new confidence. In wartime, many virologists worked on vaccine production—the army was especially worried about another outbreak of influenza virus following the pandemic of 1918–19. Those virologists gained experience with a relatively new instrument, the ultracentrifuge. The ultracentrifuge generated higher forces than regular centrifuges, allowing the separation of particles based on very small differences in mass. The new technology could be used not only as a means of isolating but also as a means of purifying and classifying different virus strains. Virologists used this capability to separate viruses from other cellular material. In turn, it encouraged a shift in the presentation of the problem of viral transmission and detection from a qualitative question of "masked" or "latent" virus forms to a quantitative question of virus purification.[57] Flush with this confidence, proponents of the ultracentrifuge contended that the best solution to the difficulty of satisfying Koch's postulates was to seek pure samples of virus through the ultracentrifugation of tumor samples—a means of avoiding the complicated immune reactions inherent in animal studies.[58]

After the Second World War, the National Foundation for Infantile Paralysis (NFIP), whose reasoning is discussed further in chapter 3, decided to make an aggressive investment in virology, following from the conviction that polio was caused by viral infection. Whatever the motives for the foundation's investment, its decision energized virology as a whole. The foundation underwrote a substantial amount of research activity in virology during the late 1940s and early 1950s, expanding virology into a robust community with numerous ties to genetics, biochemistry, and biophysics. The lack of the ability to grow and study viruses outside of their host organisms had proved to be one of the most persistent obstacles for virology. Accordingly, the NFIP invested generously in the creation of reliable in vitro systems for the culture of viruses, drawing on antibiotics, cell lines, and other technologies, which provided a new means of studying the world of viruses. By some estimates the foundation trained one-third of the virologists working in the United States by 1956.[59]

New methods of virus culture suggested that the infection of humans by viruses was the rule rather than the exception. In 1948 virologists acknowledged 20 viruses specific to humans; by 1958 the total stood at 70.[60]

In a study reported by the *New York Times*, a single virologist working for the National Institute for Allergy and Infectious Disease, Robert Huebner, used tissue culture and antibody tests to isolate 25 unique viruses from just two kinds of human and monkey tissue. The sheer number of previously unknown viruses reversed the question faced by advocates of viral disease causation; the challenge was not what virus went with what disease but what disease went with what virus.[61] Revelations concerning the ubiquity of viruses and their diverse modes of infection suggested that most individuals already carried latent viral infections that might give rise to illness, either alone or in combination with some chemical or hereditary cause. "There is ample reason to believe that neither the recent viruses nor their diseases are actually new but only our recognition of them," another set of reviewers commented.[62]

The final technical innovation to shape midcentury virology was the electron microscope. Of all the new techniques, the electron microscope offered the most direct means of hunting for viruses—the electron micrographs researchers produced fit within the visual tradition in pathology of inspecting tissues with the optical microscope. The promotion of the electron microscope, which had been developed for commercial use by the Radio Corporation of America during the Second World War, resulted in widespread enthusiasm for creating images of the subcellular world.[63] Leon Dmochowski, a pathologist at the MD Anderson Hospital and Tumor Institute in Texas, emerged as the most prominent hunter of virus particles using the electron microscope. Dmochowski had started his career in the Department of Experimental Pathology at the University of Leeds, where he adopted the ultrathin sectioning techniques employed for studies of subcellular structures to the search for virus particles in both breast cancer cells and milk.[64] After his move to the United States, Dmochowski continued to use the electron microscope to seek out cancer virus particles in tumorous tissue, authoring numerous reviews on the topic.[65] In early 1957 Dmochowski announced to newspapers that he had placed viruses at the "scene of the crime" for human leukemia.[66] Later that year Dmochowski claimed that he had identified virus particles in human leukemia patients visiting the MD Anderson Hospital.[67] Cancer virologists, such as Peyton Rous, were encouraged by the rhetorical power of electron microscope images. Solicited for his thoughts on the design of an exhibit on cancer viruses for the 1958 World's Fair, Rous responded that he thought enlarged micrographs of viruses would "make a prodigious hit."[68]

The growing faith in viruses as molecules that the laboratory could identify and study created conditions for new cancer virus candidates to receive a more sympathetic hearing than they had before. A wide range of potential tumor viruses had been identified in the late 1930s and 1940s. Yet, with the exception of Bittner's "milk factor," none of those viruses appeared to exist in mammals.[69] This changed with the identification of the first mouse leukemia virus by Ludwik Gross in 1953. Gross arrived in the United States and at the study of murine leukemia after a long journey. Born in Poland, he trained as a doctor in Krakow. He traveled to Paris in 1931 to conduct postgraduate research at the Pasteur Institute, a bastion of microbiology and vaccine development. At that institute, Gross sought to immunize mice against the tumor cell transplants. During his research Gross tried and failed to cause cancers in mice using filtered extracts of tumors (an experimental approach borrowed from Rous). In 1938 Gross visited the National Cancer Institute, Yale Medical School, the Rockefeller Institute for Medical Research, and other American institutions in an unsuccessful effort to find a position to continue his studies. Frustrated, he returned to Krakow, only to flee to Romania after the Nazi invasion of Poland in September 1939. As Nazi power spread across Europe, Gross fled further to Italy, France, and finally the United States. He managed to leave France only two weeks before Paris fell, but not without retrieving a few samples of the tumor tissue he had studied for years from the Pasteur Institute. In 1943 Gross gained American citizenship and entered the Army Medical Corps.[70]

Over the course of the Second World War, Gross became an increasingly vocal proponent of viral carcinogenesis. He reinterpreted hereditary data on the appearance of cancer in alternating generations of families (both mouse and human) as evidence of a "temporarily latent factor" transmitted in the milk of nursing mothers, "which is probably a virus." Citing Bittner's work on foster nursing in mice, Gross warned that cancer was a "vertical epidemic" and that "persons with demonstrable tumors represent but a fraction of those carrying the disease." While acknowledging that experimental evidence was slight and "speaking with all reserve," Gross concluded, "There is reason to anticipate that the incidence of certain tumors, at least, such as breast cancer in man, could be substantially reduced if women with families with any tumors in their ancestry were to refrain from nursing . . . *from birth*."[71] Gross had no reservations about taking his own advice to heart. When his daughter was born in 1944, he was adamant that she be fed formula rather than breast milk, an apparent effort to curtail the transmission of any hidden virus.[72]

At the end of the war, Gross obtained a permanent position at the Cancer Research Unit of the Bronx Veterans Administration hospital. However, he found it difficult to continue his studies of breast cancer transmission: there were no strains of mice with an especially low incidence of this cancer—a prerequisite for demonstrating the transmission of a breast cancer virus. Leukemia, however, presented a more promising target. Gross possessed one purebred strain of mice, "C3H," that rarely developed leukemia, and another, "AK," that often did. He established breeding colonies of each in a small room (an upgrade over his car trunk, which had hosted earlier colonies) and set out to inoculate adult C3H mice with filtrates of AK tumors. In the ensuing four years of research, Gross enjoyed little success. The inoculated mice did not develop leukemia. Gross could carry out only a limited number of experiments because of uncertainty over when leukemia would emerge in his mice, if at all. Each generation of mice had to be observed until death (about eighteen to twenty-four months), and the cost of keeping these mice alive strained his modest resources. A promising new approach arrived in 1950. Gross attended a lecture on the coxsackie virus family, where he learned that that virus produced paralysis in mice, but only if they were inoculated before they were forty-eight hours old, when their immune systems were not yet developed. He repeated his experiments with cells from AK mice and recently born C3H mice and found that a high fraction developed leukemia by their "middle age" of eight to eleven months.[73]

Gross struggled with the weak activity of the leukemia agent he derived from filtrates, as the relatively low occurrence of cancer attested. His results still required considerable confidence in the idea of viral transmission to be plausible. Like Rous's results, his findings encountered critical reception from oncologists at New York's Memorial Hospital.[74] However, in other quarters Gross found more interest. The New York Academy of Sciences invited him to speak at a conference on cancer viruses in 1952.[75] His work gained wider exposure as well; the *Journal of the American Medical Association* published sympathetic accounts of his results in both 1951 and 1953.[76] Crucially, several other sets of investigators took up the idea of searching for viruses in infant mice, revealing a growing number of murine leukemia viruses.[77] Tumor virus studies received further impetus in 1958, when two researchers at the National Cancer Institute seeking to replicate Gross's results, Bernice Eddy and Sara Stewart, found a virus named polyoma that was capable of producing tumors in a wide variety of adult rodent species.[78] Adding to this excitement, in 1958 Charlotte

Friend, a virologist at the Sloan-Kettering Institute, announced that she had not only isolated a mouse leukemia virus but also had developed a crude vaccine against the virus, suggesting a promising model for treating human leukemia.[79]

The Nobel Prize–winning virologist Wendell Stanley assured the Third National Cancer Conference, "Viruses . . . provide [a] rational experimental approach to the human cancer problem."[80] However, Stanley was not a specialist in cancer. His declaration, "The experimental evidence is consistent with the idea that viruses are the etiological agents of most, if not all, cancer, including cancer in man," was grounded not in experimental work that he carried out in his University of California, Berkeley laboratory but in his sense that such viruses could be identified.[81] In fact, Stanley's embrace of cancer virus research reflected the aftermath of the deployment of the Salk vaccine against polio in 1955, when public concern for polio dwindled and the virologists trained by the National Foundation for Infantile Paralysis sought other projects and funding sources.[82]

Cancer in the "Virus Age"

The science journalist Greer Williams wrote in 1959, "The microbe age is past. . . . Today we live in the virus age."[83] At the close of the decade, the question of human cancer viruses and their prevention by vaccination could now be considered in the light of the existence of numerous animal tumor viruses and the successful experience of developing vaccines against other viral diseases in humans, particularly polio. The number of tumor viruses observed in rodents, frogs, and birds was so great that when Gross published a textbook for the field in 1960, *Oncogenic Viruses*, it ran to 393 pages. The second edition, published in 1970, had 991 pages.[84] The American Cancer Society, the successor to the American Society for the Control of Cancer, dropped its previously circumspect attitude toward viral theories of cancer, writing in a 1960 pamphlet, "A virus has been isolated from leukemia mice and a vaccine has been developed which protects mice against the virus. Although this does not apply to man as yet, it offers a promising avenue of research."[85]

Unlike the skepticism of two generations earlier, widespread belief in viruses as molecular entities that might cause disease prevailed. However, the process of molecularizing cancer through virus studies was uneven. Even as the effort advanced with new technologies and new viruses, there

FIGURE 2.1. "Unauthorized" advertising appearing in western Pennsylvania in 1963, preserved in the correspondence of the National Foundation for Infantile Paralysis and the American Cancer Society. The hope for vaccination against cancer was buoyed by the advertising and publicity around the polio vaccine. Image courtesy of March of Dimes Archives.

were still many skeptics. Presented with images claiming to show viruses associated with human leukemia, a virologist pointedly retorted: "Particles are unlabeled in the electron microscope."[86] Many clinicians still did not trust reports of cancer viruses, deeming them inconsistent with their own experience. Even the newfound ubiquity of viruses created challenges. According to one scientist, an invited speaker at a tumor virus symposium, "The considerable number of nononcogenic viruses in man, which might be present by pure coincidence in some cancer cells, complicate" the task of linking tumors to viruses "tremendously."[87] Even cancer virologists reviewing the literature were prone to make comments such as "Out of a massive flow of words and ideas it is difficult to synthesize a coherent picture and to sift out the meaningful from the imaginary."[88]

Troublingly, it appeared as if the standards of causation developed in bacteriology might never prove a link between viruses and human cancer. Prominent virologists in the polio vaccination effort now seeking carcinogenic viruses, such as Albert Sabin, found the experimental hurdles to demonstrating the existence of cancer viruses by conventional methods nearly insurmountable. Even if the analogy between murine and human leukemia held, Sabin cautioned, "final proof" of a link, following Koch's postulates, required transferring the virus to "newborns" to see if they developed leukemia. This step to show that "virus particles in tissue culture cells actually cause leukemia in man obviously is not possible," he concluded.[89] In light of these ongoing uncertainties, new laboratory findings only went so far in resolving the question of how to confront cancer as a viral disease. Probing the molecular roots of contagious cancer said little about who would become responsible for making the therapeutic promise of cancer research a reality.

CHAPTER THREE

Policymakers and Philanthropists Define the Cancer Problem

In 1925, citizens of the Commonwealth of Massachusetts were alarmed to learn from new census data that their state's cancer death rate was the highest in the nation. In response, the Massachusetts House of Representatives considered an ambitious bill for a statewide system of dedicated cancer hospitals. Such hospitals, which would be the first of their kind in the United States, would not only treat cancer, but also track the incidence of cancer throughout the state. For this system to function, the house proposed a law expanding the state's framework for the surveillance of infectious diseases to include cancer, designating cancer a legally "reportable" disease alongside measles, influenza, and other hazards to public health. Such improved data, in conjunction with the new hospitals, would provide the basis for the state's cancer control effort.[1]

However, the doctors of Massachusetts united against the plan. Unlike epidemic diseases, they argued, cancer was not a concrete "clinical entity." State-mandated cancer reporting would only serve "statistical purposes" rather than aiding public health efforts. Since cancer was incurable in its advanced stages, collecting statistics would be little more than "a brief anticipation of the death returns." It was better, the doctors concluded, for measures against cancer to continue on a "voluntary" basis.[2] The proposed system was modified to defend against charges of "state medicine" by requesting approval from local physicians before opening clinics and dropping the reporting requirement. However, without the ability to know where cases of cancer occurred, the original ambition of the system was reduced to a single clinic on the outskirts of Boston.[3]

The experience of Massachusetts highlights some of the principal obstacles facing those who wished to expand the government's role in addressing cancer throughout the nation. Cancer did not fit easily within the system of government public health measures developed to confront infectious diseases.[4] The focus of the US Public Health Service, established in 1887, was the prevention of epidemic disease threats from outside the nation, exemplified by its surveillance and quarantine of sailors and recent immigrants at Staten Island's Marine Hospital or Angel Island's Immigration Station. Only after the Spanish-American War did the organization turn its attention to domestic public health, and in this sphere it struggled until the First World War. Amid its expansion, the Public Health Service maintained its focus on acute infectious diseases such as plague, malaria, yellow fever, and typhoid.[5] That the most detailed source of statistics on cancer—a prerequisite for any public health intervention—came from private life insurance companies was a sign of the limits of state capacity to confront cancer.[6] Given this context and the reluctance of doctors to accept infringement on their professional autonomy in all but the greatest of emergencies, it seemed unlikely that the federal government would come to bear responsibility for addressing the cancer problem as its role in other parts of American society grew during the New Deal, even after the creation of the National Cancer Institute in 1937.

As in the case of other social welfare activities, the eventual mobilization of the American state against cancer initially followed the avenues defined by a "submerged" sector of philanthropies and professional organizations rather than carving out new pathways into the problem.[7] Federal approaches to the cancer problem at first hewed to the structures of the "associative state," a way of making policy developed in the 1920s that viewed the government as a coordinator among private groups established to address particular issues rather than as an independent actor in its own right.[8] The cancer problem was defined by a combination of shifting ideas about the nature of the disease and about the appropriate organizational relationships among state agencies, philanthropies, and doctors. Only after changes on both of these registers did large-scale federal intervention against cancer enter the realm of political possibility. Chapter 4 discusses the process by which cancer-control activists moved away from the associative state as they encouraged federal anticancer efforts after the Second World War; this chapter traces the actors and ideas that sustained resistance toward federal cancer control efforts in the first half of the twentieth century.

Cancer and the Social Question

At the opening of the twentieth century, cancer fit uneasily between the power of state public health authorities to fight epidemic diseases and the work of voluntary social welfare organizations. Cancer's complexity stymied the usual mechanisms developed for public health threats. As the case of Massachusetts suggests, physicians generally did not accept it as a contagious disease, so its control and prevention by public health officials appeared unlikely, even as evidence of its rising incidence mounted. The long-term suffering and economic harm caused by cancer and other chronic diseases fell outside the scope of government and into the realm of social welfare organizations. This provided the inspiration for the formation of the American Society for the Control of Cancer. However, the doctors who guided these efforts were not convinced that cancer could be cured or that they should sacrifice their autonomy to provide cancer care. Those views limited the avenues of intervention that any organization—private or public—could follow against disease.

Care for those struck by debilitating long-term illnesses was the domain of voluntary social welfare organizations, some of the central institutions of the Progressive Era in US history. Those organizations proliferated during the economic depression of the 1890s, which starkly revealed the range of social problems intensified by industrialization and urbanization. The founders of such organizations highlighted the danger of instability if these problems were not resolved. Their efforts resulted in a growing network of voluntary philanthropic organizations, all of which sought to answer various components of what was known as the "social question." While these Progressive Era organizations differed in important respects, their sources of support tended to reflect the impulses of the emerging professional middle class and its "search for order," especially its faith in expertise and its enthusiasm for education as moral uplift.[9]

For such organizations, tuberculosis provided a pivot point from social work to what came to be called chronic diseases. Tuberculosis, or "consumption," was a leading cause of death in the nineteenth century and the focus of many public health campaigns. Like cancer, tuberculosis was a disease that killed slowly, creating a population of sick individuals in a state of convalescence—neither acutely sick nor capable of supporting themselves through work. But medical and scientific authorities were confident that the cause of tuberculosis, in contrast to that of cancer, had been identified. In 1882 Robert Koch famously articulated the foundational principles of infectious disease causation as he demonstrated the microbial cause of tuberculosis. Belief in tuberculosis microbes inspired new public health efforts to separate individuals with the disease from the general population or to prohibit behaviors (such as spitting) associated with its transmission.[10]

However, even after these discoveries, tuberculosis continued to be addressed as a problem of social welfare. It still appeared to be a disease that medicine could resolve through attention to individual constitutions rather than large-scale public health works. Infection with tuberculosis bacteria, many doctors were quick to point out, did not necessarily cause the development of tuberculosis as a disease; susceptibility to tuberculosis and recovery were specific to the individual, not the illness. Many anti-tuberculosis activists continued to argue that the causes of tuberculosis were to be found in social factors such as poor housing, nutrition, or behavior. Within the United States tuberculosis campaigners were more likely to be suspicious of biomedical interventions, such as Koch's tuberculin therapy, and in favor of social reform and welfare programs that focused on the living habits and conditions of those seen at risk of tuberculosis, such as the urban poor. In individual cities, those programs tended to foster cooperation between charities and doctors at sites such as tuberculosis dispensaries.[11]

When an alliance of urban reformers and physicians established the National Tuberculosis Association (NTA) in 1904, a new chapter began in medical philanthropy. The NTA aimed to address tuberculosis at a national level. However, it possessed neither a large membership nor wealthy patrons. The association therefore sought out new ways of supporting its work. In 1907, the Danish-American urban reformer Jacob Riis, whose six brothers had died of tuberculosis, published a glowing report of the Danish practice of selling Christmas stamps, or seals, to fund tuberculosis sanitariums. Although each seal represented only a small contribution, the sum of all these contributions provided impressive resources. In 1908 the Red Cross started to sell similar "Christmas Seals" in the United States on behalf of the NTA. Rather than relying on medical professionals or elite donors, the Christmas Seals campaign demonstrated that a voluntary organization could marshal substantial resources against disease by tapping the wealth of the public as a whole. After a brief lull in its activities during the First World War, the NTA managed to draw a half million volunteers as well as support from insurance companies, businesses, and labor groups. Reflecting skeptical attitudes toward tuberculin therapy and the local roots of its fund-raising efforts, the NTA directed most of the money raised through these measures to the treatment and social service efforts of its local chapters rather than to tuberculosis research.[12]

The success of the NTA galvanized philanthropic efforts against cancer. When the American Society for the Control of Cancer was founded in 1913, the *New York Times* reported that it aspired to "a campaign of education against cancer similar to that which has been carried on against tuberculosis."[13] The steady creation of hospitals dedicated to cancer treatment in the first decade of the twentieth century testified to growing local concern about cancer, but until the ASCC was formed, this had not taken national form.[14] Following the mold of other Progressive Era organizations, the ASCC framed its overall program as an attack on ignorance about cancer. Its motto, "Fight cancer with knowledge," sought to defuse the tension between hope and pessimism in presenting the cancer problem. Educating the public about cancer detection and treatment placed physicians at the center of work against cancer without jeopardizing their professional credibility by promising that the disease was curable. Within this frame, awareness of cancer and an individual's willingness to seek treatment were as much an element in avoiding death from cancer as the limited therapeutic offerings of the medical profession. By arguing for the need for individual self-monitoring and regular contact with doctors, the ASCC's public education efforts harmonized with the overall effort of groups such as the American Medical Association to extend the legitimacy of professional medicine.[15] These same motives, as chapter 1 explored, also contributed to the ASCC's unwillingness to discuss cancer as an infectious or contagious illness.

In line with other professional medical organizations, and unlike the NTA, the ASCC was unwilling to go further than providing information. It not only avoided research in the 1920s, as discussed in chapter 2, but it also hesitated to provide treatment. Since cancer treatment required the services of surgeons and radiologists often not found in normal medical practices, establishing clinics on the model of tuberculosis dispensaries appeared to be a reasonable extension of the ASCC's activities. Moreover, as its first president noted, because of the expense of cancer care, "a sure way to drive the public to the quacks who infest the country is to advise them to go to specialists whose charges are beyond their means." However, only a handful of cancer clinics operated in the United States, whereas there were 800 clinics for the treatment of sexually transmitted infections, 600 clinics for the treatment of tuberculosis, and 100 clinics for heart disease. The operation of clinics was certainly expensive compared to public education, but the primary reason for their limited growth, the ASCC grudgingly acknowledged, was that most doctors opposed the charitable provision of cancer care by the ASCC or any other organization.[16] Their opposition reflected the medical profession's hardening resistance to the work of any organization in the early twentieth century—governmental, corporate, or charitable—that attempted to infringe on the discretion of doctors to charge for care in their private practice as they saw fit.[17]

The threat of cancer, in the eyes of doctors, did not warrant abandoning their professional autonomy, even to organizations such as the ASCC. That conviction remained in the face of the crisis that the Great Depression marked for the voluntary philanthropic approach to social problems. Faced with rising need and declining support, many voluntary organizations turned to cooperative relationships with the federal government during the New Deal, accelerating the involvement of the federal government in areas of social and economic life that had previously been beyond its reach. However, the physician leadership of the ASCC viewed New Deal efforts, such as the use of Social Security funds to expand cancer care for the poor, as potential harbingers of socialism.[18] Uncommitted to understanding the causes of cancer and unable to enlist physicians in the provision of cancer treatment, the ASCC's efforts to "fight cancer with knowledge" were unlikely to foster the expansion of a federal role in either research or therapy.

The National Cancer Institute and the Associational Approach to Cancer

Whatever approach might be taken to addressing the disease, during the 1920s cancer emerged as a problem for national public health efforts. Stoked by the education efforts of the ASCC, the number of articles devoted to cancer in popular publications outstripped the ones on tuberculosis for the first time.[19] Municipal and state public health authorities started to take action addressing chronic illnesses, including cancer, but like their voluntary counterparts, they emphasized public education rather than research or treatment. Nor did these local efforts supply a pathway for federal intervention.[20] In 1922 the Public Health Service (PHS) established a modest laboratory outpost for cancer research at Harvard, but its operations were geographically and institutionally peripheral to PHS operations. It was only the extraordinary fear associated with cancer that prompted any action at all. Indeed, after establishing the laboratory, the PHS discovered that it did not possess the legal authority to share laboratory space with a university![21]

However, calls mounted for a dedicated federal response to cancer. In 1928, Senator Matthew Neely (D-WVa) issued a call to action. Relying on statistics from the ASCC, Neely warned that the challenge to public health posed by cancer was of a different kind than that of the infectious diseases that had been targeted by the government's previous public health campaigns. It was nonetheless a challenge that the government should rise to meet. "Medical science has conquered yellow fever, diphtheria, typhoid, and small pox. Medical science has robbed even leprosy and tuberculosis of their terrors. But in spite of all that . . . cancer remains the unconquered, the unconquerable, and defiant foe of the human race." Neely exhorted Congress to increase federal spending on cancer from the "ridiculously inadequate" annual sum of four hundred thousand dollars to a more robust $5 million a year.[22] However, Neely's rhetoric and statistics did not move his colleagues. His bill failed to pass, and Neely lost his reelection campaign that fall, leaving federal cancer research with no strong advocate in Congress.[23]

Neely's campaign did manage to raise the political profile of cancer. During the 1928 elections, the Democratic Party platform endorsed doing "all things possible" to "ascertain preventive means and remedies for . . . diseases, such as cancer [and] infantile paralysis" that had "largely defied the skill of physicians." In his 1929 inaugural address, Republican president Herbert Hoover endorsed the principle that the "Public Health Service should be as fully organized and as universally incorporated into our governmental system as is public education."[24] In 1930, building on its success in controlling yellow fever in the American South, advocates for the PHS proposed establishing a National Institute of Health as a federal center for research against disease.[25] Although the bill envisioned this institute as a center for research on infectious diseases, discussion of the proposed institute in Congress very quickly turned to cancer. One witness, a respected surgeon from Johns Hopkins University, surprised his audience when he asserted that there had been no progress against cancer in the previous two generations and urged further inquiry into the nature of the disease. "We will never wipe out cancer until we have found a cause of cancer," the surgeon concluded.[26]

With the creation of the National Institute of Health in 1930 and the expansion of federal intervention in social and economic life under the aegis of New Deal liberalism, calls mounted for an institute specifically dedicated to cancer. In 1937 several different proposals in Congress merged into a plan to establish the National Cancer Institute (NCI).[27] According to Thomas Parran, the surgeon general, the creation of the NCI marked an important shift from infectious diseases to chronic diseases as problems for public health and welfare. Since 1900, he reported, deaths from tuberculosis and other infectious diseases had decreased by 80 percent, but cases of cancer were on the rise. "It is inevitable, therefore, that public health should concern itself more with these chronic diseases. . . . Any disease, whether communicable or not, which has a wide prevalence . . . which is costly to treat, is a proper concern to the public health. Cancer fulfills all those definitions."[28]

Although the NCI emerged in the midst of the New Deal, which is often understood as a moment of unprecedented federal expansion, its aims and organization recalled the more modest voluntary structures of the Progressive Era. The NCI was envisioned at the time as a small research operation located just outside Washington, DC, in Bethesda, Maryland, not as the massive organization that emerged later in the twentieth century. It did not seek an activist role in planning cancer research on a national scale but was structured to favor an "associative" relationship between private organizations and the state, a role embraced by the Republican Hoover and Warren G. Harding administrations in the 1920s.[29]

This relationship reflected the wariness that doctors and biologists at established research centers maintained toward state intervention. Cancer research, as an activity, was limited to a handful of hospitals and centers, largely in the northeastern United States, and most of these were controlled by oncologists and surgeons. Those professionals were very skeptical of government intervention. The oncologist and ASCC consultant James Ewing cautioned that the government should venture "no further into the field of pure cancer research," in "another futile effort to discover the ultimate cause of cancer." Articulating a rebuttal to the calls for federal anticancer mobilization heard often in the chambers of Congress for the rest of the century, Ewing maintained that a cure could not be "hastened by pouring money into the effort." Not only was the principle of addressing cancer as a public health problem through research flawed,

Ewing continued, but a large government research effort would not "find 20 men competent to take up cancer work. They do not exist, and the Government has no place to educate them." At best, Ewing thought that a central cancer hospital might be established near Washington, where the government could treat the limited number of "patients for whom it is responsible."[30]

The American Medical Association cautioned against the prospect of "putting the government in the dominant position in relation to medical research."[31] C. C. Little, director of the ASCC and head of the Jackson Laboratory of mouse genetics, urged Congress to focus the new institute's resources on infrastructure for research—specifically, "controlled animal material"—a subtle promotion of the mice his laboratory produced. Drawing on anti–New Deal rhetoric, he warned that the institute should not attempt to establish a "dictatorship" or "hamp[er] the direction" of cancer researchers. At best, the new institute could provide a loose form of "coordination," such as running an informational "clearing house" or hosting "informal meetings." Even this limited vision of government intervention marked a break with "older men in the field," Little warned.[32] Peyton Rous, who maintained an interest in cancer research despite the frustration of his efforts to isolate a tumor virus, feared that government intervention would bring about "regimented direction" that favored industrial firms rather than universities. "The guns of the powerful laboratories run by the electric and engineering companies at once began to say Boom! Boom!" he wrote to a colleague.[33]

The final form of the NCI reflected the scientific and medical community's wariness of state intrusion. The NCI received guidance from a newly created body, the National Advisory Cancer Council, whose members were closely aligned with the ASCC.[34] Little, one of the council's founding members, reiterated his warning that its members should be wary of assigning too much power to "government agencies." Instead, the "coordination" of research should avoid the risk that federal efforts might "regiment" research and also should "convince the press that an effort to organize the [cancer] problem along established lines of mutually interested business concerns" was more desirable than government planning.[35] The council was unwilling to encourage work on particular topics by announcing grants; it preferred to await proposals from researchers.[36] The experts commissioned to advise the NCI on "fundamental research areas" cautioned that the "limitations of cancer therapy" required "a conservative and critical attitude" toward research into the nature of cancer

FIGURE 3.1. Meeting of the National Advisory Cancer Council, 1949. The council represented the approach to the cancer problem adopted by the associational state. These medical and professional experts were more likely to reflect the interests of their communities than to embrace aggressive federal intervention against cancer. Image courtesy of the US National Library of Medicine.

cells. "In any program of cancer research, patience and the adoption of a long-time point of view are absolutely essential."[37]

The Second World War appeared to expand the horizons of what the federal government could accomplish through its support of research. The Office of Scientific Research and Development coordinated the work of academic and industrial researchers through a central office. Individual efforts, such as the development of radar systems by MIT's Radiation Lab and the construction of the atomic bomb by the Manhattan Project, created vast organizations in which scientists served on cooperative, goal-oriented teams. Academic scientists were often stunned by the degree of material support that they could obtain for their research—instruments that would have occupied a year's budget became commonplace objects— and impressed with industry's ability to manage research. DuPont famously served as a major contractor for the Manhattan Project's uranium production facility at Oak Ridge, Tennessee.[38] In biology the mass production of penicillin, blood plasma, and vaccines provided biologists with a similar set of experiences.[39]

However, these experiences did not shift elite attitudes toward federal cancer research. The body of medical experts installed at the NCI were capable of deflecting even the most forceful demands for federal action. In 1946, 60 percent of Americans named cancer as their most feared disease—heart disease, at 5 percent, was a distant second. A majority also supported increased taxes to fund a national "Manhattan Project" to cure cancer.[40] Senator Neely, who had returned to Congress as a representative from West Virginia, proposed another "Mobilization of the World's Cancer Experts" that would provide $100 million to aid cancer research. Congressional backing seemed assured, and even members of the ASCC's successor organization, the American Cancer Society, expressed support. However, the surgeon general testified that the National Institutes of Health could spend only 2 percent of the proposed sum on research, given how "few trained persons in the country" were working on cancer.[41] While professing sympathy for the national interest in curing cancer, he maintained that unlike wartime engineering projects, anticancer research would be limited by the incomplete nature of "fundamental knowledge" about the disease.[42] The director of the NCI concurred that the institute could not expand its research to absorb the funding windfall. This opposition slowed and eventually defeated the bill in a thicket of procedural objections as the 1946 legislative session wound to a close.[43]

The defeat of Neely's second bill highlighted an emerging tension between those who sought to combat cancer and those seen as experts on the disease. For the elite members of the academic medical community who served as advisers to the federal government, the threat of a backlash from dashed public hopes seemed more dangerous than forgoing federal financial support. For these researchers, discussing the biological complexity of cancer was therefore not only a plea for more knowledge, but also a means to restrain the kinds of action that the government, or any other organization, might take in the name of curing cancer. As long as these experts remained the dominant source of advice to the government through their position on the National Advisory Cancer Council, their assiduous pessimism would limit options for federal intervention.

Chemotherapy, a New Frame for Cancer

Although the Second World War itself did not directly translate into a wider role for the federal government, it did instigate a shift in how some philanthropies understood the management of medical research. Progressive Era philanthropies had sought to foster scientific research through a rational management process. The National Research Council (founded as a joint effort of corporations and the National Academy of Sciences in 1916), the Carnegie Institute of Washington, and especially the Rockefeller Foundation sought to support and guide the growth of particular areas of academic research through means such as postdoctoral fellowships. These efforts tended to focus on efficiency in terms of making sure that money was spent well instead of on the urgency of solving any particular problem.[44] As a result, those efforts had remained deferential to the preferences and interests of individual scientists. The Second World War energized those who saw the potential of biomedical research organized on a more widespread and aggressive scale. Two campaigns, to develop a polio vaccine and to identify promising chemotherapies for cancer, illustrate how the template of wartime urgency started to shift expectations regarding the organization of peacetime biomedical research.

In the aftermath of the Second World War, radiation and surgery remained the principal orthodox treatments for cancer. Neither of these two areas of therapy were congenial to large-scale planning, and the clinical skill required for either one did not admit a leading role for research. The NCI had made one notable interventionist effort in cancer care after its establishment, spending half of its annual budget on radium for distribution to hospitals, but this effort was quickly curtailed by oncologists on the National Advisory Cancer Council.[45] However, during the late 1940s the emergence of a new means of treating cancer, chemotherapy, inspired hopes that cancer research could be managed and coordinated. Chemotherapy, the idea of synthesizing chemical compounds to treat disease, arose out of the nineteenth-century German dye industry, which produced numerous potential therapeutic agents derived from coal tar. The development of the anti-syphilis drug Salvarsan (arsphenamine) in 1909 inspired hope that similar "magic bullets" could target cancer and other diseases. However, none of the compounds tested in the 1920s or 1930s offered promising therapeutic results.[46]

The Second World War invigorated the prospects of cancer chemotherapy at a practical and a philosophical level. During the war, the Committee of Medical Research of the Office of Scientific Research and Development used contracts to speed the development of chemotherapies for several illnesses, most notably malaria and bacterial infections.[47] The most striking success of this federal effort was the mass production of the antibiotic penicillin, discovered by happenstance in 1928 when a microbiologist in England found that a mold was killing his bacteria cultures. Pharmaceutical companies built on these wartime federal programs, screening thousands of molds gathered from around the world to identify more antibiotic compounds. The rise of antibiotics, a "magic bullet" against infection, exerted a strong hold on the imagination of others planning biomedical research: major therapeutic advances could happen as a result of scale and serendipity even when a deep intellectual understanding of disease was absent.[48]

Cornelius P. Rhoads, former leader of the US Army's Chemical Warfare Service, helped forge critical connections between military studies and charitable cancer research. From his supervision of chemical warfare research, Rhoads was aware that researchers at Yale Medical School, working under secret contracts, had determined that nitrogen mustard, or "mustard gas," could produce startling, albeit temporary, remissions of leukemia and lymphoma.[49] Rhoads emerged from the war as a staunch advocate of the idea that cancer cells could be attacked by chemotherapy just as antibiotics attacked bacteria.[50] He gained the resources to pursue his ambition for large-scale cancer chemotherapy when he accepted an invitation from General Motors executive Albert Sloan to direct an institution with the intent of adapting industrial research strategies to cancer. While universities and philanthropies had eschewed research planning, American industry had not. Starting with railroads in the late nineteenth century, corporations developed a distinct "managerial class" to coordinate their complex operations across vast expanses of territory.[51] Westinghouse, DuPont, Bell Telephone, and General Electric, rather than relying on the genius of individual inventors or academic consultants, established laboratories to satisfy their need for technical expertise and innovation.[52]

In 1945 Rhoads became the director of the newly founded Sloan-Kettering Institute in New York. At a time when the NCI received five hundred thousand dollars annually, the Sloan-Kettering Institute opened with a $4 million endowment. Like General Motors, the institute was organized according to a problem-oriented, interdisciplinary divisional structure, not by conventional academic disciplines. The process of identifying and studying possible chemotherapy agents for cancer, as Rhoads advocated, was a project perfectly suited to the institute's scale and organizational ethos. Along with the Institute for Cancer Research in Philadelphia, the Sloan-Kettering Institute launched an ambitious effort to screen compounds as potential anticancer chemotherapy agents. However, few promising compounds emerged. Various candidates either produced very short remissions or proved too toxic for use in patients.[53]

The "screening" approach to chemotherapy using animal models, patterned on the success of screening for antibiotic compounds, also ran afoul of the emphasis that many academic biologists placed on developing "fundamental" knowledge of cancer that would ground a "rational" therapy in humans.[54] Critics pointedly noted that the "volume of research" discussed in the scientific and popular press ran the risk of creating a "feeling that successful control of malignant disease through chemotherapy was imminent," even though screening remained beset by "gaps" and "lags." Compounds identified as anticancer agents in animals appeared to have dubious clinical use.[55] "One cannot witness the kaleidoscopic appearance and disappearance" of potential compounds, an observer wrote, "without becoming pessimistic."[56] Nonetheless, interest in chemotherapy continued to grow.[57]

The successor to the ASCC, the American Cancer Society (ACS), struggled with how it would support chemotherapy work. Reflecting the views of its academic scientific advisers, the ACS remained committed to the view that cancer research was a "highly individualistic enterprise" and that "discoveries of great practical significance" were "more dependent upon the quality than upon the quantity of scientific investigators."[58] The industrial-scale screening of many compounds to gain prosaic information such as toxicity and dosage levels did not appear to meet its definition of research. Beyond the substantial expense associated with screening tens of thousands of compounds, the next stage of drug development, clinical trials, required support for hospital facilities. Here too, the ACS wrestled with how its research mission related to the sponsorship of clinical trials. Subsidizing the care of terminal cancer patients, a population of individuals who were the possible recruits for chemotherapy trials, remained controversial for the ACS's physician members.[59] Barring exceptional circumstances, the principal arm of the ACS research program—the Committee on Growth supervised by the National Research Council—declined to assume the costs associated with "hospital beds, nursing care, or related services."[60] For these reasons, screening for chemotherapy remained the domain of a limited number of institutions that integrated laboratory and hospital facilities, chiefly the Sloan-Kettering Institute. However, as is discussed in chapter 4, public expectations soon outstripped the capacity of these voluntary organizations and created new pressure for federal intervention.

Polio Vaccination and New Urgency for Biomedical Research

The development of the Salk polio vaccine by the National Foundation for Infantile Paralysis (NFIP) during the 1950s is an iconic medical breakthrough in the modern American imagination.[61] The polio vaccine represented a milestone for biomedical research, not only for what it did but for the way that the NFIP framed its fund-raising for research as a matter of combating childhood disease, which provided a potent template for future advocates of other biomedical research efforts. After a notorious outbreak among children in New York City, polio seized the national imagination in 1916. While the number of deaths that summertime polio outbreaks claimed was small compared to other threats to children's lives, such as pneumonia, the painful paralysis that it inflicted on survivors made polio a target of particular fear.[62]

The most famous survivor of polio, however, caught the disease in adulthood: future president Franklin Delano Roosevelt. After he became paralyzed, Roosevelt became involved in charitable efforts to provide rehabilitation services to polio survivors at a resort in Warm Springs, Georgia, a town known for its mineral springs. When he returned to political life, he turned leadership of these activities over to his law partner, Basil O'Connor. Both men sought to use Roosevelt's ascent to the presidency as a way to rally support for polio fund-raising. In what became an annual event, on January 29, 1933—Roosevelt's birthday—O'Connor organized thousands of "Birthday Balls" to raise money for the Warm Springs Foundation. Like the National Tuberculosis Association's campaign, this effort focused on the public as a whole rather than on a few wealthy donors. The President's Birthday Ball Commission used a full range of aggressive public relations techniques—including radio broadcasts, appeals by celebrities in movie theaters, and airplane barnstorming—to "sell" polio fund-raising not to affluent donors but to millions of modest ones. The success of this strategy was evident. Even as charitable giving slumped during the Great Depression, giving to the Birthday Balls rose.[63]

This fund-raising strategy, while initially successful, hit an impasse in the late 1930s that forced the polio campaign to take a new direction. Rising partisan rancor surrounding the New Deal blunted the fund-raising appeal of Roosevelt's image. Moreover, the main beneficiary of the Birthday Balls—the Warm Springs Foundation—operated on a social services model, helping children who visited the springs. Critics of Roosevelt and the Birthday Balls questioned whether a national organization was necessary, arguing that the money sent to Warm Springs might be better spent within local communities.[64] Beset by these challenges, O'Connor and Roosevelt chose to inaugurate a new strategy for medical philanthropy, establishing the NFIP in 1938. Under O'Connor's leadership, the NFIP cut its overt ties to Roosevelt and emphasized the research work that donations would support. From the scientific advisers of the Birthday Balls, O'Connor recruited a staff of respected virologists to supervise the NFIP's Committee on Scientific Research, which underwrote generous multiyear grants to a broad range of virus studies on the hope that they would give insight into the nature of polio.[65]

O'Connor refashioned the public relations strategies developed by the fund-raising arm of the NFIP, the March of Dimes. The advertisements run by the March of Dimes pioneered the use of unabashedly sentimental images of children and confidently promised its donors imminent break-throughs. The March of Dimes used images of children to underline the moral imperative of preserving childhood life, elevating the effort to cure polio to the level of what a later chronicler called "a *holy* quest." Buoyed by this campaign, the NFIP raised an average of $25 million annually between 1938 and 1962—far beyond what any medical philanthropy had raised before.[66]

This new fund-raising style created new tensions between the NFIP and the scientists that it supported. One virologist recalled, "Most of us were motivated mainly by curiosity and by the challenges of the many unsolved problems concerning the interaction of virus and host, rather than by the hope of a practical solution in our lifetime."[67] Yet the NFIP operated without a reserve endowment or wealthy donors capable of ameliorating a fund-raising shortfall. One unsuccessful fund-raising year risked stalling its research program as a whole. As an organization the NFIP "could not appear sluggish or over-cautious. It was trapped within its own image of dynamic optimism."[68] A memo to members of the NFIP Speaker's Bureau cautioned, "The word 'soon' can be variously interpreted; it could mean tomorrow or next year. In fact, we do not believe it will be either tomorrow or next year, but probably will entail at least several years." Therefore, "we must not mislead the public into thinking that all of these problems can be licked overnight."[69] However, a pamphlet published within the same year as this admonition foretold, "The conquest of polio is now in sight."[70] O'Connor became increasingly frustrated by the diverging aims of scientists and the NFIP. He scrawled in the margins of a report in 1945: "*No—No*—let's have a *new* philosophy of *doing* things in medicine. . . . Let [us] see how *quickly* we can do it . . . and not get lost on how we can *study* it. . . . We get money to spend—and from the people."[71]

To address the dissonance between the rhetoric of its successful fund-raising campaigns and the attitudes of its scientists, the NFIP launched a wholesale reorganization of its research efforts in 1946, under the

FIGURE 3.2. The "poster child" images of the March of Dimes established a new and powerful rhetoric of urgency around childhood disease research. Image courtesy of March of Dimes Archives.

leadership of a new director of research, Harry Weaver. Weaver was inspired by the success of the development of penicillin during the Second World War to refocus the NFIP's resources on the development of a vaccine, a task that had been abandoned by virologists after several public failures in the late 1930s. He wrote that the solution to the problem of polio required "group planning and the pooling of ideas and resources . . . a master plan which would permit a cheaper and quicker solution than is possible by the individual approach."[72]

Many grantees of the NFIP protested Weaver's restructuring of the research program. In the face of this dissent, Weaver did not waver in his ambition. Instead, he recruited younger, less established researchers who were willing to pursue the projects he deemed necessary. Famously, Weaver appointed Jonas Salk to oversee an expensive and repetitive effort to determine how many strains of polio virus existed, a critical step toward developing a vaccine. The testing and production of the Salk vaccine as a whole presented a monumental testament to the power of coordinated and well-funded biomedical research.[73]

While the case of polio might appear to rebut the argument that biomedical research could not be organized, it did nothing to create opportunities for increased federal intervention. Given its reliance on voluntary philanthropy, the NFIP was a steadfast opponent of federal intervention against polio. As an epidemic disease menacing the health of children, polio was an area of public health research meriting federal attention twice over. Within social welfare and public health policy, child welfare was one of the few areas where the federal government had a long tradition of action.[74] During the late 1940s and early 1950s, legislators introduced several bills seeking to use federal resources to fight polio. The NFIP opposed these bills, pointing to its own "vast research program" and averring, "There is no real need, or necessity" for federal intervention against polio. Federal involvement would do "great damage to the cause of battling this dread disease."[75] The foundation painted a grim portrait, redolent of the anticommunism of the early Cold War, of the active harm that federal involvement would bring about. Would volunteers "evince the same enthusiasm, the same interest—these 30,000 people . . . if the word goes out that the Government is taking over? . . . You have the Government taking care of the personal diseases of our people, a completely totalitarian idea that never was intended in America."[76]

The NFIP's stark rhetoric rested on the assumption, widely held before the Great Depression, that there was a zero-sum relationship between federal action and contributions to philanthropies.[77] A congressional representative sympathetic to the NFIP echoed its anxiety that government intervention would "interfere with the making available of large funds by private individuals and agencies." If the public thought an "all powerful Federal Government" would handle these efforts, then they would have no motive to continue to contribute to voluntary organizations.[78] Because of its strenuous lobbying, the NFIP's research effort vastly exceeded that of the federal government against polio, and the release of the Salk Vaccine in 1955 was a public relations triumph for the NFIP and the voluntary sector, not for federally supported biomedical research.[79]

Contesting the Rule of Experts

The structure of the associative state left cancer specialists as the gatekeepers for defining anticancer policy and, in a broader sense, in control of setting the boundaries of what progress society could hope for in combating cancer. In the 1920s, voluntary and professional organizations framed the cancer problem in ways that sought to dampen hope for a cure on the part of the public. Cancer was a feared disease, likely incurable unless detected very early. This message addressed what these organizations regarded as the primary danger posed by the disease: overzealous calls for intervention and challenges to the authority of the medical community. Thus, their gatekeeping status succeeded in deflecting or minimizing calls for federal mobilization against cancer.

The voluntary sector's understanding of the cancer problem played a critical role in constraining federal intervention, but its understanding was not static. Vaccination and chemotherapy produced changes in the understanding of cancer that opened new avenues for the organization of biomedical research and offered the possibility for a new federal role. However, growing optimism about biomedical solutions to the cancer problem did not yet suggest that the federal government should take a leading role in fostering these solutions. Many medical philanthropies, such as the NFIP, remained stalwart opponents of federal intervention in research. The growth of the NCI into the flagship organization for biomedical approaches to cancer required a further shift in how some ambitious medical philanthropists envisioned their relationship to the state, one driven by the frustration of their campaigns on behalf of national health insurance after the Second World War and the emergence of the biomedical settlement.

CHAPTER FOUR

The Biomedical Settlement and the
Federalization of the Cancer Problem

A t midcentury the restraints on federal participation in cancer research weakened and then broke. Twenty years after the National
Cancer Institute (NCI) was founded in 1937, its budget exceeded, and
soon thereafter vastly surpassed, the budget of any voluntary agency pursuing anticancer research, a status that it maintains to the present day.
The pressing question was no longer whether the federal government
would play a role in anticancer research, but how large that role would
be. However, the federal government did not mobilize against cancer because the established cancer control community shifted its stance. This
chapter explores how the federal government became involved in anticancer research despite the challenges and obstacles discussed in chapter 3.
It might seem that the American government's expanding role in cancer
research was part of its larger expansion as it confronted the twin challenges of the Great Depression and the Second World War. But it was
not. Health and medicine were one sector of society where the federal
government found its expansion checked during the New Deal and its
aftermath. Medical and business groups succeeded in defeating several
efforts to provide health security through a program of national health
insurance, most notably President Truman's effort during the "Fair Deal."
Even as confidence in the power of scientific research grew overall, the
surgeon general, the American Cancer Society, and prominent cancer researchers doubted that a large-scale federal campaign against cancer was
desirable or even possible.

The expansion of the NCI reflected the rise of new forces in the realms
of medicine and philanthropy. Though it followed a different path than

better-known efforts aimed at economic and social security, the growth of federal support for biomedical research raised similar questions regarding the role that the government would assume in American society in the second half of the twentieth century.[1] The growth of the NCI highlights the fact that the boundaries of state intervention in American life were not set at the end of the New Deal but underwent a continuous process of revision that played out through legislation, administrative politics, and backroom lobbying.[2] The federal government became involved in cancer research through a set of conflicts, negotiations, and compromises among a growing cast of participants from the public health, legislative, philanthropic, and medical communities.

Mary Woodward Lasker, an advertiser, art dealer, and New Dealer who became a powerful advocate for health research after the Second World War, was one of the most adroit of a new group of activists that sought to draw the government into the cancer problem. After the failure of Truman's national health insurance proposal, Lasker and others made a fateful decision to focus on biomedical research as the best way to use federal resources to protect and advance the health of the American people. Lasker's role outlines how shifts in the private sector facilitated the emergence of the "biomedical settlement," creating the conditions for federal efforts against cancer on a previously unimaginable scale. Whereas the "submerged" private sector had at first hindered the expansion of the federal government into cancer research, by the 1950s it served to promote it.[3]

Lasker's efforts on behalf of the NCI also demonstrate the political character of claims about the nature of cancer as a disease. The growth of the NCI took place over opposition from the very experts it aimed to support. Members of the clinical and scientific communities continued to describe cancer as complex and mysterious—and therefore a matter beyond state intervention. Faced with this pessimism, Lasker and her allies looked for new ways to frame cancer alongside new channels of political influence. Finding and highlighting ways to present cancer as a knowable, and curable, disease—particularly through the new field of chemotherapy for childhood cancer—provided political and moral leverage for federalizing cancer research. The consequence of this strategy, however, was a persistent tension between the way activists encouraged the public to understand the problem of cancer and the way cancer experts approached the disease. Lasker and her allies encouraged increasing federal spending on cancer research to discover a cure for the disease, while the recipients of this support did not envision that their biomedical inquiries would result in therapeutic progress. Although cancer viruses were not initially part of the biomedical settlement, it came to include them later in the 1950s, when vaccination promised a means of addressing this growing tension between scientific inquiry and therapeutic goals.

Mary Lasker and the Operation of the Biomedical Settlement

While federal cancer spending benefited from growing public concern for cancer, public concern alone could not account for the expanding role of the federal government in light of the political headwinds against federal intervention in both health and research after the Second World War. Rather, this dramatic expansion owed its origins to the creation by activists and legislators of a new political culture around cancer research. Confronting the pessimistic stance of prevailing medical opinion toward the potential of research, advocates for federal involvement developed an alternative set of experts and provided them with political access to shape cancer research on a national scale.[4] This new culture was not the product of a mass movement, but the work of a small group of advocates with a commitment to a vision of federal cancer research initially not shared by legislators or cancer experts. Following Lasker's political and philanthropic activities helps illuminate how expertise and power were drawn together in new ways to shape federal cancer policy after the Second World War.

Born Mary Woodward in Watertown, Wisconsin, in 1900, Lasker graduated from Radcliffe and studied art at Oxford. Arriving in New York in 1926, she established herself as a successful gallery owner and later as a designer of textiles. Through her work, she met Albert Lasker, a wealthy and well-connected advertising executive best known for his work for the Lucky Strike cigarette company, such as the notorious "Reach for a Lucky instead of a Sweet" campaign.[5] Mary and Albert married in 1940. The two shared a commitment to party politics, although for different parties. The summer after their wedding Albert and Mary celebrated with a "honeymoon of conventions." Mary, an ardent New Dealer, was an advocate for national health insurance and birth control. Albert was a moderate Republican who had provided public relations advice to Warren Harding's 1920 presidential campaign.[6]

Whatever their political differences, Mary and Albert both had great enthusiasm for the potential of medical research to improve human health. Mary attributed her interest to her own family's experience with illness.

FIGURE 4.1. Mary Woodward Lasker (1900–1994), about 1940. Lasker was one of the most significant activists on behalf of biomedical research after the Second World War. Photograph by Dorothy Wilding. Courtesy of John Hustler and the UK National Portrait Gallery.

Both were confounded by the general indifference they encountered when they spoke to doctors around New York about ways of increasing the quantity of medical research. In 1942 they established the Albert and Mary Lasker Foundation, a philanthropic effort to encourage biomedical research.[7] After the death of a member of her household staff from cancer in 1943,

Lasker sought to understand why doctors had been unable to offer better treatment. She was struck by a claim made by a member of the American Society for the Control of Cancer that with more money—around five hundred thousand dollars annually—a cure for cancer would be possible. While this figure roughly matched the annual budget for the NCI at the time, it seemed to Lasker that the sum was small in comparison to the good that it might accomplish.[8]

Lasker recalled in later years that as she became an active fund-raiser for the ASCC, she was stunned to learn that "they had been in business for 36 years and not raised a cent for research." Such indifference to research reflected the dominance of the ASCC by surgeons, who saw little use for biomedical approaches to cancer. Albert and Mary used their own fortune to subsidize the ASCC's fund-raising efforts—extracting a promise from the society that it would devote one-quarter of the funds raised to research activities.[9] Despite these steps, the Laskers were concerned that the ASCC would remain too restrained in its anticancer efforts. In 1944 they conspired with their allies to wrest leadership of the ASCC from physicians and install health activists drawn from the New York business community on the board of the renamed American Cancer Society (ACS).[10] The ACS placed research at the center of its annual fund-raising drives, invoking the new motto "research and progress" instead of "fight cancer with knowledge."[11] While the ACS public affairs staff acknowledged that many of the stories featured in these campaigns held no "direct bearing upon cancer in humans," the ACS saw very real value, supported by rising donations, in telling the public about the "minor and major battles waged in winning" the "war on cancer."[12]

Refocusing the mission of the ACS on research also drew power away from local chapters of the organization, which were under the leadership of doctors, and toward the central leadership, which was aligned with the Laskers. Funds raised for cancer education or social services were often distributed locally, but the ACS channeled the evaluation of research grants through its national office in New York, which claimed the scientific expertise necessary to review the merits of these projects. The embodiment of this impulse was the creation of the "Committee on Growth" in 1946, an initiative funded by the ACS and supervised by members of the National Research Council of the National Academy of Sciences to distribute financial support for long-term research on the nature of cancer among the "basic sciences."[13]

As the Laskers remade voluntary cancer philanthropy, Mary was also at work as an advocate for national cancer policy. Her choices as an activist illustrate the important role that philanthropic advocacy played in forming the biomedical settlement. The mobilization of the federal government in anticancer research during these years was not the result of popular pressure as much as of the efforts of activists to outmaneuver the influence that the medical profession wielded in Washington's political arena. Mary's marriage to Albert brought her into contact with other well-connected individuals who shared her New Deal views, such as Florence Mahoney, the wife of a Miami newspaper publisher with connections to several senators. These connections provided the nucleus of Mary's first lobbying efforts in Washington. Through a political ally of Mahoney, Democratic Florida senator Claude Pepper, Lasker became one of the advocates of Representative Matthew Neely's "Manhattan Project" bill to cure cancer in 1946.[14]

Just after this bill failed to pass Mary's most ambitious legislative cause, national health insurance, also floundered. For scholars of American political history, the seating of the 80th Congress in 1947, after the Republican Party succeeded in regaining control of the House of Representatives in the elections of 1946, is notorious for its efforts to diminish or reverse the most ambitious elements of the New Deal amid the anticommunist political environment of the early Cold War. This Congress, with the urging of the American Medical Association, helped defeat President Truman's proposal for national health insurance and placed a check on the further growth of the New Deal's social welfare agenda. It appeared likely that the attitude of the National Foundation for Infantile Paralysis and other voluntary medical philanthropies would succeed in erecting a permanent barrier to the expansion of the federal government into biomedical research.[15] The 80th Congress also witnessed the beginning of contentious debates over a "National Science Foundation," pitting the plans of New Deal Democratic senator Henry Kilgore for state-directed scientific planning against a much more restrained organization envisioned by Roosevelt's science adviser Vannevar Bush and his conservative allies in Congress. The report that Bush produced to buttress his position, *Science: The Endless Frontier* (1945), subsequently became a touchstone for advocates of the position that the federal government should support "basic" research—but not direct it.[16]

These setbacks did not dissuade Mary from continuing to pursue "a really big scale dynamic research program," but they did cause her to adjust her focus.[17] Mary's new lobbying effort began from the premise that if the government would not provide health security directly through insurance, then it could indirectly promote health by sponsoring medical advances. In this effort, she was moved by Albert's earlier service as a member of the US Shipping Board during the First World War and the enduring impression that the power of "federal money" had left upon him.[18] Unlike the leadership of the NFIP, Mary envisioned the federal government not as a competitor in fund-raising efforts but as a supplier of research and training services, while the ACS would focus on public education and innovative science. "There's so much to be done in the field of cancer," an ACS annual report commented, "that there is no room to quibble over the line between the responsibility of government and enterprise."[19]

Lasker turned her resources and energy to understanding and mastering the levers of power that moved medical research policy in the nation's capital. As a Democratic donor and activist, she was already aligned with the party that controlled both chambers of Congress, with a few lapses, from the 1930s through the 1990s. Building on these connections, Lasker retained an experienced lobbyist, Colonel Luke Quinn, as her personal representative in Washington. Their correspondence gives a sense of the subtle routes of influence that Lasker developed within the machinery of government in the service of her effort to expand support for cancer research.[20] Lasker often sought to use her social connections in New York as a source of influence. In one instance, Quinn wrote to inform Lasker that the new assistant director of the Budget Office might be within the social orbit of one of her friends, given his previous position at the National City Bank in New York. The budget process was complete for that year, but Quinn hoped to "cultivate him at as early a date as possible" so as to be ready for the next round of appropriations.[21]

When issues related to cancer appeared in other hearings, such as the possible carcinogenicity of some dyes used in food, Quinn sought to find witnesses sympathetic to cancer research to appear before the committees involved.[22] Lasker could also be more direct. Enclosed with a one-thousand-dollar donation to one senatorial candidate was her note, "I understand . . . that you have always been sympathetic to Federal appropriation for federal research in cancer. . . . I do hope you will be elected to the Senate and will be helpful to the various voluntary groups who are dedicated to assuring funds are available to keep people alive and well."[23] Friendly legislators also provided Lasker with invitations to social events in Washington where she could speak with administrators in the NCI and the National Institutes of Health.[24]

Lasker soon determined that the organization of the House and the Senate worked in favor of her political connections. Congressional rules during these years gave the leaders of committees and subcommittees broad discretion over the progress of legislation and the setting of budgets for federal agencies. The defeat of Neely's bill to cure cancer, or of initial efforts to establish the National Science Foundation in the late 1940s, suggested that seeking to increase federal cancer research through stand-alone legislation was risky. A lone opponent could derail a bill's progress through the House or the Senate. Yet the power of individual legislators in the appropriations process could work to Lasker's advantage. A few well-placed allies in the governing party could direct substantial support to favored projects, especially in an area with few other well-established constituencies.[25] With the cooperation of these allies, Lasker sought to increase federal support for cancer research not through new legislation but through the budget process. She soon fashioned alliances with Representative John Fogarty (D-RI) and Senator Lister Hill (D-AL), who rose to chair the committees overseeing the Public Health Service by the early 1950s. Both of these legislators might have already been inclined to back federal funding for health research, but their contact with Lasker gave them the impetus and scientific backing to dramatically increase appropriations for the NCI in the 1950s.[26]

While Lasker's efforts have long been acknowledged by political observers as critical to the NCI's growth, they have often been unaware of the labor that went into the moments of public testimony. Her preparations underline the importance not only of political influence, but also of new cultures of expertise for allowing the federal government to enter cancer research. At the beginning of Lasker's lobbying campaign, the cancer experts who traveled to Washington on behalf of the ASCC or the American Medical Association (AMA) had been uniformly skeptical of an expanded role for the government. Lasker responded not by trying to change the opinion of these experts, but by recruiting a new cohort of specialists to lobby Congress on biomedical issues that arose during the appropriations process. Months before the process started, Lasker and Quinn started to discuss possible witnesses, with an eye toward making their testimony as engaging as possible. The aim of presenting these witnesses in early hearings, held in subcommittees under the eyes of friendly legislators, was to "put ammunition in the hands" of allies seeking increased funding of the NIH and NCI during general debate on the floor of Congress.[27] Before one hearing, Quinn asked Lasker for her opinion on a lineup of doctors, commenting that the speakers the previous year had been "a little bit dry."[28] Reporting on the appropriations hearings in 1955, Quinn praised their staging. Senators and witnesses worked together to "underscore" remarks for the record, even though one of the doctors was "somewhat long and tiring," spending too much time on a discussion of "research itself."[29] Quinn was especially pleased with one physician-witness, who had brought along several live rats and mice, creating an effect that was "quite dramatic."[30]

NIH administrators sympathetic to Lasker's goal, such as longtime NIH director James Shannon, also became a part of the theater. In prepared exchanges, legislators would appear to intensively question administrators about whether the "official" budget adequately supported cancer research. The answer from the administrators was invariably that more money could be usefully spent, and the committee would have its evidence to justify increasing the budget of the NIH.[31] By the mid-1950s, Lasker exercised substantial influence over both the expert advice supplied to Congress about cancer and the machinery of appropriations for federal spending on biomedical research. In Quinn's eyes, Lasker's efforts could occasionally be too successful. On the eve of the appropriations process in 1954, Quinn groused that the Public Health Service had submitted an unusually low official budget because, he said, "it is obvious that the PHS counts on us to get increases."[32]

Lasker also benefited from the indifference toward cancer research on the part of other equally politically savvy constituencies. Since cancer research did not appear to impinge on the economic prerogatives of doctors, the AMA, preoccupied with the threat of national health insurance, remained neutral on its expansion.[33] The cancer research community itself had limited means of advancing its interests on the national stage, and Lasker controlled the ACS, the most prominent of these. The absence of other constituencies magnified the power of Lasker's new experts. In one instance, a skeptical legislator dispatched a staff member to check some of the statistics Lasker presented to him during a meeting, only to find that the only other sources of information were from organizations allied with Lasker.[34] However, the power of Lasker's alliance was still limited by the broader antistatist views of doctors. Once, when she could not muster ACS support for a hospital construction bill, one of her oncologist allies wrote to her of his regret that the physician "rank and file" of the ACS remained in agreement with the AMA's opposition.[35]

Lasker and the ACS proved, nonetheless, that cooperation between philanthropy and the federal government could increase the resources available to both organizations.[36] The ACS became the first medical philanthropy to adopt the relationship to the federal government that other social welfare groups had embraced during the Great Depression. While the budget for research at the ACS continued to grow, it was rapidly surpassed by appropriations to the NCI. In 1949 the budgets of the two organizations had been comparable, but by 1959 the NCI's budget was double that of the ACS, and they diverged even more dramatically in the 1960s.[37]

Cancer as a Children's Disease

With enthusiasm for scientific research rising, the cancer research community had ample opportunity to emphasize the incremental nature of their work as they sought to temper hope for a cure. These customary responses to public demands for a cancer cure, however, were rooted in a set of assumptions about the populations that cancer struck. The statistics collected and the educational programs produced during the 1920s and 1930s encouraged the assumption that cancer struck adults. Until the late 1940s, the assumed victim of cancer was an adult, more often than not a middle-class, educated, white woman. The gynecological and breast cancers that struck this population were among the cancers that were relatively amenable to surgical intervention, and the ASCC had shaped its education campaigns accordingly. Its message—early detection, self-monitoring, and seeking out medical authority—reflected the assumption that fearful women were reluctant to talk to their doctors about possible cancer symptoms.[38] Meanwhile, the causes of cancer were often sought in mood, diet, or other behaviors—doctors continued to minimize the potential of heredity and contagion to cause cancer. The sum total of these messages was that, though cancer might be a feared disease capable of causing great suffering and death, the individuals affected by cancer bore responsibility for the course of the illness by virtue of their behavior.

    After the Second World War, a new set of associations arose between cancer and children. Before this decade, the possibility that cancer would become visible as a children's disease seemed slim. Childhood cancer cases, which compounded the general taboo associated with childhood death, did not appear in the education materials of the ASCC. Initially, this silence also reflected a genuine lack of awareness as to how many cases of cancer existed among children. With the decline of other childhood diseases in the 1920s, cases of cancer became more noticeable among the causes of death in childhood. By the late 1930s, childhood cancers started to draw more attention at a few hospitals. Memorial Hospital in New York began to maintain a separate registry of childhood cancer cases, which formed the basis of *Cancer in Childhood* (1940), the first textbook on pediatric cancer. The connection between children and cancer gained further publicity with the publication of *Death Be Not Proud* (1947), a memoir by the noted journalist John Gunther of his young son, who died of brain cancer. Media coverage captured the realization that cancer might be a disease of the young as well as of adults. "Cancer Kills Children Too!" announced an article in the *Women's Home Companion*.[39]

The ACS and Memorial Hospital cautiously started to feature children with cancer as a part of their fund-raising appeals in the 1940s. As with polio, the public relations staff of these groups faced the dilemma of balancing hope for progress with therapeutic pessimism. Physicians and surgeons could promise very little in terms of improved treatment for the featured children. With no breakthrough treatment to offer, such appeals emphasized the duty of mothers to detect the early warning signs of cancer in their children, much as they should in themselves.[40]

The prospects of childhood cancer treatment improved dramatically in 1948, when Sidney Farber, the chief pathologist at the Children's Hospital in Boston, reported that he had produced dramatic remissions in children with leukemia by treating them with a set of compounds known as antifolates. Leukemia itself was an unlikely site for promising news about cancer treatment. Unlike tumor-forming cancers of tissues or organs, leukemia involves the rapid growth of immature white blood cells in the blood stream, which thicken the blood, causing pain, fever, and death. Given its association with blood, oncologists and hematologists had only agreed on leukemia's identity as a cancer just before Farber's first results. Moreover, leukemia was especially challenging to diagnose in children, where the common form of leukemia, acute lymphatic leukemia, could easily remain undiagnosed during its rapid and fatal course—it was often misdiagnosed as fever. Even if leukemia was correctly diagnosed, the two major therapies for cancer, surgery and radiation, were of little use because they were designed to attack localized cancers. The same property of leukemia that made it difficult to treat—its spread throughout the blood rather than its localization in compact tumors—also made this form of cancer particularly susceptible to chemotherapy, which also distributed compounds throughout the body.[41] Although the remissions

FIGURE 4.2. The fund-raising activities of the "Jimmy Fund" on behalf of Sidney Farber's leukemia chemotherapy research deliberately quoted the visual rhetoric of the child-centered advertisements pioneered by the March of Dimes. Note the iconography on this coin-collection case, which features a wheel-chair-bound child wistfully watching baseball. Image courtesy of Dana Farber Cancer Institute.

Farber attained were measured in months, the occurrence of any remission at all inspired immense excitement—these were not only the first remissions reported for leukemia, but were also among the first remissions from chemotherapy for any kind of cancer. Farber's success in Boston provided new impetus for all chemotherapy efforts.[42]

Farber's decision to embrace the link between children and cancer arose from the challenges he faced at Children's Hospital. As a pathologist he was an outsider to the oncology community, which was dominated by surgeons, the professional group most likely to view chemotherapy with skepticism. Farber responded to this skepticism by reaching outside the hospital for support. He fashioned an alliance with several charities committed to child welfare writ large to create the Children's Cancer Research Foundation (CCRF). In doing so, Farber focused on a particular subpopulation of people at risk of leukemia, which according to contemporary epidemiology was a disease that largely struck adults.[43] The fund-raising campaigns of the CCRF drew heavily on the iconography pioneered by the March of Dimes and, in a departure from earlier childhood cancer publicity, left no doubt that cancer should be thought of as a dread— and curable—disease in children. Famously, the CCRF introduced a child known as "Jimmy." Unlike most treated children, Jimmy had seen his leukemia go into complete remission following chemotherapy. An all-American name meshed with the venues where the CCRF distributed its appeal. At baseball games, in movie theaters, and on radio broadcasts, the Jimmy campaign created an enduring icon of hope against cancer that cemented both the presence of cancer as a threat to children and the hope of a cure. This appeal was enormously successful. In four years the CCRF was able to finance the construction of the Jimmy Fund Building, a center for pediatric chemotherapy research, at the Children's Hospital.[44]

The Cancer Chemotherapy National Service Center

The association of chemotherapy with the protection of childhood life proved a potent means of mobilizing popular support for chemotherapy research. Diseases of childhood did not exist within the same moral frame as adult illness. Children did not choose their circumstances, nor did they possess the education and awareness that adults might acquire about cancer. The cultural value attached to children and the horror inspired by childhood suffering rose rapidly during the 1930s and 1940s. Following the deprivations of the Great Depression and the Second World War, the birth rate nearly doubled. During the most intense period of the "baby boom," 75 million children were born. The care, housing, feeding, and entertainment of these children shaped postwar American culture.[45] Those developments completed a longer-running reevaluation of the worth of the lives of children, one that had begun with their removal from the labor force and incorporation into the domestic lives of middle-class families at the end of the nineteenth century.[46]

The mismatch between philanthropic capacity and public expectations for chemotherapy threatened to overwhelm the resources of voluntary cancer groups. The power of children to infuse biomedical research efforts with moral urgency had already been suggested by the successful fund-raising campaigns of the March of Dimes, which used anticipated threats and anticipated cures to elevate the status of childhood disease research.[47] The promise of chemotherapy soon placed burdens on cancer research organizations that they could not bear. Just as polio fund-raising had boosted hopes for a polio cure, appeals for the support of childhood chemotherapy raised expectations for the tempo of progress against cancer that even ardent advocates of cancer research funding struggled to sustain. Research into cancer cures had never moved quickly, but the association of chemotherapy with childhood illness encouraged by Farber and others introduced the expectation that therapeutic advances should arrive not eventually, but rapidly.

Charitable organizations appeared unable to bear these new expectations alone. In the United States, only the CCRF in Boston, the Columbia College of Physicians and Surgeons in New York, and the Sloan-Kettering Institute maintained the mixture of laboratory space and hospital beds that allowed for the screening of promising chemotherapy compounds and their clinical evaluation. Sloan-Kettering carried out the bulk of this work, accounting for roughly three-quarters of all chemotherapy agents screened in the early 1950s.[48] The leadership of the Sloan-Kettering Institute was keenly aware that it could not keep pace with the growing influx of possible compounds. In 1956 the director of Sloan-Kettering noted that the institute had screened twenty thousand compounds, at a cost of twenty dollars for each "primary" test. Against this accomplishment, the American pharmaceutical industry produced 1 million potential compounds every year.[49] Researchers faced an "almost unlimited number statistically of organic compounds" that could be tested, according to the chairman of the National Research Council's

Medical Sciences Division.[50] As early as 1950, one of the major patrons of Sloan-Kettering, the multimillionaire Laurance Rockefeller, wrote to a senator that while "private resources can and will continue to press forward," more federal aid was necessary to maintain the "tempo of the attack."[51]

The newfound sense of urgency associated with childhood chemotherapy, coupled with a growing awareness of the scale of chemotherapy screening, provided Lasker with grounds for imagining a new relationship between cancer philanthropies and the federal government. Chemotherapy not only provided a problem on a scale demanding federal assistance; it also provided a community of medical experts willing to break with the restrained attitude favored by surgeons and to make bold promises that cancer was curable. In the early 1950s, Lasker and the ACS turned their political resources toward lobbying for chemotherapy. When the ACS issued its first pleas for federal chemotherapy aid, some members of Congress were skeptical about the need for government intervention, quizzing witnesses as to the contributions of state governments or charitable foundations.[52] The promise of protecting childhood life helped quell this skepticism, guided by Lasker and Quinn's keen sense of legislative theater. Older associations of cancer with the threat of epidemic disease were one element of this appeal. Speaking at a hearing, the ACS's representative invited legislators to imagine that a new illness that struck one in five people and claimed more lives among children five to fourteen years old than any other disease had just appeared in New York. Soon, this same disease was in California. "This, you would agree, is an epidemic . . . such as we have never seen in this country . . . [a] national calamity." This disease, of course, was cancer, which at that moment presented the same threat to the country, "at a somewhat slower rate."[53]

A physician from Cornell Medical School underscored the special moral status of children when he assured Congress, "We feel that if a widespread and well organized and well supported program of investigation continued that it seems very probable that a real cure for this disease and for many other types of cancer can be found in the not too distant future. Our present goal . . . is to keep these children, as many of them as possible, alive in the hope that some of them may be here alive and kicking when that discovery is made."[54] Farber appeared at a Senate hearing to guide his audience through images of people with cancer, including a young child treated for leukemia. He was able to assure a senator that the child's present condition four years after treatment was "indistinguishable from normal. . . . We can only wish that we could duplicate the set of circumstances in the treatment routinely."[55]

Congress responded with growing sums of money for chemotherapy research, which in turn provided a means of increasing overall federal spending on cancer. In 1953 Congress awarded $1 million to chemotherapy research at the NCI. The "full research potential" of chemotherapy was "not being exploited as fully as it is warranted," its advocates concluded, even though the "serial examination of clinical agents . . . could suitably be engineered."[56] In 1955 Congress, under further urging by Farber and Lasker, directed the NCI to establish the Cancer Chemotherapy National Service Center (CCNSC). Rapidly growing federal support for chemotherapy research—$1 million in 1953, $5 million in 1955, and $20 million in 1957—quickly outstripped private resources. Farber claimed that the chemotherapy effort constituted "the greatest mobilization of resources . . . ever undertaken to conquer a disease."[57]

The administrators of the NCI, who might have been expected to respond to this largesse with excitement, granted these initiatives a frosty reception. Like their colleagues in the voluntary sector, the leadership of the NCI doubted that the testing and screening of chemical compounds rose to the level of scientific investigation. After the decision of Congress to award $1 million to chemotherapy in 1953, the director of the NCI sought to diminish the sense that the extra money had been awarded with the "express intent" that the NCI devote it to chemotherapy. He maintained that the bill showed only that "Congress had indicated a special interest in . . . chemotherapy . . . with a special emphasis on leukemia."[58] The director of the NIH sought to place the best face on the congressional demand, stating during the following year's budget hearings that "the stimulus provided by the Congress may result in a truly national effort toward a cooperative and integrated search for agents useful in the treatment of leukemia."[59] Here, the legislative power marshaled by the ACS proved critical to overpowering the more conservative role for the federal government envisioned by the leadership of the NCI.[60]

Reflecting the NCI's misgivings about the intellectual worth of screening, the CCNSC opened as a semiautonomous part of the NCI. Through a new funding instrument for cancer research, federal contracts, the CCNSC coordinated closely with pharmaceutical companies to start testing tens of thousands of compounds in animals for potential antitumor activity and later oversaw extensive human clinical trials, especially of antileukemia compounds. Its efforts rapidly became the single largest budgetary item at the NCI, eventually including cooperative clinical trials that spanned dozens of institutions under the direction of officers based in Bethesda.[61] Kenneth Endicott, the first director of the CCNSC, explained that the "spectacular temporary remissions in acute leukemia" made chemotherapy research a promising area for the NCI to coordinate cooperative work: "With the impetus given this field by Congress, it has been possible to bring together the pharmaceutical industry, research organizations, private investigators, and the United States Government, each contributing their varied skills and resources to implement an effective cooperative national program."[62]

Once established, the CCNSC provided a beachhead for the further expansion of the federal government into other fields of cancer research. In the early 1960s, its staff pioneered the successful use of aggressive multidrug therapies for leukemia. Appropriations for the CCNSC grew much faster than any voluntary screening effort, and before long the contracts issued by the center sustained large-scale work at several centers across the country. It provided a new model for cancer research, with the government actively supporting and coordinating efforts among pharmaceutical companies, hospitals, university laboratories, and the NCI. With the development of the CCNSC, the locus of experimental cancer research shifted to the federal government in other subtle ways. The scale of the CCNSC's work prompted numerous innovations in the application of biostatistics to clinical trials and forged new links between oncologists at different institutions. Although the federal government did not possess the capacity to direct research in a top-down fashion through the CCNSC, the NCI was integral to the development of chemotherapy investigations after 1955.[63]

When it came to the fledgling field of cancer virus research, the CCNSC's choice of model organisms went far toward resolving the reservations regarding the relevance of Gross's mouse leukemia findings to human cancer—less by addressing earlier critiques than by creating a new community of researchers who implicitly accepted the relevance of mouse research to human cancer. The CCNSC's choice of two inbred mouse tumors and one strand of murine leukemia as the standard screen for anticancer activity helped elevate purebred mice as the "gold standard" model organism for cancer research.[64] Endicott later confidently asserted that "in the mouse, leukemia is essentially a virus disease. . . . It can be prevented by several techniques of vaccination. . . . I, for one, do not believe there's that much difference between a man and a mouse!"[65]

Future Pathways for Federal Action

With the political alliances that the ACS and Mary Lasker had fashioned in place, annual appropriations for the NCI quintupled between 1952 and 1962. The scale of federal spending on cancer outstripped the spending of voluntary organizations by orders of magnitude.[66] This expansion came about as a result of a convergence of factors, but primary among them was the decision made by anticancer advocates like Lasker to pursue federal biomedical research funding as a pragmatic response to the difficulties she and others faced in promoting federal health insurance—what I have called the biomedical settlement. The elite cancer specialists of the prewar ASCC, whose assiduous pessimism had restrained the expansion of research against cancer, no longer controlled the machinery of federal policymaking. In their place, a new political culture of biomedical research had arisen. Its participants, combining the moral urgency of childhood disease philanthropy with a commitment to advancing health through government action, sought to identify and promote particular ways of approaching cancer that might result in cures. Chemotherapy, a new approach that required urgency and planning, was the first such ground for the biomedical settlement, but it was not the last. Its therapeutic promise and demand for resources charted pathways that future federal interventions would follow.

Federal support for chemotherapy fostered the dramatic expansion of a new community of cancer researchers who had a very different understanding of the cancer problem than surgeons and new ways of creating knowledge about the disease—most notably, drug screening and clinical trials.[67] Chemotherapy fostered excitement that translated into a new willingness to consider other neglected approaches to the cancer problem and created a community of experts willing to speak in favor of further federal intervention. With these new institutional tools and public expectations for addressing the cancer problem, administrators, legislators, and activists turned their attention to cancer viruses. In 1958 Congress, buoyed by testimony from the director of the NCI that "the stage was set" for "major breakthroughs" against cancer, allocated $1 million to study leukemia viruses—a sum triple what the ACS spent on this field annually.[68] The dilemma of spending this money in a way that met the expectations fostered by the biomedical settlement, however, remained to be resolved.

CHAPTER FIVE

Managing the Future at the Special Virus Leukemia Program

In 1964 the National Cancer Institute unveiled the Special Virus Leuke-mia Program (SVLP), an ambitious effort to identify a human leuke-mia virus and develop a vaccine. Befitting the urgency associated with the prevention of childhood leukemia, the program promised to do more than award grants to cancer researchers and wait. Instead, *Life* magazine en-thused, the management of the program would "*plan* research and *make* results."[1] Speaking to a group of cancer researchers, Carl Baker, the assis-tant director of the National Cancer Institute and the program's architect, reminded his audience that leukemia killed dozens of people daily, many of them children. Organizing cancer research to meet its goals as quickly as possible was as critical a "methodological" question as any point of labo-ratory technique, he concluded.[2] The scale of the SVLP matched its ambi-tious organization. It devoted to leukemia virus research alone upwards of twenty-five times what the American Cancer Society spent annually on the whole of cancer virology.[3]

At a time when biology and medicine were still "small science," the creation of the SVLP forced questions of "big science"—organization, centralization, and state intervention—to the fore.[4] The creation of the program marked a further development in the federal government's grow-ing role in cancer research. Earlier, the leadership of the NCI had dis-avowed the idea that it might attempt to manage research outside of lim-ited contexts, despite the growing hope fostered by Mary Lasker and her allies that biomedical research could produce a cure for cancer. "Scientific research is largely a cumulative process," the director of the NCI had writ-ten in 1957, in response to a senator concerned about progress against cancer. Even growing budgets could not distract from the NCI's dedication to the "freedom of scientists to pursue freely and independently avenues of investigation which their studies indicate are most promising."[5] The SVLP promised to change this relaxed attitude. Further confounding traditional research arrangements, the rapid growth of the SVLP occurred despite a persistent dearth of evidence that its central target, a human cancer virus, existed. In fact, numerous experts in virology and oncology forcefully maintained that viruses played no role in human cancer. For the conventional process of vaccine development, these protestations would have presented a fatal obstacle. The expansion of the program for so many years in the face of such challenges is a sign that something different animated its operations.

The central innovation of the SVLP was a new way of thinking about human cancer viruses, not as objects in the laboratory but as *administrative objects* that served as the basis for planning and infrastructure. Even as they remained stubbornly elusive in the laboratory, human cancer viruses were the object of thousands of pages of description and elaboration in planning documents and memoranda within the SVLP.[6] As scientists at the bench sought proof of human cancer viruses, Baker and others at the NCI drew on Cold War managerial theory to create a system that fixed cancer viruses as objects for administrative action. As administrative objects, cancer viruses possessed a different type of reality than that associated with objects studied in the laboratory. Their existence was as much a matter of process as of their physicality.[7] This managerial framework provided a means of building infrastructure to facilitate the emergence of the viruses—and to develop a vaccine—rather than passively awaiting their discovery.[8]

The SVLP's exercise in future-oriented management also addressed problems that the NCI faced in the midst of its rapid expansion during the 1950s and 1960s. Lasker's political alliances had proved adept at marshaling federal support for cancer research, but that funding would be jeopardized if the NCI appeared to have no means of producing research advances. Developing a managerial framework for cancer viruses addressed both internal and external threats to the NCI's legitimacy. The "reform" of the NCI's organization prescribed by management theory allowed a new group, the administrators of the NCI, to claim a greater share of control over the direction of scientific research at the expense of academic physicians and scientists. Administrative objects performed a dual function: they made both cancer viruses and the community of cancer virologists more manageable for the staff of the NCI.[9]

Setting the Pace of Research at the NCI

During the 1950s, Mary Lasker and her allies succeeded in creating a political culture that generously funded the NCI. Cancer chemotherapy and cancer vaccination emerged as promising areas for the application of biological studies to the cancer problem. As the expansion of funding for cancer virus studies by Congress in 1958 attested, the political will could be mustered to support a cancer vaccination campaign. However, when this first round of funding arrived, the leadership of the NCI was neither capable of directing virus research on a large scale nor willing to do so. Despite intense public anticipation, the leadership of the NCI adopted an unhurried approach. The spirit of the approach, as one NCI administrator noted, was very much in accord with the novelist H. G. Wells's description of cancer research in *Meanwhile* (1929): "The disease of cancer will be banished from life by calm, unhurrying, persistent men and women, working, with every shiver of feeling controlled and suppressed. . . . The motive that will conquer cancer will not be pity or horror, it will be curiosity to know how and why. . . . Desire for service never made a discovery."[10]

Presenting cancer research as a gradual, individual effort spoke to the political foundations of grant research at the NCI. While support for cancer research expanded at a breathtaking pace during the 1950s, the administrative mechanisms for dispensing that support had not. After its creation, the NCI, like other branches of the National Institutes of Health, remained circumspect about the use of federal money to support research in medical schools and universities. That arrangement initially suited the interests of both the government and the biomedical research community. Academic physicians and biologists sought to protect their autonomy, and federal administrators sought to preserve scant resources for "intramural" research at the NIH's headquarters in Bethesda.[11] The National Advisory Cancer Council—a body populated by doctors and scientists at universities and medical schools rather than officials from the NCI—oversaw the distribution of what funds the NCI did possess for "extramural" grants.[12] Suspicion of federal involvement in cancer research persisted even after the excitement of mobilizing science and technology during the Second World War. As chapter 4 discusses, the leadership of the NCI and the NIH raised their voices in 1946 against aggressive efforts to confront cancer on the model of the Manhattan Project.

Though outwardly optimistic about the prospects of identifying a human leukemia virus, privately the leaders of the NCI were uncertain that there would be rapid progress toward a vaccine. The director, fresh from lauding the prospects of cancer virus research to Congress, wrote to a colleague that "work on human tumors would be a long, arduous, and often unrewarding task."[13] Deliberating over how to spend the NCI's legislative windfall, the National Advisory Cancer Council endorsed a traditional, sedate approach. The NCI should award grants to individual scientists on the basis of their individual merit, not on the capacity of their studies to advance a cancer vaccine. Moreover, the council recommended that decisions about how to allocate money should remain the responsibility of academic specialists, not federal administrators.[14] When consulted by the council, the Virology and Rickettsiology Study Section of the National Institutes of Health endorsed support for individual research projects of up to ten years at a time, cautioning that the "exploration of the possible role of viruses in malignancy may include long periods in which positive progress may not be apparent."[15]

The NCI shared the reluctance of the leadership of the NIH to treat peacetime biomedical research as matter of national urgency. Even as challenges mounted in the mid-1950s, the National Advisory Cancer Council maintained that the administrative architecture of scientific freedom, especially the review of grants by academic peers rather than by federal officials, should be preserved. In the future, the council emphasized, "there should be greater confidence in men and less emphasis on specific detailed single research studies to be undertaken." Extramural grants distributed by the NCI should allow even more time between evaluation sessions, so as not to interrupt the work of "proven investigators."[16] Scientists reading the fine print of their grants from the NCI would find that they contained a "Scientific Freedom" clause, guaranteeing, "The grantee is not required to follow the specific project submitted for review."[17]

The terms of the biomedical settlement designed by Lasker and her allies placed this commitment to scientific freedom in tension with the rationale for funding biological research. By 1954 the Department of Health, Education, and Welfare provided 48 percent of all federal support for the life sciences and nearly two-thirds of all support in biomedical sciences, such as molecular biology, developmental biology, and pathology.[18] As the scope of the settlement expanded, the connection between many supported projects and advancing human health became increasingly difficult to justify—unlike the National Science Foundation, the leadership of the NIH could not claim that they were advancing knowledge for its own sake. In 1958, "The Advancement of Medical Research and Education," a review of the NIH by former surgeon general Stanhope Bayne-Jones, brought these issues to the surface. Given the immense national importance of health research and the "inevitable" expansion for its support at the federal level, the review concluded that the NIH should expect to account for the "effectiveness" of its research administration efforts and their relevance to preserving and promoting the nation's health.[19]

Challenges to Planning for a Cancer Vaccine

The clash between autonomy and accountability bedeviled the NCI's first flagship effort to develop a cancer vaccine: the Viruses and Cancer Panel. This panel came into being as a means of managing the $1 million allocated by Congress to leukemia virus research in 1958. The panel held a strong commitment to vaccine development. Its advisers included prominent polio researchers, such as Jonas Salk, whose experience developing the polio vaccine provided a potential template for how to proceed against cancer.[20] Indeed, the polio research shaped attitudes toward a cancer vaccine more deeply. The research effort of the National Foundation for Infantile Paralysis (NFIP) had sustained a large community of virologists, most of whom lost funding once the polio vaccine was developed. While not all polio virologists moved from polio to cancer, those who did brought with them the confidence of having developed a vaccine against apparently long odds. The ACS hired Harry Weaver, who had reoriented the NFIP toward vaccine development, as its new research director in 1955.[21]

The accomplishments of polio research refashioned the NCI's approach to the development of a leukemia vaccine.[22] The influx of polio virologists into the advisory panels of the NCI created an alternate source of authority for assessing the potential of cancer virus research, aside from the one found among the surgeons who dominated the oncology community. A report drafted for the National Advisory Cancer Council by a committee composed largely of polio specialists indicated the excitement of these conversations, commenting, "In view of . . . recent discoveries, it is now fully evident that there are excellent opportunities for intensive investigations of virus tumors and tumor-viruses . . . and of possible prevention by vaccines."[23] Informed by these views, the NCI's Viruses and Cancer Panel sketched an ambitious mobilization plan. It envisioned standardizing and distributing reagents, laboratory animals, virus strains, and cell lines that would be beyond the capacity of any individual laboratory to develop or maintain.[24]

Despite this ambition, the panel's members soon found that they did not have the ability to pursue their vision. The panel possessed authority to "stimulate" communication and consider training or resource needs— hallmarks of the loose management of extramural research at the NCI.[25] But because it lacked the mandate to sponsor specific projects, the panel's plans for the construction of infrastructure were often frustrated. Efforts at standardization, for example, were stymied by the unwillingness of many virologists to voluntarily consign responsibility for culturing cells to an organization outside their own laboratories, thus hampering any effort to mass-produce resources for research on cancer viruses.[26] Moreover, the panel was unwilling to move forward with projects that individual scientists were hesitant to pursue. While the panel did see a pressing need to collect and bank tumor cell lines, it feared that starting this undertaking under "government auspices" would create too many problems.[27] After working for a year to set up a tumor cell bank, the panel was informed by the NCI that its structure did not "permit entry into long-term commitments of money or staff that would presumably be required for the complete development" of many of the basic resources for cancer virus research.[28] Indeed, two years after its creation, the panel still lacked authority to fund the construction of special facilities for cancer virus research.[29]

Meanwhile, external calls for the effective management of biomedical research at the NIH as a whole continued to mount. Opponents of federal spending on medical research, such as Representative Lawrence Fountain (D-NC), jumped upon the challenges of administering the NIH's growing roster of grants. For Fountain and others vexed by the expanding federal budget, challenging the political consensus in favor of spending on research that Lasker had created in Congress proved too difficult. However, Fountain found alternate grounds for his critique in the rhetoric of government efficiency and taxpayer rights.[30] From the fall of 1961 to the spring of 1962, the investigations of his committee brought to light a series of scandals in the administration of grant funds. Fountain managed to maneuver the usually adroit head of the NIH, James Shannon, into a series of statements that appeared to dismiss the need for vigorous oversight of grants.[31] Press coverage, including headlines such as "Overspending on U.S. Medical Research Cited," "Stanford Denies Building Pool with U.S.

Grant for Research," and "Lack of Proper Policing Called Chief Fault in NIH Grant Plans," were not only an embarrassment to those at the NIH responsible for managing grants, but also a threat to the core idea that research sponsored by the NIH could promote American welfare.[32]

The crux of Fountain's attack was that no measures existed to ensure that the NIH was accountable for the tax dollars it channeled into biomedical research. In front of his committee and the press, he castigated Shannon: "We certainly ought to expect reasonable management procedures designed to ensure that the money is being spent prudently and for the purposes for which it was intended by Congress. . . . You are spending many hundreds of millions of dollars to support research for which we cannot see and measure the results."[33] Even a sympathetic congressional ally to the NCI complained about the "apparent lack of success of the Cancer Institute to date . . . in systematic development of this new knowledge with specific orientation to cancer."[34] While outwardly assuring legislators that the NIH and NCI were up to the task of managing research, administrators inwardly betrayed acute anxiety that the NIH lacked the capacity to supervise research in a way that would consistently yield benefits to human health.[35]

For the NCI's new director, Kenneth Endicott, these were critical questions to address after he assumed office in 1960. Endicott came to cancer research from his post as head of the NCI's Cancer Chemotherapy National Service Center, the only part of the NCI where industrial-style organization and contracts predominated over grants—a function of both the scope and the urgency instilled in this work by anticancer activists. Endicott stated with certainty, "We are on the verge of a major breakthrough in the cancer area." He took it for granted that "the Institute has within its own resources the capacity to make the break . . . through with practical programs of cancer prevention and control."[36]

Cancer viruses were the first targets of Endicott's attention. In his understanding, the slow progress of the Viruses and Cancer Panel was due to organizational bottlenecks rather than scientific challenges. The structure of both research and training adopted by the Viruses and Cancer Panel was too "permissive." Without the unpredictable initiative of individual researchers, "not very much would have happened." Endicott urged an "aggressive" approach by the panel to any problem that "holds back research expansion."[37] As a first step, Endicott established both the Laboratory of Viral Oncology and the Virology Research Resources Branch. The latter was intended to "provide essential materials and services to all scientists working in the field," through "contracting with a commercial concern so that in as short a time as possible" the necessary cell cultures, animals, and virus strains would be available "in mass quantities."[38] However, neither of these two units moved quickly enough to satisfy Endicott's desire for progress. In late 1961, he gave his assistant director, Carl Baker, a wide-ranging mandate to refashion the NCI's cancer virus research structures with the aim of realizing the goal of a vaccine.

Carl Baker Learns to Manage the Future

While nothing in his early career suggested it, by 1961 Baker was well on his way to becoming a leading advocate of biomedical research management—a position he solidified during his leadership of the NCI during the start of the War on Cancer in the 1970s. Born in 1920 and raised in Louisville, Kentucky, Baker studied zoology and then medicine. During his medical residency in Milwaukee, he became interested in cancer. However, upon his graduation from medical school in 1944, the navy drafted Baker to serve as a combat physician. Baker's wartime service was uneventful—he went through basic training but never saw battle. With his discharge, Baker returned to his interest in cancer, pursuing a doctorate in biochemistry at the University of California, Berkeley. His research sought to identify differences in the metabolic rates of normal and cancerous cells, extending an approach to the biochemistry of cancer that was popular before the war.[39] Instead of completing his doctorate, he accepted an offer in 1948 to move to Bethesda and start work in the NCI's intramural biochemistry section.[40] Baker's career as an intramural researcher at the NCI was short-lived. In an era before cell culture or other *in vitro* systems for studying cancer, Baker's work required frequent contact with laboratory animals, which severely aggravated his asthma.[41] With such severe allergic reactions, it seemed as if he would need to leave the NCI entirely. However, he remained interested in cancer research and found work as a grant reviewer outside of the institute's laboratories. While his research career was frustrated, Baker's work reviewing grants brought him a temporary appointment as the director of all of the NCI's intramural scientific research. This appointment resulted in an offer to serve as the assistant director of the NCI in 1958. He remained second in command at the NCI until rising to director in 1969.[42]

As his administrative duties grew, Baker took an interest in expanding his knowledge of the management of science as a whole. His exploration soon

FIGURE 5.1. Carl Baker played a major role in planning and operating the SVLP as associate director of the NCI in 1960–69. He served as director of the NCI from 1970 to 1972. Photograph by Edward Hubbard, courtesy of NCI.

acquainted him with wider-ranging discussions concerning the relationship between scientists and the federal government. During the late 1940s many scientists heatedly debated the question of whether, or how, the increase in federal support for research would change the kinds of work carried out by the scientific community. The debate over the structure and mission of the National Science Foundation provided a touchstone for the definition of this new relationship. Many voices on the political left advocated aggressive government intervention in the planning of scientific research, while those on the right and center championed scientific autonomy and "fundamental" research as quintessential values of democracy. Physicists, among the most prominent voices in the debate, remained dedicated to what one historian has termed "laissez-faire communitarianism," or the capacity to regulate themselves through peer review rather than being subject to outside management. This view praised the judgment of scientific ideas among peers as freedom and conflated state planning with the dangers of communism.[43] The Soviet Union's suppression of classical Mendelian genetics and its embrace of the theories of Russian geneticist Trofim Lysenko in the late 1940s hardened these rhetorical divisions. The ideology of scientific autonomy deeply shaped the final version of the National Science Foundation, which enshrined peer review and shunned state control.[44]

The administrators of the NIH were intensely concerned with how they might foster scientific freedom while also directing research. This topic was the first taken up by a discussion group Baker joined in January of 1957 for the purpose of exploring questions of "research administration." Notes of their discussions, preserved in Baker's archives, offer a window into how the administrators of the NIH grappled with the ideological and practical aspects of research management. Unlike the National Science Foundation, the NIH had adopted advancing human welfare as part of its mission. This emphasis also set the management efforts of the NIH apart from those of private philanthropies such as the Rockefeller Foundation, which had sought to foster the growth of molecular biology in the 1930s.[45] For the NIH, the tension between "relating the problems society needs solved (e.g., the cure of disease) on the one hand, and the activities of scientists (many of whom believe that they can serve society best if they are left to . . . their own topic of work)" featured prominently in the group's conversation. Their first discussion ended firmly on the side of advocates in the tradition of laissez-faire communitarianism, such as Anglo-Hungarian physical chemist and philosopher of science Michael Polanyi and Harvard president James Conant. The main challenge for administrators in Bethesda, as Baker understood it before attending this first meeting, was how their organizations would accommodate the "individual freedom" due to scientists.[46]

Nonetheless, Baker remained open to understanding scientific inquiry as a process amenable to organization, coordination, and planning. The opening discussion of the gathering considered the proposition that

scientific research was of a different kind than other activities that organizations dealt with. The group rejected what the meeting's notes record
as the "19th century," "mechanistic," and "materialist" view that scientific research's main goal was "delineating the real and absolute nature
of the universe in objective and immutable terms." The group preferred
Conant's view, formulated in the wake of quantum mechanics, that science was an open-ended, dynamic, intellectual process. The conditions
for conducting scientific research were both "managerial and political."[47]
Moreover, Baker added, he thought the management of science could not
exist apart from accommodating public expectations. He was emphatic,
however, that the principal question was one of maintaining scientific
independence — of "encompass[ing] an essentially individual and autonomous activity within a purposive arrangement of forms and goals."[48]

Beyond any theoretical concern for scientific freedom, the NIH administrators gathered for these discussions were concerned with the smooth
operation of their institutes as organizations. At the second gathering,
the group considered, and ultimately dismissed, Polanyi's assertion that
"the pursuit of science can be organized . . . in no other manner than by
granting complete independence to all mature scientists." One member
of the group recorded the rejoinder, "[The] question is not if scientists
are 'guided ultimately by the authorities.' They are. The key point is the
nature of the freedom left to them, and the nature of the guidance."[49]
Managers of research might act primarily as "buffers" between a restive
public and the "scientist at the bench."[50] In May 1957, the group's final
meeting arrived at the question of what kinds of administrative systems the
NIH could fashion to meet its goals. Whatever system they might follow,
this group felt that the status quo could not be maintained. "In 1948," the
minutes read, "the NIH was still a comparatively small organization. Decisions could be arrived at upon very simple principles with considerable
assurance that they were the right decisions." A decade later, however, the
organization had grown so large that its ability to communicate with itself
was "jammed."[51]

These discussions provided Baker with a point of departure as he set
out to explore other frameworks for the management of scientific research. His journey took him into areas not often visited by biologists and
physicians. He attended seminars at the Brookings Institution for managers of research-and-development laboratories. The works on his catho-
lic reading list encompassed the philosophy of science and management
theory, including *The Dynamics of Bureaucracy* (1957), *Effecting Change*

in Large Organizations (1958), the facetious critique of management *Parkinson's Law* (1957), and Peter Drucker's *The Practice of Management* (1955).[52] The Cold War politics of science planning were never far from his work. Preparing for a session of the "Management Institute for Leaders in Scientific and Professional Programs," hosted by the US Civil Service Commission, Baker read the blunt assessment that the Soviet Union had exceeded the United States in understanding that various fields of biological research were "vehicles for social as well as individual integration." Yet, productive as the Soviet system of science might be, Baker's reading cautioned, their research-and-development efforts were hobbled by the assumption that it was important "not *to know* the world, but *to change* it." This approach risked reducing "science to technique." The government needed to provide "a model to which highly creative scientists will consistently be attracted" rather than overbearing scientific direction.[53]

The framework of greatest consequence for Baker's future approach to cancer virus research was systems analysis.[54] Whereas chemotherapy screening efforts had taken inspiration from the research-and-development divisions of large corporations, such as General Motors and the pharmaceutical industry's antibiotics screening programs, Baker drew on a new set of management ideas.[55] The 1950s and 1960s witnessed the spread of systems analysis and its cousin, cybernetics, into many corners of American thought, including architecture, psychology, computing, neurology, ecology, and molecular biology.[56] While not as well known as cybernetics, systems analysis exerted a profound impact on federal bureaucracy. Starting from the techniques of operations research developed during the Second World War, systems analysis gained prominence through the Santa Monica–based RAND Corporation, a think tank created by the US Air Force to help solve challenges associated with aerospace production. Using the mathematical language of operations research and game theory, the engineers and statisticians at RAND offered the Air Force, recently established as an independent military branch, a useful set of tools for justifying its production needs when it wrestled for resources within the armed forces. The flexibility of its methods allowed RAND to expand the scope of its activity during the 1950s, including applying systems analysis to nuclear grand strategy. Notoriously, reports by RAND stoked fears of an illusory "missile gap" between the United States and the Soviet Union during the 1960 presidential election campaign between John F. Kennedy and Richard Nixon.[57]

After Kennedy assumed the presidency, he appointed the president of the Ford Motor Company, Robert McNamara, to serve as his secretary of defense. McNamara enjoyed a reputation as an excellent manager, but he had no military background. His appointment reflected Kennedy's desire to exert greater civilian control over the Pentagon's weapons procurement process. Distrustful of the advice of military leaders, McNamara was attracted to the apparently objective criteria that the RAND reports provided for coordinating Cold War grand strategy with the process of weapons design and procurement. Drawing from his prior experience as an executive at Ford, McNamara formed a task force that combined accounting techniques with systems analysis to create a management method for defense projects known as the Planning Program Budgeting System (PPBS). Though the intellectual merits of PPBS were fiercely contested during early debates over its application to weapons procurement decisions, its econometric approach to defense policy succeeded in the ultimate aim of placing the different branches of the military under the rubric of civilian-controlled "rational defense" policymaking.[58] McNamara subsequently incorporated these methods into a broader, computer-based cybernetic vision of command-and-control war-fighting, which was embraced by air defense specialists contemplating nuclear war and counterinsurgency efforts in Vietnam.[59]

McNamara's high-profile reorganization of the Pentagon also drew the attention of civilian agencies interested in the "rationalization" of their efforts through systems analysis. Federal administrators involved with the War on Poverty turned to program planning on a routine basis later in the 1960s.[60] However, program planning gained its most visible civilian applications in space exploration, an arena of intense Cold War rivalry. Stung by the first successful launch of an astronaut into space by the Soviet Union, Kennedy committed the United States to placing a person on the moon by the end of the 1960s. The "Space Race" brought the National Aeronautics and Space Administration (NASA) both generous resources and a sense of great urgency. In a situation similar to that of the NCI in the late 1950s, the director of NASA, James Webb, sought to achieve balance between what his technical staff deemed possible and what Congress expected. Webb embraced systems analysis as a means both of problem solving and of navigating the shoals of legislative appropriations.[61] He cultivated an aura of managerial acumen. As NASA came closer to its goal, his book *Space Age Management* (1969) became a touchstone for management seminars.[62]

While it is commonly assumed that the NCI turned to NASA for inspiration during the War on Cancer of the early 1970s, a decade earlier

Baker had already begun to consider how similar methods could be used for cancer research. To Baker, systems analysis suggested that the resistance of biological systems to planning was a matter of degree, not of kind. Program planning was powerful, Baker noted, because it supplied a vocabulary for describing the management of biomedical research that could encompass different scales, ranging from individual laboratories, to departments, to the whole of the NCI. The concept of a cancer research "program," Baker wrote in his reading notes, offered a "useful word" precisely for its multiple meanings, most of which implied movement in an "organized way" toward a goal.[63] Echoing the terms that he had absorbed from management seminars, Baker argued that progress in the physical sciences "resulted from the intensive application of planned applied and developmental research efforts . . . the selection of important targets, the sense of urgency, [and] the careful delineation of plans continually modified as necessary." Through focusing on "higher priority items in the target pathways," systems planning could attain "objectives that only a few years ago were thought impossible." Such impressive methods should also be used in the NCI's "war against disease," Baker concluded.[64]

Baker's first opportunity to apply systems planning to cancer research came when Endicott placed him in charge of restructuring the Viruses and Cancer Panel. As an *ex officio* member of the panel since its inception, Baker had witnessed its organizational frustrations. The logistical challenges of preparing standard reagents, collecting tissue, and designing laboratory space seemed to be tasks ideally suited for the application of systems analysis methods. With Endicott's endorsement, Baker first created the Human Cancer Virus Task Force, which, Baker explained to its participants, aimed to elucidate the role of viruses in human cancer "more rapidly than would be done by a less concentrated and cooperative venture." Reflecting his faith, Baker expected the task force to continue its work for five years at most.[65] The staff of the NCI optimistically concluded that the "the slow progress in getting on with the human cancer virus problem" was due to logistical challenges "comparable to . . . the early days of polio research."[66] A special report for Endicott, likely prepared by Baker, promised that the discovery of a human cancer virus was imminent, based on an "avalanche of evidence from an unprecedented number of scientific disciplines."[67]

Despite this confident assessment, however, Baker's planning effort struggled with the stubborn absence of laboratory proof that an infectious agent caused human cancer. The ongoing expansion of cancer vaccine research at the NCI, an expensive matter of building infrastructure and recruiting scientists, seemed difficult to justify without a known infectious agent. Even ambitious program planning in the physical sciences usually proceeded with the confidence that the objects of its action existed. Under intense scrutiny for its use of research funding, the NCI seemed unlikely to commit more resources to a field that appeared to have such a slim and distant hope of therapeutic payoff.

Cancer Viruses as Administrative Objects

Baker's response to the impasse he faced was to approach cancer viruses not as objects of laboratory research but as objects of a process of administrative planning. He understood that language was a critical part of his effort to make human cancer viruses tractable for the bureaucracy of the NCI. Systems analysis and program planning offered Baker a vocabulary for discussing the future of cancer vaccine research that foregrounded the process of searching rather than current experimental findings. He took pains to train his staff in the discursive aspects of these methods. The management system that Baker envisioned required "a wholly new way of approaching the description" of research. The NCI needed "new philosophies, concepts, terminologies, and procedures." While moving into planning might seem "unique or even strange" for biomedical researchers, Baker acknowledged, it was "necessary" to meet the challenges involved in multidisciplinary research devoted to "the needs of society and matters related to health." In this effort, "the concepts and philosophies, the terminologies and procedures satisfactory for the description" of contemporary biological research were "not suitable for describing the newer developments proposed."[68] Those new methods were important given persistent doubts that human leukemia viruses existed. Summarizing the results of research into the causes of leukemia in 1958, the Nobel Prize–winning immunologist MacFarlane Burnet belittled viral leukemia as "a laboratory curiosity."[69] Later high-profile epidemiologic reports questioned the viral etiology of leukemia or placed emphasis on environmental and genetic factors, such as those entailed by exposure to radiation.[70]

Drawing on his study of management theory, Baker offered a specific definition of program planning for use at the NCI. Seeking to bridge a divide that he perceived between "basic" and "programmatic" research, Baker emphasized that program planning must advance on the basis of

"common procedural elements," which in turn required common concepts and definitions. "Terms such as *program, program planning, programming*," Baker explained, "are used as *mechanistic* terms to describe
a particular approach to work performance; they are used as *adminis-
trative/management* terms to describe a method for the coordination of
many activities." Baker elaborated that program planning was "a broad
conceptual and philosophical approach." "Program plans" were intentionally agnostic on the question of particular scientific outcomes; plans
were not to "require or involve . . . the detailing of specific research efforts,
the establishment of priorities, the selection of the mode of operation, or
the insertion of a time-frame." The paramount advantage of program planning was that it provided "a mode of approach to the planning of research
efforts which are oriented and focused to the achievement of the end result
or product."[71]

To hone this program plan, Baker brought in personnel fluent in the
vocabulary of management rather than that of medicine. In particular, he
recruited Louis Carrese, an industrial psychologist with experience as a
contract systems analyst for the Department of Defense.[72] While cancer
research administrators had offered optimistic projections of progress
before, Carrese emphasized that the chief aim of planning was to deliver
results within a specified time period. Writing to the NCI's staff, Carrese
explained that "the projection period should present a reasonably work-
able timespan (5 to 10 years) suitable for the framing of critical questions
in research in a manner which will permit the evaluation of progress." This
was a way of discussing biomedical research that, as Carrese underscored,
provided "*the basis for action decisions—not further discussions.*"[73]

Charting a Course for Acceleration

In particular, the management process envisioned by Baker and Carrese
emphasized the tempo of cancer research as an important area of intervention. The importance of tempo arose from the associations between
leukemia vaccine research and childhood disease through polio and chemotherapy. The means of integrating time into their process came from
a variant of systems analysis known as the Program Evaluation Review
Technique, or PERT. While systems analysis as a whole offered a framework for setting goals and drafting budgets accordingly, it did not necessarily address the question of how to meet these goals quickly. Baker was

not the first administrator to encounter this problem of urgency in the context of systems planning. In the late 1950s, members of the Polaris Missile Program—a crash effort to build a submarine-launched ballistic missile—had developed PERT in an effort to accelerate the research-and-development process. The construction of the missile in record time gave PERT immediate cachet.[74] PERT promised to account for the uncertainties of new research and development even as it kept projects on a planned timeline.[75] Its advocates praised its focus on attaining goals in a minimum amount of time for "new," "untried," and "non-routine" activities.[76] Baker was so taken with this approach that he commissioned translations on its process from francophone operations research journals.[77]

As Baker and Carrese delved into program planning, they left paper traces of their process that provide an index of the development of cancer viruses as administrative objects, which culminated in their proposal of a new technique for planning biomedical research, the "convergence technique," in 1965. Baker and Carrese adopted a central icon of the PERT planning process: the graphical representation of research pathways. Charts, graphs, and other visual representations of production processes were well established in management practice by the middle of the twentieth century, a prominent example of the "visible hand" of corporate management.[78] Indeed, some of these methods could be found in previous efforts to transfer antibiotic screening techniques to chemotherapy research.[79] PERT broke with these representations in its emphasis on a network of many pathways rather than a single line of development. A PERT network claimed to represent all of the events and tasks required to complete a project as well as the unknown contingencies that might arise. In creating a field of events and activities, known and unknown, PERT offered planners the capacity to envision multiple paths to their goal and to specify with confidence how long each path would take. Indeed, rather than moving forward from what was known to a program's goal, PERT planning handbooks often recommended working backward from the goal itself, dispensing with what was known, to design a network based on these potential developments.[80]

As the SVLP developed, its focus on the time to complete research emerged as a unique element of its planning process. As Baker considered the expansion of cancer virus research at the NCI in the middle of 1961, he sketched a chart showing how to arrive at a vaccine. The chart mapped out a series of steps that would take place between identifying a link between a virus and leukemia and preparatory research for a vaccine.

However, Baker saw only one path forward to a vaccine.[81] The emphasis on time characteristic of the PERT planning process soon started to appear in the program's iconography. One innocuous chart appeared, at first glance, to compare the state of progress in developing a leukemia vaccine with progress in developing vaccines for other human viral diseases and viruses linked to cancer in animals. Beneath the apparently neutral description, the chart presented a powerful rhetorical argument about time and progress. Unlike previous diagrams, which showed a set of particular steps for developing a leukemia vaccine, this chart posited eight steps through which every viral vaccine passed—from the acquisition of materials to the industrial production of the identified virus and vaccination. Moreover, the progress vector for each virus was labeled with the number of years the viruses had been studied.

The identification of a leukemia virus in mice had been followed by the development of a vaccine in fifteen years. It had taken fifty-eight years to develop a polio vaccine, but only thirty-four to develop a measles vaccine. Human leukemia viruses had barely progressed beyond the "detection" and "identification" stages (the use of these terms themselves an optimistic assertion). Most beguiling, however, were the dotted lines included with the human leukemia viruses: evidence of discoveries that were anticipated.[82] Echoing the universal stages of economic development posited by modernization theory, the power of this representation for planners and managers rested in its assertion that the steps of vaccine development were uniform and might be amenable to acceleration with the appropriate intervention.[83]

The ultimate result of Baker's thought was the "convergence technique," a method of research management tailored to biomedicine. Planning biomedical research, as Baker and Carrese understood it, presented two challenges. Biomedical research was much more unpredictable than industrial production, for which straightforward timetables could be constructed. Moreover, biomedical concepts appeared to be resistant to the quantification favored by systems analysis. Baker and Carrese's technique, and its associated convergence chart, enfolded these challenges in a new set of planning concepts: flows and arrays. For example, in the case of developing a human cancer vaccine, the overall "flow" of research outcomes or resource development followed a set of conceptual phases. However, each phase specified particular experiments as "tactical elements" within this larger flow, with each experimental outcome indicating a forward move into a different stage of the subsequent phase. The convergence

ACTIVITY	LEUKEMIA/LYMPHOMA							POLIO	MEASLES
	AVIAN	MURINE	FELINE	CANINE	BOVINE	AMPHIB.	MAN	MAN	
ACQUISITION OF MATERIALS	1969	1961	1966	1965	1960	1965	1967 1963	1900 1930	
DETECTION									
ISOLATION							"H"		
REPLICATION									
IDENTIFICATION									
CHARACTERIZATION							"C"		
INDUSTRIAL REPLICATION							"H"		
CONTROL									

	AVIAN	MURINE	FELINE	CANINE	BOVINE	AMPHIB.	MAN	MAN	
ANIMAL:	AVIAN	MURINE	FELINE	CANINE	BOVINE	AMPHIB.	MAN	MAN	
YEARS:	57	15	4	1	6	1	9 3	58 34	
NO. "VIRUSES":	3	16	1	-	-	-	-	3 1	
DISEASE CONTROL:	None	Vaccine Genetic Therapy Fomite	None	None	Sacrifice	None	Therapy	Vaccines	

FIGURE 5.2. Imagining the future of research. This chart, from the 1967 progress report of the SVLP, demonstrates the contours of thinking about human leukemia viruses as administrative objects—emphasizing time, comparison between species, and uniform steps from the identification of a virus to a vaccine.

chart, which traced flows horizontally through a set of phases—an information flow, a resources flow, and a research flow—offered a graphical summary of its planning assumptions. This way of presenting research planning provided a portrait of the research process as a *concurrent* process, in which a negative or absent result from one experiment at one point did not necessarily prevent the pursuit of goals in other phases. "Convergence" of these different activities on the overall goal represented the net result of the "research package" designed by planners at the NCI.[84]

The convergence technique allowed the SVLP to displace uncertainties about the existence of human cancer viruses in its present moment in favor of confident statements about the future process of vaccine development. As Carrese explained in a radio interview, the term "convergence" conveyed the idea that the research plan was designed so that "the results of the discrete phases and steps will, in fact, converge on a point in scientific development which represents the achievement of the objective."[85] Like PERT, the convergence technique "started with the ultimate end objective" and worked backward from it rather than focusing on present capabilities. Instead of waiting for flashes of inspiration or moments of serendipity in the laboratory, the convergence technique offered managers "monitoring" and "decision" points at which they could decide whether a line of investigation warranted further resources. The "linear array" of each flow expressed a rational ordering of events, but it did not prescribe that each event had to take place in that sequence. The technique would permit administrators to anticipate the "lead-time" for developing important materials or the infrastructure for mass-producing a cancer vaccine before a human cancer virus was identified.[86] The logic of its organization suggested what divisions or units should bear responsibility for particular tasks, and it offered a cognitive tool for the overall managers of the program, allowing them to "simultaneously" consider the "many complex interrelationships" of biomedical research.[87]

The convergence technique embraced a full range of actions directed by the NCI toward a vaccine—from the design of institutions to the design of experiments. The technique's capacity to address these different scales arose from Baker's conviction that the binary between "fundamental" and "applied or developmental" research was misleading. Rather, the technique treated all forms of scientific research as a "continuum." Each program demanded the "full spectrum" of research forms. Baker saw his emphasis on planning theory as a redress for the lack of attention given to planning in biomedical research; he hoped the convergence technique would fill this gap. Academic scientists were trained only in "tactical research," and their research projects might lack "strategic significance." The SVLP offered a model of how to bridge the chasm between benchtop science and the diseases that it was expected to address. Baker envisioned that the convergence technique would have its "greatest utility" at the interface of these two.[88] Baker and Endicott included a $30 million program of cancer virus research based on this technique in the NCI's proposed budget for 1964.[89]

While the convergence technique may have focused on the future, its initial uses addressed the present, especially the difficulties faced by the leadership of the NCI during the congressional appropriations process for cancer vaccine research. Presenting the outline of the SVLP to Congress, Endicott invoked polio and defense research as examples of the potential of management methods to accelerate research: "With the availability of adequate funds and the attraction to the problem of large numbers of qualified investigators, the solutions to the problems could begin to come." According to Endicott, the polio vaccine effort had gathered the resources to fund the extensive use of primates for virus typing and had followed this typing project with the development of a tissue culture system for the virus that allowed rapid vaccine production. Although no human cancer virus had yet been identified, Endicott argued that similar ambiguities had plagued polio research. Invoking the idea of developmental stages, Endicott maintained that despite these present challenges, leukemia virus research was where polio research had been twenty years before a vaccine was found.[90]

In many instances, congressional allies of the NCI would add to the institute's budget requests after these appearances. Their response to a special request for vaccine research had been generous in 1958. However, in early 1964 Endicott's appeal for the SVLP to the House of Representatives was unsuccessful. Rather than accept this setback, Endicott drew upon the climate of expectation and crisis surrounding leukemia viruses to organize an end run around the normal NIH budget-approval process, obtaining a special appropriation from the Senate to get the SVLP started.[91] This path was risky. Definite public commitments to progress against cancer were controversial—most members of the cancer research community feared public backlash when expectations were dashed. This time, however, Endicott had a system that appeared capable of meeting those expectations.

The following year Endicott needed to justify the SVLP's budget to Congress as a whole. His changed tone indicates the importance of the idea of a manageable future to marshaling support for cancer vaccine research. The NCI, he promised, was poised to deploy a system of organization suitable for an accelerated push to develop a leukemia vaccine. Congress could not afford to delay this effort:

> We believe that the developments in the research areas mentioned are so important and so opportune that we must do everything possible to push forward now. . . . We cannot await full understanding of the nature of cancer. . . . We have learned to work with mechanisms for a planned approach. . . . We await with impatience developments in the human virus area which will demonstrate that a virus is a causative factor in at least one type of human cancer. Our organization is being made ready for such a development because we know that when this happens we must exploit this lead with all possible speed.[92]

Unlike the previous year's appeal, this one succeeded. His presentation did more than promise cancer prevention. In directing the attention of legislators to the need for resources and organization to *accelerate* vaccine development, Endicott neatly foreclosed debate regarding the viral causes of human leukemia itself. The urgency of acting quickly to exploit a breakthrough reframed the question as not *whether* leukemia had a viral cause but *how* to speed vaccine development.

Press coverage responded with enthusiasm to this new presentation of cancer viruses. *Life* magazine elaborated that in the SVLP, "the plans called for starting far in advance to work out the specifications, devise instruments, put up buildings, train personnel, breed animals—everything to ensure that all systems would be Go and A-OK as if for a countdown at Cape Kennedy."[93] Headlines such as "Leukemia Cure Near?," "Hope Raised for Cure of Leukemia," and "Virus Link Found: Huge Leukemia Project Pushed" indicate the success of Endicott's narrative in framing the program as a response to an acute crisis, as opposed to a long-term search for cancer control.[94] An interview with Frank Rauscher, staff coordinator of the SVLP, appeared in the *Baltimore Sun* under the headline "Leukemia's Cure Seen: Scientist Predicts Vaccine within 5 to 7 Years."[95] "In devising any experimental attack on a biological problem," Rauscher, the future director of the NCI, explained, "one is invariably faced with the need not only to design the experiment, but the need for many materials with which to conduct the experiment." While he admitted that he had initially been skeptical of the uses of program planning, the creation of the convergence chart had convinced him that dealing with resources first was necessary, lest important experiments be forced to wait "months, or perhaps years" for lack of prior planning.[96]

As they emerged from the chambers of the NCI into public discussion of cancer, the administrative methods produced by the SVLP, especially its charts, served as icons of the power of systems thinking to bring a future free of cancer into being.[97] Baker understood the public relations value of these methods, recalling that the charts and calculations of the convergence technique were extremely helpful for defending the particular amounts of money the NCI sought from Congress.[98] Variants of the SVLP chart were reproduced in public relations materials, reprinted in *Life* magazine, and even appear in President Nixon's papers. Generations later, they remain the most prominent traces of the new way that the NCI talked about the future of biomedical research. Produced at a moment when both the existence of human cancer viruses and the capacity of the NCI to effectively produce a vaccine were in doubt, the charts were the most stable representatives of human cancer viruses available—supplanting the ambiguities of laboratory representations with the permanence of an administrative process.

FIGURE 5.3. Charts demonstrating the convergence technique became prominent visual representations of the NCI's approach to human cancer viruses in the late 1960s. Here (*left to right*) Carl Baker, Louis Carrese, and Frank Rauscher, each of whom played a large role in the War on Cancer, pose in front of a chart. Image from National Cancer Institute, copy provided by Frank J. Rauscher III.

Future Objects and Present Power

The tradition of invoking hope for a cancer cure during appeals to the public was not unique to the SVLP. Unlike other appeals, however, the statements of the SVLP went beyond rhetoric. The planning methods of the program had a powerful impact on the practice and politics of cancer virus research. Judged as a political and administrative performance, the SVLP succeeded in realigning control among activists, researchers, and managers and summoning infrastructure for biomedical research that likely would not have come into being otherwise.

In 1964, as Baker and Carrese's planning for the SVLP neared completion, another prominent review of the NIH's overall administrative capacity appeared. The Wooldridge Report, as this review was known, issued a pointed critique of the NIH's system of peer-review study sections for managing biomedical research. "No matter how much the NIH may wish to leave the development of basic fields of science to be guided by the independent judgments of scientists in their private laboratories," the report concluded, "a quite different set of policy questions is now forced on the NIH by the very magnitude of its program."[99]

The Wooldridge Committee reserved its sternest criticism for the NCI. It expressed skepticism about the organization of the NCI's flagship effort, the Cancer Chemotherapy National Service Center, and questioned the wisdom of placing cancer above other diseases. It was not clear, the report noted, that the NCI possessed the ability to manage large-scale research or compare the results of this approach to traditional alternatives.[100] While not dismissing out of hand the idea that the NIH could manage larger projects, the report cautioned that the right "technical management" was essential. Without a "full-time program management team" featuring an "unusual combination of scientific and administrative capabilities," the inherent "inefficiency" of large-scale work would reduce the productivity of researchers below "conventional scientific standards."[101] The Wooldridge Report caused consternation within the NIH and the NCI. To rebut the report's criticisms, the director of the NIH demanded proof that the institutes were capable of planning on a large scale.[102]

From Baker's perspective, the Wooldridge Report offered an ideal opportunity to promote his approach to the administration of biomedical research. On behalf of the NCI, Baker offered a detailed response, explaining that he had already been working to devise planning methods, particularly where contracts gave administrators the power to coordinate research and "combine the talents in science and in medicine and in administration in effective organizational arrangements." Following his review, Baker concluded that the SVLP represented "one type of planning that . . . needs to be introduced into biomedical research," a step that was necessary as the NIH moved into "the large budget era."[103] Endicott, in his own meetings with the leadership of the NIH, argued that the planning efforts of the NCI, and especially the SVLP, provided a "prototype" effort for the management of biomedical research by the NIH.[104] "No one investigator, indeed no one institution, can expect to carry out a research program which would be wide-ranging enough to make use of all new knowledge and techniques," while still remaining "focused" enough to produce results, Endicott wrote.[105]

The NCI explained to the Wooldridge Committee that the SVLP was designed not only to determine whether viruses were a cause of human cancer, but also to "activate measures for prevention and control of the disease as rapidly as possible." Within this objective, the contracts administered by the program were intended to "take the burden" of logistical work from academic scientists, who were not capable of dealing with the care of "many thousands of test animals of various species," the manufacture of viruses in substantial quantities, or the purchase of equipment too large to be kept on academic campuses. Not only were these capabilities beyond the reach of academic scientists; these "routine" but vital tasks were assignments that few academics wished to perform.[106]

The Wooldridge Report's critique of peer review continued a long-running conflict over the use of grants versus contracts to fund research. The system of peer-reviewed grants ensured that as the federal government provided growing support for cancer research, the flow of those resources would follow the interests of the existing cancer research community rather than supplanting it. The SVLP, however, used contracts on a large scale to recruit researchers to work on the problems its management deemed important. Unlike grants, the contracts were rarely approved by peer reviewers in the NIH's study sections or the National Advisory Cancer Council, bodies filled with physicians, academic scientists, and other leaders who were deferential to the idea of scientific autonomy. The use of contracts transferred power over research priorities to the leadership of the NCI. Baker's newfound fluency with program planning served a useful role as Endicott asserted greater control over other areas of cancer research.[107] In 1965, Carrese and Baker used the contract planning of the SVLP as a template for a far-ranging reorganization of the Cancer Chemotherapy National Service Center.[108] In 1966, the newly established Breast Cancer Task Force drew up its own convergence chart, and similar programs followed in chemical carcinogenesis testing and tumor immunology.[109]

Baker and Endicott may have been satisfied with this result, but their ally, the ACS, was not. While it was willing to be aggressive in promoting federal grants for cancer research, the ACS remained an organization of physicians favoring a minimalist role for the government. Moreover, by the early 1960s, Lasker had succeeded in recruiting her own allies within the peer-review system and the National Advisory Cancer Council, previously bastions of opposition to the biomedical settlement. Contracts reduced the influence that the ACS had enjoyed over the machinery of federal cancer policy. The ACS protested vigorously to the NIH, casting doubt on the legal authority of the NCI to administer so much money without the sanction of the National Advisory Cancer Council.[110] The managerial expertise promised by the SVLP played a critical role in the NCI's victory over the ACS in this bureaucratic clash. The Wooldridge Report had convinced Endicott that "advanced salesmanship" would be necessary to defend the NCI's administration of contracts for chemotherapy research.[111] Facing the ACS's challenge, Endicott converted this planned defense of the NCI's administration into a brief in favor of both the contract mechanism and the discretion of the NCI's leadership to set their own agenda for biomedical research.[112]

Chaired by the president of the Institute for Defense Analysis, the committee that was established to adjudicate this dispute over contracting returned a verdict that went beyond vindicating the NCI on the statutory question of contract oversight; it cast doubt on the credibility of peer review as a whole. The committee bemoaned the arrangement of priorities at the NIH; it meant that "many of the staff members who have the technical competence for program management are motivated, by the attitudes of their professional community, to apply this competence in individual research rather than to 'waste' it in management affairs."[113] Developing administrative capacity was critical because "biomedical research is now yielding, with increasing frequency, results which promise major benefits in health and longevity, but which call for large-scale *directed research* or for *development* before they can be put to use." The report urged the NIH to embrace "new methods of organization and procedures," rather than focusing on peer review. It singled out the NCI as a notable success in developing a managerial cadre and urged that the rest of the NIH emulate the example that the NCI set in its chemotherapy and vaccination programs by seeking administrative personnel with "top flight managerial talent."[114]

Cancer Research as Cold War Science

A decade after the first call to mobilize for a cancer vaccine, cancer viruses were firmly embedded in the managerial culture of the NCI as well as in the laboratories of academic scientists. Administrators at the NCI came to think and talk about cancer viruses in a way that gave them the conceptual tools for action in the absence of approval from academic cancer researchers. This mode of operation demanded different management styles and created different forms of scientific life for biomedical researchers. It also produced a conflict between the administrative legitimacy of

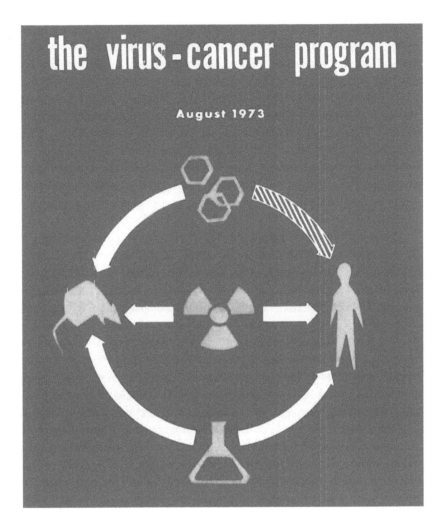

FIGURE 5.4. The logo of the SVLP, retained for the Virus Cancer Program of the 1970s, was designed to underline the importance of the inevitable discovery of human cancer viruses. Its designer explained that of the three causes of cancer in the center column (chemicals [represented by the flask], radiation, and viruses), all had been connected to animals. "It is possible that most of man's leukemias can be prevented by a vaccine directed against specific human viruses. This attack and the objective of the SVLP are illustrated by the hatched arrow — the one remaining link, in this diagram, to be closed." Courtesy of National Institutes of Health Office of History.

the NCI and the reticent and independent personae cultivated by many biologists, whose style of research disavowed connections to advancing medical therapy for cancer. Indeed, when filling out a questionnaire for scientist-administrators about the skills essential to his job, Baker selected "knowledge of managerial technique," a category from the survey, and then penciled in: "an optimistic attitude and commitment to [the] mission."[115] Since possible human cancer viruses often existed at the limit point of the experimental and epidemiological criteria used by virus researchers, the *performance* of managerial claims became an integral part of guiding the interventions that Baker and other administrators made in the process of biomedical research during a period of unprecedented expansion.[116]

The history of the SVLP adds to those histories of molecular biology, ecology, genetics, and other fields, which highlight the fact that biomedicine grappled with questions of big science throughout the Cold War, not just at its end.[117] The NCI had not yet articulated the idea of a "War on Cancer," but the patterns of the Cold War already permeated its approach to the disease in the late 1960s. The discussion of cancer viruses within the SVLP corresponded, in this sense, to the efforts in the same decade of systems planners and nuclear strategists to create "closed worlds." Closed worlds substituted techniques and scenarios drawn up in the language of engineering for actual events and outcomes. The aim of all of these practices of planning, language, and technique was to create a stable conceptual space that anticipated any event that might occur.[118] The question of human cancer viruses, reframed through the convergence technique, provided stability for future planning. The growing count of lives lost to leukemia and the possibility that these were signs of an emerging epidemic provided the sense of urgency to animate the closed world of vaccine development. This new world gave the SVLP the power to create changes in the social and material infrastructure of cancer virology and the emerging field of molecular biology.[119]

CHAPTER SIX

Administrative Objects and the Infrastructure of Cancer Virus Research

The Special Virus Leukemia Program marked the emergence of an ambitious vision for government-sponsored "big science" in peacetime biomedical research. While the Second World War had witnessed the mobilization of doctors and biologists, and the chemotherapy program had encouraged federal support for clinical trials, the SVLP envisioned channeling immense resources under the command of a single organization into the study of cancer at a biological level. After their successful defense of its contracting authority against the challenge of the American Cancer Society, the administrators of the SVLP enjoyed an unprecedented degree of freedom from the usual constraints of biomedical expertise in spending money for cancer virology. As Kenneth Endicott, the director of the National Cancer Institute, had emphasized in his testimony to Congress, the program did not ask whether a human cancer virus existed, but when it could be found. The effort of the SVLP was predicated on the assumption that "at least one virus" was "an indispensable element" in human cancer.[1] This orientation removed control over the boundaries of federal cancer virus research from the academic biomedical research community writ large, who lambasted the program's questionable "intellectual underpinnings."[2]

However, this criticism did not diminish the SVLP's capacity to shape cancer virology. Philanthropic organizations, especially the American Cancer Society, continued to support cancer virus studies, but those organizations focused on academic laboratories and typically offered less than $350,000 (2017 USD) annually.[3] By contrast, the National Cancer Institute would devote more than $6.5 billion (2017 USD) to cancer virus studies carried out by the SVLP and its successors, the Special Virus Cancer Program (SVCP) and the Virus Cancer Program (VCP)—spending as much as $700 million annually (2017 USD) later in the 1970s. At the beginning of the War on Cancer in 1971, its operations accounted for 10 percent of the NCI's overall budget.[4] The magnitude of support the SVLP enjoyed allowed its systems planning approach, which had initially satisfied the NCI's need to confront skepticism from Congress, to reshape the field of cancer virology. The NCI's cancer virus programs later formed the core of expansive, and controversial, plans to manage cancer research as a whole during the War on Cancer.[5]

The SVLP and its successors were controversial because they treated human cancer viruses as administrative objects rather than as objects of scientific research. The leadership of the SVLP understood their efforts as more than a means of studying cancer viruses. Rather, they saw their program as a way to hasten the inevitable emergence of a human cancer virus. The same challenges that had bedeviled cancer virology a decade before still existed, and no decisive observation of a human cancer virus had yet taken place. However, treating human cancer viruses as administrative objects rather than as laboratory objects resolved this challenge and allowed the federal government to act in advance of scientific consensus. Administrative objects were more than simply rhetorical pawns for intra-Washington political maneuvering; they also suggested novel ways of practicing cancer virus research, practices that underline the importance of infrastructure for biomedicine.

The SVLP approached the process of cancer research as one of joining together elements of a system for studying viruses—enzymes, scientists, tissue samples, and instruments—that had not been united before. The process of building this infrastructure helped constitute cancer viruses as scientific objects. This interpretation implies that the relationship between the scientific study of ontology, what is in nature, is far more closely tied to epistemology, the process of how we know it, than many accounts of scientific discoveries assume.[6] For the SVLP, objects such as viruses did not exist in nature awaiting scientific observation; rather, they would emerge at the center of a system of materials and practices arranged to reveal them. The existence of human cancer viruses became closely tied to the social and material process through which the SVLP conducted its search.

By examining three of the SVLP's projects in the late 1960s, this chapter highlights the ways in which the administrative nature of cancer viruses allowed the pursuit of questions that would not have otherwise been socially or materially feasible. The first two of the examples—the construction of industries for the study of leukemia viruses in the United States and the support for epidemiological searches for cancer viruses overseas—illustrate how the SVLP was able to sponsor work that would not have been pursued through the traditional channels of peer review. The third example—the development of the viral oncogene theory and the SVLP's embrace of molecular biology—is a more complex case. The oncogene theory was proposed at a moment when the program had failed to meet its own goals, and it suggests that administrative thinking was not an incidental, but a generative, element of the experimental systems that arose to study cancer at the molecular level in the 1970s.

Creating a "New Sort of Industry"

The SVLP succeeded in removing the medical and biological professions from the gatekeeping role they had previously enjoyed in the conduct of federal cancer research. The administrators of the program drew on physicians and biologists for advice and followed the results of their work enthusiastically, but their options for action were not limited to the opinions expressed by this group, as the peer-review system would have demanded. This independence opened up many more areas for potential intervention and permitted the creation of new communities of biomedical researchers. Carl Baker highlighted the importance of this active management for the SVLP. Thinking of viruses as administrative objects, he argued, demonstrated that the important organizational question was not whether individual investigators would ever find proof of human cancer viruses, but whether they even had the capacity to do so.

Human cancer viruses and other carcinogens, Baker explained, could be identified only through "greatly expanded" search efforts. For academic scientists, this kind of screening work represented "mere data collection," too "pedestrian" to merit funding. It fell to the NCI to handle the "large and complex" logistical effort of maintaining the "great numbers of animals . . . over long periods of time" necessary to generate statistically useful findings. These efforts, Baker concluded, would "not be done in the laboratories of individual investigators." The scale of the SVLP's operations was necessary not only for speed, but also for capturing phenomena that individual laboratories would miss. Its "complex, integrated, system-atic" efforts were not only required; they were "overdue."[7]

Kenneth Endicott, who had experience with this kind of approach in chemotherapy, presented contracting as integral to the NCI's new opera-tions. For projects on which academic investigators were unwilling to work, the NCI would create a new set of research institutions. In that effort, mission-oriented contracts allowed coordination between private industry and the government. Testifying to Congress, Endicott explained that the NCI was "trying to develop . . . a new sort of industry to do the types of technical jobs that need to be done here." The Department of Defense, Endicott explained, "reached this scale a long while ago. . . . A whole series of consulting laboratories . . . developed around the country."[8] He was more expansive regarding his ambitions in an interview: "We'll build some indus-tries. What the Hell! We'll build them because we've got to have them."[9]

The contracts issued by the SVLP reflected its emphasis on construct-ing this new infrastructure as quickly as possible. The contracts awarded in its first year, for example, succeeded in spending $9.9 million on forty-eight projects in eight months—far faster than an equivalent amount could have been disbursed through grants. True to the process-based ap-proach that its planning embraced, the contracts distributed by the SVLP were aggressively targeted at the logistical obstacles that cancer virus research grappled with. Some of the most expensive contracts were for the creation of animal models for the study of leukemia, making use of defense-related contractors and agricultural schools across the nation. Bionetics Research Labs in Falls Church, Virginia, received $853,033 to start screening different primate species for their susceptibility to poten-tial human leukemia viruses. Contracts worth approximately $470,000 went to both the University of California's School of Veterinary Medicine and the University of Pennsylvania to test the transmissibility of bovine leukemia in specially constructed isolation facilities. Large contracts were also awarded to other agricultural and medical schools to develop systems for studying the transmission of leukemia in cats and dogs. Another line of animal resource development was to create "germ-free" populations of mice, hamsters, and quails for further research—which Germfree Prod-ucts Inc. of Tampa, Florida, promised to do for $720,335.[10]

Another group of contracts focused on assembling the materials that cancer virology required. Defense contractor Melpar Inc. received $244,380 for the production of 3,473 ml of purified mouse leukemia vi-rus, to spare laboratories interested in working with leukemia viruses the

FIGURE 6.1. Building 41 on the NIH Campus in Bethesda, MD, completed in 1970 at a cost of nearly $10 million, contained flexible laboratory space to allow work with biohazards such as human cancer viruses. Image courtesy of National Institutes of Health Office of History.

expensive burden of maintaining their own mouse colonies. The SVLP did not wait for promising results from animal studies—it sent $328,399 to Baylor University to use electron microscopy and other methods to reveal the presence of viruses in blood samples taken from children with leukemia, collected by the Texas Children's Hospital Hematology Service. Reflecting the SVLP's confidence that the identification of a human leukemia virus was imminent, the largest contract it awarded in its first year was for $989,000 to the Dow Chemical Corporation for the design of a "state of the art" leukemia virus research laboratory incorporating the latest biohazard control measures.[11] The following year, the National Institutes of Health broke ground on the laboratory—Building 41 on the NIH's Bethesda campus. The building eschewed "conventional laboratory design," favoring a system of containment "modules" that could be reconfigured to suit future developments in research, especially work with human cancer viruses.[12]

The same administrative logic could manifest itself in even larger infrastructural forms. In 1969 President Nixon signed the Biological Warfare Convention, which committed the United States to ceasing any research on the use of microbiological agents in war. Fort Detrick, located in Frederick, Maryland, had been the center of the US Army's research into biological warfare—ostensibly for defensive purposes, although the line between defense and offense was easily blurred. Sensing that a dramatic example of turning "swords into plowshares" would both feed his détente-era diplomacy with the Soviet Union and position himself as an anticancer crusader, Nixon announced that Fort Detrick would be converted into a cancer research center in the summer of 1971, a first salvo in the War on Cancer. The very traits that associated Fort Detrick with biological warfare—its ability to handle infectious viruses and bacteria, its large facilities for growing viruses in tissue culture, and its colonies of experimental animals—were exactly the features necessary for studying cancer viruses. "It can be expected that additional new, unknown, potentially hazardous human tumor viruses will be discovered," an NCI press release warned, and it therefore made sense to conduct this research with "all possible precautions" to protect laboratory workers.[13] Under Baker's direction, the NCI brought in the defense contracting company Litton Bionetics to manage the complicated logistics of such a large biomedical research facility and to supervise a program carried out by Litton scientists and staff from the NCI, who were quickly finding their own research quarters at the NCI headquarters overcrowded.[14] In the late 1970s, the contract to operate Fort Detrick as the Frederick Cancer Research Facility reached a value of $25 million annually. Two hundred researchers, among a scientific staff of more than six hundred, were at work in its laboratories on viruses.[15]

A Global Hunt for Cancer Viruses

Alongside its efforts to create laboratory infrastructure for virus studies, the leadership of the SVLP sought to identify a virus causing human cancer "from nature."[16] For the SVLP, these studies were an opportunity to demonstrate through natural experiments, such as the immunological comparison of otherwise similar healthy and sick individuals, that viruses could cause a human cancer. Such a finding promised to resolve one of the major frustrations of cancer virus research—its inability to fulfill Koch's postulates by inducing disease in a healthy individual.[17] The effort first focused on collecting samples of blood and tissue from domestic leukemia "outbreaks" such as the one in Niles, Illinois, mentioned in the introduction, but the effort soon expanded further afield. The SVLP maintained ongoing contact with the Centers for Disease Control in Atlanta to investigate and collect tissue from "unusual epidemiologic or genetic situations," as well as with hospitals across the country.[18]

The most promising candidate in the mid-1960s was found not in North America but in sub-Saharan Africa: a rare lymphoma reported by the Irish missionary surgeon Denis Burkitt in 1958. The geographic distribution of this cancer bore a suggestive relationship to mosquito-borne viral illnesses such as yellow fever.[19] In 1964 Michael Epstein and Yvonne Barr, an electron microscopist and a pathologist at Middlesex Hospital in London, created excitement when they announced that they had identified virus particles—called Epstein-Barr Virus (EBV)—in tissues from individuals with Burkitt lymphoma.[20] This discovery promised to advance the search for leukemia viruses. Given that the United States and Uganda, where Burkitt lymphoma was first observed, appeared to have mirror-image incidence rates of leukemia and lymphoma, some epidemiologists speculated that Burkitt lymphoma might be a kind of leukemia itself.[21] Understanding its viral causes would thus directly advance the SVLP's mission of finding a leukemia virus.

However, virologists and cancer researchers in the United States reacted with skepticism toward undertaking the large-scale epidemiological fieldwork that would be required to demonstrate a link between EBV and Burkitt lymphoma. Writing to a Rockefeller University colleague seeking to survey cases of Burkitt lymphoma in Africa and South America, the dean of American tumor virology, Peyton Rous, argued, "Your idea of going here and there in the tropics and becoming essentially a medical ecologist instead of the trenchant discoverer through experimentation that you have over and over shown yourself to be, is a greatly disturbing idea. You would not be coming to grips with the vital problem of the lymphoma but merely skirmishing on its fringes."[22]

The structures of the SVLP sought to surmount precisely that kind of academic disinterest. Faced with doubt from American cancer virus researchers, the NCI and the SVLP used contracts as a means of spurring research that would otherwise not have occurred. Their effort built upon the interest of the NCI's Cancer Chemotherapy National Service Center (CCNSC) in the treatment of Burkitt lymphoma. In the 1950s, driven by a similar desire to screen the maximum number of compounds as quickly as possible, the CCNSC had issued contracts for the testing and collection of potential chemotherapeutic agents in countries around the world, including India, Japan, Poland, Egypt, and Hungary. The NIH was prohibited by law from providing grants to researchers outside the United States, but there were no similar constraints on contract funds.[23]

Those interested in chemotherapy found a perversely favorable setting for clinical testing amid the "underdevelopment" of cancer treatment in newly independent African nations such as Uganda.[24] At midcentury, the principal therapy for solid tumors was some combination of radiation and surgery. Access to both forms of this treatment, even in North America and Europe, was still limited. Burkitt, after his initial observations, had chosen to explore the chemotherapy of lymphoma rather than its infectious etiology, on the grounds that East Africa's lack of radiation treatment facilities or trained surgeons made it possible to test chemotherapy, still an unproven treatment for solid tumors, as a stand-alone treatment rather than an experimental supplement to established therapies in the United Kingdom. Fortunately, some of the compounds Burkitt tested did indeed produce striking remissions in the children he treated.[25]

The CCNSC envisioned Uganda as a laboratory for the testing of new chemotherapy regimens. In 1967 it issued a contract to establish the Lymphoma Treatment Center at Makerere University in Kampala.[26] This location ensured access to a large number of Burkitt lymphoma patients to serve as subjects in clinical trials for new chemotherapeutic agents.[27] The treatment center operated from 1967 through 1976, supported entirely by contract funds in addition to a free supply of experimental chemotherapy compounds.[28] Its principal mission was to obtain information that would be "of direct interest to the treatment of cancer in the United States." Rapid growth was possible since its Ugandan medical staff, who grew to forty in number, were more affordable to hire than equivalent personnel for clinical trials in the United States.[29]

Meanwhile, the SVLP sought to resolve several bottlenecks facing further studies of the relationship between EBV and Burkitt lymphoma. Given the rarity of Burkitt lymphoma cases, shortages of biological material posed a serious challenge. In one early study, thirty-eight of ninety-six serum samples collected from patients in Kenya were so small that they could not be subjected to the full range of tests that investigators sought to carry out for antigens and DNA fragments.[30] From its access to Burkitt lymphoma cases through the Lymphoma Treatment Center, the NCI supplied samples of tissue and blood to researchers working in the United States and Europe. This effort included maintaining its own deep-freeze blood serum bank on site, containing sera collected "from African patients with a diagnosis of Burkitt lymphoma . . . frozen as soon as feasible and sent in dry-ice by air express" to laboratories in the United States.[31] As part of its Human Tissue Procurement Program, the SVLP also issued contracts to Makerere University in Kampala, Korle Bue Hospital in Accra, Ghana, and the University of Ibadan, Nigeria, for samples of Burkitt lymphoma, which were processed and distributed by Virginia-based defense contractor Melpar.[32]

As further information concerning the association of EBV with Burkitt lymphoma accumulated, the SVLP convened a meeting of American virologists in 1968 to discuss the organization of a much larger survey, in conjunction with the International Agency for Research on Cancer (IARC), of EBV infection in the blood of residents of East Africa.[33] The IARC's president, John Higginson, hailed the proposed survey as a new template for large-scale biomedical research. He saw large-scale studies of the pathogenesis of cancer and other degenerative diseases as a key opportunity for "multidisciplinary investigations in several different populations . . . at the international level." Diseases, he argued, "do not respect national boundaries." Public health officials, epidemiologists, and laboratory workers were traditionally separated, Higginson explained, but research into diseases such as cancer demanded that they be brought together within integrated research organizations. The value of such a project was evident in the collaboration of public health officials, virologists, and epidemiologists in the analysis of Burkitt lymphoma, an example of an "environmental biology" capable of uniting laboratory and field studies of carcinogenesis through long-term observations that "many workers prefer to avoid but which are essential to any rational control program."[34]

The leadership of the SVLP foresaw that it would be "necessary to develop a capability of making field studies in human populations and the associated viruses in their natural ecology plus the expansion of related laboratory capabilities." This capability would be "costly" because of the need for "massive numbers of tests" and the extensive record-keeping work necessary to correlate laboratory data with epidemiological observations.[35] Indeed, given the rarity of Burkitt lymphoma, it seemed likely that only a survey of tens of thousands of individuals would reveal a close association—work that greatly exceeded the dozens of individuals that doctors in clinics treated.[36] The IARC proposed working as a prime contractor for the SVLP, expanding the program's reach around the world through its ability to subcontract with health agencies connected to the World Health Organization.[37] Soon after receiving this request in January 1969, the program chairs of the SVLP voted unanimously to award $350,000 to the IARC to start preparations for its survey.[38] With the grudging comment, "Since fulfillment of Koch's postulates in humans is unethical, the most direct evidence for etiology can be obtained by combining epidemiological and virological expertise," the SVLP issued a contract to the IARC for a full-scale survey in the middle of 1970.[39]

The effort to carry out a large-scale survey of EBV created new communities of experts trained in techniques for identifying viral infections. In 1965 Werner and Gertrude Henle, virologists at the Children's Hospital of Philadelphia, developed the first immunofluorescent antibody test for the presence of EBV, using a collection of EBV cultures supplied from Kampala, London, the NCI, and the Sloan-Kettering Institute in New York City.[40] The Henles became major proponents of immunofluorescence tests to track EBV infection. In 1968 they identified a causal link between EBV and infectious mononucleosis, based on the observation that one of the technicians in their laboratory fell ill with the disease at the same time that her blood sera showed a rise in EBV antibodies, a link that was substantiated by subsequent examinations of sera from other mononucleosis cases.[41] The SVLP generously supported the Henles' laboratory on the grounds that it served a "major program interest" in seeking to associate EBV particles with tumor cells.[42]

The IARC survey depended on the testing of blood samples using standardized immunological tests, and while this work was initially decentralized, it eventually concentrated in the Henles' laboratory. With NCI sponsorship the laboratory grew into a service center for screening lymphomas from around the world for the presence of EBV.[43] From 1968 through 1974, their laboratory performed screening tests on biopsied tissue sent to them from the United States, the United Kingdom, Germany, Israel, and Uganda.[44] That transition accelerated when the IARC and the SVLP sought to incorporate molecular technologies such as DNA hybridization into the tests for the presence of EBV—those tests could be carried out only by a limited number of researchers trained in the technique. In Germany, for example, a former member of the Henles' laboratory, Harald zur Hausen, started to use DNA hybridization to construct probes with the theoretical capacity to detect fragments of EBV DNA within cells.[45] In July of 1970, Werner Henle requested that the SVLP send four thousand dollars' worth of radioisotopes to zur Hausen, who collaborated with the Henles' Philadelphia laboratory to produce radioactively labeled DNA to test for the presence of EBV DNA in lymphoma tissue.[46] Zur Hausen later, in the 1980s, used similar hybridization probes to demonstrate the relationship between strains of human papillomavirus and cervical cancer.[47]

This sketch cannot fully cover the range of actors and issues raised by the conduct of biomedical research in decolonized nations, but it suggests that the infrastructure of the NCI was essential to fostering the movement of materials and the formation of communities around particular research problems that would otherwise have been passed over by established communities of virologists.[48]

From Administrative Contradictions to Oncogenes

The true potency of treating cancer viruses as administrative objects revealed itself in the late 1960s. This development came about, ironically, as efforts to identify human cancer viruses using the SVLP's considerable resources floundered. While a program predicated on the discovery of these scientific objects might have faced an insoluble problem at this point, the institutions that Baker had created based on his administrative vision of cancer viruses provided the opportunity for the program's dramatic reorientation toward molecular biology. Managerial imperatives shaped the study of cancer biology rather than biological discoveries determining the management of cancer research. The managerial theories embraced by the SVLP allowed it to turn to new iterations of the process of hunting for viruses in place of the viruses themselves, and this shift drove new visions of biological carcinogenesis.

From the moment of its formation, the principal methods of the SVLP's search reflected its mission of developing a vaccine. Drawing on the traditions of immunology and microbiology, the research projects sponsored by the program aimed to associate viruses with cancer rather than to understand the mechanism by which viruses induced cancer.[49] This non-mechanistic approach had a strong track record in the context of vaccine development. Ignorance of the genetic or biochemical aspects of polio, flu, measles, rubella, and countless other viruses had not prevented the successful production of vaccines, and there was no reason to think that cancer would be an exception.[50] In one sense, immunological searches had been extremely productive. Projects sponsored by the SVLP around the country succeeded in associating viruses with cancers in eighty-five species of animals, including chickens, frogs, mice, cats, and nonhuman primates. The sheer ubiquity of those viruses made the discovery of human cancer viruses seem imminent.[51] Yet as the NCI's intensive search for human cancer viruses reached the end of its first decade, the administrators of the SVLP faced the challenging truth that it had not produced any concrete evidence that viruses played a role in human cancers.

The tension between the SVLP's administrative commitment to the role of viruses in human cancer and the absence of immunological evidence for this claim drew the attention of Robert Huebner. Before transferring to the NCI in 1968, Huebner had achieved a remarkable string of successes at the National Institute of Allergy and Infectious Diseases. Commissioned into the US Public Health Service during the Second World War, he continued to serve at the NIH after the war, studying respiratory diseases common among army recruits. He established a reputation as an innovative virologist for his adoption of new immunological tests to reveal a previously unsuspected world of latent, or hidden, human viral infections.[52] Huebner could not say how these viruses caused disease, but his immunological detection methods implicated viruses as the causes of many kinds of illnesses. In the late 1950s, he turned his expertise in the detection of latent viruses to the detection of viruses in the tumors of mice infected with polyoma virus. His laboratory developed an immunological test for the virus rather than waiting months for tumors to appear in test animals.[53] Frustrated by his laboratory's first attempt to isolate viruses from human leukemia samples, Huebner returned to study a common respiratory virus, adenovirus, as a possible carcinogen.[54] He found evidence that adenoviruses were oncogenic in hamsters—the first instance of a human virus that caused cancer, albeit in a different species.[55] In 1968 Huebner moved to the NCI and started working full-time within the SVLP.

Huebner was initially confident that these methods could solve the riddle of cancer cells without cancer viruses.[56] However, after several years of developing tests, Huebner was also confounded by the absence of any trace of viruses in human tumors. Rather than discarding the premise that viruses were associated with human cancers, Huebner decided to formulate a new mechanism of carcinogenesis: the oncogene theory. Virologists studying cancer in mice had noticed a chain of apparent "vertical" transmission of cancer from one generation to the next. Most, like the mouse virologist Ludwik Gross, proposed that this chain of transmission rested

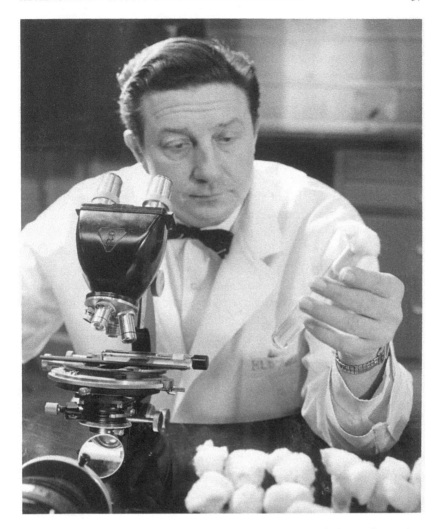

FIGURE 6.2. Robert Huebner, about 1965. After an initial career developing vaccines against respiratory illnesses in the army and at the National Institute for Allergy and Infectious Diseases, Huebner became one of the NCI's most ardent virus hunters. Image courtesy of National Institutes of Health Office of History.

on the infection of embryos by viruses in the womb before their immune systems had developed.[57] Huebner suggested that vertical transmission was a feature not of infection, but of the inheritance of "oncogenes"— genetic elements from a virus that were capable of giving rise to cancer generations after a virus infected a cell.

Huebner drew upon some of the latest findings in the regulation of genes in bacteria, particularly the theory of French biologists François Jacob and Jacques Monod that genes could exist in a "repressed" state in bacteria. However, the impetus for his new theory arose not from his enthusiasm for molecular biology, but from his sense that "a simple contradiction . . . needed to be solved." "When viewed in relation to all the data," he continued, "the inherited oncogene theory is the only concept which covers all the bases." Huebner presented his new theory not only as a mechanism of how viruses caused cancer but as an explanation for the biological process of carcinogenesis as a whole.[58]

The vectors of Huebner's oncogene theory were C-type RNA tumor viruses, or retroviruses, which virologists inside and outside the SVLP frequently observed in animal tumors. Drawing an analogy between theories of gene regulation in bacteria and the causation of cancer by viral oncogenes, Huebner and his colleague George Todaro speculated that retroviruses had two types of genes. The first kind, "virogenes," were responsible for making the protein capsule of the virus. The second kind, "oncogenes," were responsible for transforming a normal cell into a cancerous cell after infection. After infection, the genetic components of a retrovirus—virogenes and oncogenes—remained in a "repressed state" within the genome of a normal cell, passing from "animal to progeny animal and cell to progeny cell." In this state, a retrovirus would be invisible to immunological detection and even electron microscopy. Occasionally, if a cell were exposed to radiation or chemicals, the repression of the oncogene would cease, causing cancer to seem to appear spontaneously when it had in fact been caused by viral infection generations before. A finding that supported this theory was that supposedly "virus free" colonies of cells often showed traces of viral proteins.[59]

The oncogene theory ran against prevailing opinion in the fields of medical genetics, oncology, and molecular biology in the late 1960s. Within the field of medical genetics, the physical relationship between changes in genetic material and the occurrence of disease remained opaque. In 1914 Theodor Boveri, who had identified chromosomes as carriers of heredity in 1902, suggested, with little evidence, that cells became cancerous because of changes or abnormalities in their chromosomes.[60] In the late 1950s, medical geneticists linked various conditions, such as intersex characteristics and Down syndrome, to identifiable variations in a cell's chromosomal makeup.[61] In 1960 a pair of researchers in Philadelphia announced that small changes in one chromosome were associated, although not necessarily causally connected, with a particular kind of leukemia, but

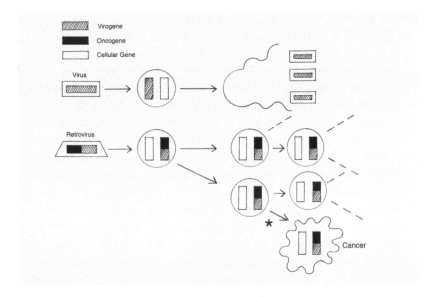

FIGURE 6.3. An illustration of the viral oncogene theory, as articulated in 1969. (1) Normal viruses contained "virogenes" that allowed the viruses to reproduce. Reproduction took place by infecting a cell, causing the cell to rupture and release new viruses. This made viral infection very easy to associate with cases of disease using immunology. (2) Retroviruses contained both virogenes and "oncogenes." The retrovirus could insert an oncogene into the genome of a cell along with its virogenes, where it would reproduce over several generations instead of killing the cell. In some cases, an event (*), such as radiation, chemical exposure, or chance mutation, could cause the oncogene to be expressed by the cell, creating a cancerous growth with none of the immunological signs of infection. Illustration by Steven Parton.

oncologists did not explore this result.[62] In the 1960s and 1970s, studies of the family history of breast and colon cancer patients established the hereditary basis of these diseases, but that work took place in Nebraska, far from the attention of experimental cancer researchers on the coasts. Moreover, from the perspective of midcentury oncologists, few therapeutic gains would ensue even if hereditary factors were identified in causing human cancer—the stigma of eugenics haunted such investigations.[63]

Meanwhile, to those familiar with the biochemistry of RNA and DNA synthesis, the oncogene theory's insistence that RNA-based viruses could create DNA copies of their genome violated the "central dogma" of molecular biology proposed by Francis Crick, the codiscoverer of DNA's structure, which held that information flowed from DNA to RNA to proteins but not in the reverse direction.[64] At the very best, the oncogene theory relied on a biochemical mechanism that had not yet been proved to exist; even if such a mechanism did exist, Huebner and Todaro had no sense of where an oncogene lay in the retroviral genome.

However, as leader of the SVCP's Solid Tumor Virus Segment, Huebner held the power and resources to move ahead with his investigation even though it ran against the opinion of other researchers. "Perhaps one of the more debilitating concepts in cancer etiology," he bemoaned to the NCI leadership, "is the notion promulgated by cancer therapists and pathologists that cancer is 100 to 200 'different diseases'; thus inferring many causes and the need for many differing modes of control." For Huebner, research into the molecular mechanisms of viral carcinogenesis promised to resolve this complicated picture by demonstrating that different kinds of cancer were the result of "essentially similar molecular events taking place in many different cells and organs of the body."[65] This apparent similarity justified his confidence in moving from animal models to human vaccination. "We can now wipe out cancer among mice. Our job is to isolate similar genes in humans," he announced at the laboratory of a major SVLP contractor in St. Louis in February 1970.[66]

The Reverse Transcriptase Boom

The nature of the power that the SVCP's infrastructure allowed Huebner to wield became evident after the identification of reverse transcriptase, an enzyme that provided a plausible mechanism for retroviruses to alter the DNA of infected cells. The SVCP could not claim credit for the discovery of reverse transcriptase, but the contrast between the career path of one of its codiscoverers, Howard Temin, and Huebner's suggests how the expansion of cancer virus research infrastructure shaped work on retroviruses.

As a doctoral student at the California Institute of Technology in the late 1950s, Temin helped to develop an *in vitro* system for studying the behavior of the tumor virus first identified by Peyton Rous, known as Rous Sarcoma Virus (RSV).[67] Temin's work fostered an initial wave of interest among molecular biologists regarding the potential of studying the genetics of cancer through viruses, which is discussed at greater length in chapter 7. In the 1960s Temin, now a professor at the University of Wisconsin–Madison, devoted much of his research to examining the behavior of RSV and the role it played in the transformation of cells. He proposed his own theory of transformation, the "provirus" theory, in 1964. Drawing on the emerging metaphor of genes as "information" carriers for protein production, Temin argued that RSV was capable of inserting into "the cell genome," the set of DNA usually carried by a cell, "sufficient information to make it a tumor cell."[68] His theory suggested that all cases of viral carcinogenesis required the "formation, activation, and expression of new genes."[69] Rather than searching for a virus itself, Temin focused his efforts on the hunt for "virus-carried information."[70]

During the second half of the 1960s, Temin sought to resolve the challenges presented by his provirus theory. In particular, that theory, like Huebner's oncogene theory, ran afoul of Crick's "central dogma" of molecular biology, according to which the information flow of molecular biology passed from DNA to RNA or DNA, and from RNA to proteins or RNA. Information could not pass from proteins to RNA and DNA or from RNA to DNA. This was a challenge for Temin's theory, because RSV was an RNA-based virus. There was no biochemical mechanism that Temin knew of that allowed RNA to create the DNA provirus that his theory proposed. Stymied at the level of mechanism, Temin attempted different methods to show that the RNA of RSV created a DNA provirus within infected cells, but none met with success. Ultimately, Temin came to rely on techniques that were unfamiliar to the molecular biologists of the phage school. He turned to a complex, but hypothetically accurate, biochemical technique for detecting the provirus, if it existed: RNA-DNA hybridization. Unfortunately, the test's hypothetical precision was matched only by the difficulty of employing it. The challenges of the hybridization reaction required an adeptness with biochemistry that Temin did not possess—the phage school had emphasized training in quantitative methods rather than biochemistry. As a result, though Temin was a respected member of the expanding molecular biology community, few embraced his theory.[71]

Temin's approach embodied the modest and restrained style of academic biology—he generally worked alone or with a few collaborators, and his principal aim was learning more about the behavior and reproduction of RSV itself. The grants that he received for his work from the NCI, following the structure endorsed by academic biologists in the 1950s, placed no obligation on him to produce particular results. Accepting the US Steel Award for Molecular Biology from the National Academy of Sciences for his discovery of reverse transcriptase, Temin reminded his audience, "It is clear that this type of research has as of yet had no immediate helpful consequences for people with cancer." Nor was it "clear whether it will have such consequences in the future. We may hope that it will, but we must not confuse this hope with the actualities."[72]

In 1970, Temin and one of his collaborators succeeded in isolating reverse transcriptase from RSV, an enzyme that allowed RNA viruses to edit the DNA of cells. This accomplishment was soon embraced as a major discovery, and Howard Temin was a corecipient of the Nobel Prize in 1976 for his contributions to virology and molecular biology. As much as his work appeared to stand on its merits, its reception benefited dramatically from the SVCP's interest in retroviruses. Temin had claimed to provide proof of the provirus in earlier papers, and his new claim could have been greeted with further skepticism.[73] Molecular biologists might have remained reluctant to set aside the central dogma, preferring to treat his discovery as an isolated curiosity rather than a finding that suggested revising the dogma itself. For example, James Watson's *Molecular Biology of the Gene* (1965) discussed the central dogma, holding that it was generally applicable with a potential exception for "certain RNA viruses."[74]

In the late spring of 1970, a young virologist named David Baltimore independently succeeded in isolating reverse transcriptase. The simultaneous publication of Temin's and Baltimore's findings in *Nature* that June did a great deal to draw attention to reverse transcriptase as a major discovery. While Baltimore first came to suspect the existence of reverse transcriptase in his studies of a viral cattle illness, vesicular stomatitis, he did not possess the ability to grow enough of the virus to isolate the enzyme. He appealed to Huebner's colleague Todaro, a fellow Swarthmore alumnus. Through this informal connection, Baltimore obtained almost $1 million worth of a purified mouse leukemia retrovirus, which the SVCP had stockpiled for further virus studies. Baltimore isolated reverse transcriptase from these materials.[75] While he could have isolated reverse transcriptase by other means, his access to the virus sped his work, allowing the appearance of his results with Temin's.[76]

Huebner and the SVCP greeted the discovery of reverse transcriptase as an opening for their administrative approach. The oncogene theory, far from providing a glib excuse for a setback, appeared to have anticipated a major discovery, and now the SVCP would help the scientific community pursue its implications. In particular, these developments led the leadership of the NCI to believe that molecular biology offered a new basis for the detection of viruses and a site for "major expansion"; they hoped that these approaches would eventually provide a "rational basis" for cancer therapy or prevention at the cellular or subcellular level.[77] SVCP contractors worked rapidly to confirm the isolation of reverse transcriptase. After Baltimore announced his findings on the Friday afternoon of a conference at Cold Spring Harbor Laboratory in Long Island, Sol Spiegelman, a SVCP-sponsored expert in RNA-DNA hybridization, hurried back to his laboratory at Columbia University. The following Monday he announced the successful replication of Baltimore's and Temin's results.[78] A representative from the SVCP invited scientists in the "packed hall" of Cold Spring Harbor Laboratory's annual tumor virus meeting "to send requests for free samples of RNA tumor viruses to Bethesda as soon as possible."[79]

The SVCP, in effect, sought to fill the sparsely populated field of retrovirus research with as many virologists and molecular biologists as possible. It launched an ambitious set of human, animal, immunologic, and molecular studies of retroviruses.[80] Ray Gilden, who worked at Flow Laboratories, said that Huebner would "run through walls" in his effort to mobilize as many scientists as quickly as possible for research on promising viruses. Murray Gardner, who oversaw the VCP's largest academic contract at the University of Southern California, remembered the "almost unreal" experience of being recruited by Huebner during a visit to Los Angeles and being left with money to "learn on the job" and assemble an interdisciplinary research team. Robert Gallo, later one of the first virologists to isolate HIV, recalled that his work on viruses at the NCI's Laboratory of Tumor Biology was supported by a contract and that it allowed him to rapidly expand his work by hiring postdocs and acquiring laboratory space.[81]

The scale of this support was generous beyond the imagination of molecular biologists. Until the late 1960s, a grant of $100,000 for molecular biology represented "a large sum of money."[82] Huebner, by contrast, controlled a budget of more than $10 million annually, and his segment was only one of several within the expanding scope of efforts under the NCI's renamed Virus Cancer Program. In 1972 the largest contracts awarded by Huebner's segment included $550,000 to the University of California for the study of two simian Type-C retroviruses, $2,410,000 to Flow Laboratories in Maryland for comprehensive immunological and molecular studies of possible human Type-C and herpes viruses, $2,080,000 to Microbiological Associates for the study of viral and chemical cocarcinogenesis, $1,200,000 to St. Louis University to use molecular methods to detect cancer virus–specific genetic material, and $2,499,040 to the University of

Southern California School of Medicine for studies regarding the etiology and epidemiology of human cancer.[83]

Many scientific commentators were sharply critical of the NCI's investment in retroviruses. The intellectual dividends of reverse transcriptase research did not seem to justify the immense investment. Their dissent provides an ironic testament to the efficacy of the SVCP in populating and sustaining activity at the juncture of virology and molecular biology. "If a prize were to be given for the most whip-cracking and numerous unnecessary duplication of experiments," a *Nature* commentator groused, retrovirus studies "would win hands down." Even if it "prov[ed] hard to find even epidemiological evidence of human RNA cancer viruses, the SVCP will be content if its pensioners manage to manufacture them in their laboratory."[84] By 1971 another writer marked the first anniversary of the enzyme's discovery by worrying that, like the Vietnam War, reverse transcriptase studies were beginning to suffer from a "credibility gap."[85]

Building Up Cancer Virology

The SVCP's investment in cancer virus research in all its forms, especially retroviruses, succeeded in creating a "boom" in these studies just as planning for the War on Cancer began. The expansion of this field was a function not only of scientific interest and federal money, but also of the infrastructure built by the SVCP. The mass production of viruses, the creation of international epidemiological projects, and the rapid exploration of reverse transcriptase would not have been possible absent the new administrative and political structures that allowed the NCI to operate independently of the peer-review system. The SVCP created experimental systems that otherwise would not have existed, prompting researchers to ask and answer very different questions about cancer and viruses than they had thought possible before.

The growth of cancer virus studies at the NCI also differs from accounts of booms in other scientific fields, which have emphasized the need to enroll scientists themselves.[86] Baker's attitude toward the motives of individual scientists recalls a point made by historians of American westward expansion. Faced with seemingly limitless "free" land, companies found that the territory was valuable only insofar as labor was available to work it. To compel workers to undertake the arduous and dangerous tasks of railroad construction, mining, or agriculture, these groups turned to labor contracts rather than attempting to entice workers by offering higher wages.[87] While the contracts employed by the SVCP were not as blunt as those used in the nineteenth century, they suggest that the expansion of cancer virus studies as a field was sustained as much by the agenda of the government as by attracting the interest of scientists. The resulting tensions came to the fore as the federal government contemplated the application of the SVCP's model to all forms of cancer research, and potentially biomedical research as a whole, during the War on Cancer.

CHAPTER SEVEN

Viruses as a Central Front in the War on Cancer

In December 1969, the White House was inundated by a volume of letters and petitions that was unprecedented in the first year of the Richard Nixon presidency. Citizens from across the country wrote to Nixon asking, and sometimes demanding, that he make the "conquest of cancer a national goal."[1] This chorus of demands heralded the beginning of a new era in the politics of biomedical research, one in which popular pressure assumed a role in shaping federal policy. Two years later, Nixon signed the National Cancer Act, marking the commencement of the government's "War on Cancer." To fight this war, Congress envisioned the most ambitious effort—up to that date or any time since—to manage and organize biological research for medical ends. The National Cancer Institute gained a lavish budget and a legislative mandate to "encompass the full range of all types of research and related activities" necessary for the prevention, control, and cure of cancer.[2] In its expansion, the National Cancer Institute drew upon the systems of command and control it had adapted from Cold War defense planning to guide cancer virus research. However, calls for a declaration of war did not reflect hubris, as it was often assumed by later commentators, but rather, deep concern that the biomedical settlement of the 1950s was unsustainable in the face of changing political and economic conditions.

In the late 1960s, the pillars of the initial biomedical settlement started to erode. Members of the Washington alliance for biomedical research that Mary Lasker and her allies had assembled in Congress faced renewed questions regarding the therapeutic payoff of biomedical research based on peer-reviewed grants. These questions joined the dual pressures of

spending on the War on Poverty and the Vietnam War to slow the me-
teoric rise of federal expenditures on biomedical research. Demands for
mission-oriented research escalated, reflecting anticancer activists' frustration with the slow pace of cancer research and their concern that taxpayers would no longer support research whose aims strayed so far from
human health. For Lasker, the National Cancer Act provided an opportunity to imagine not only an end to cancer but also a new way of organizing
biological research for the sake of public welfare.[3]

The rise of calls for mission-oriented research also augured a shift in
the relationship of another set of parties to the biomedical settlement, biological researchers, to the federal government. The War on Cancer was
especially striking for witnessing the entry of a new community of biologists, molecular biologists, into debates over the terms of the biomedical
settlement.[4] For this community, the 1950s and early 1960s had marked
their discipline's triumphant emergence from work in virology, biochemistry, biophysics, microbiology, and genetics. Evidence of molecular biology's ascendency appeared to be everywhere: its leading researchers
received a string of Nobel Prizes, new journals arose, and university biology departments were renamed to reflect the importance of molecular
studies. Molecular biologists saw themselves in the vanguard of a new way
of understanding life based on the construction of experimental systems,
such as the interaction of bacteria and viruses, which were amenable to
quantitative modes of analysis borrowed from physics.[5] In the late 1960s,
their mood changed. The pace of major discoveries slowed, and a sense of
limitless potential gave way to one of growing anxiety. Some worried that
molecular biology was on the verge of entering a routine, largely fallow,
"academic phase."[6] Many molecular biologists thought that tumor viruses
offered the best pathway for their "migration" from the stagnating study
of simple organisms to the new frontiers of complex human and animal
cells. However, animal cells and tumor viruses were far more complex and
far more expensive to work with than bacteria and their viruses.[7]

The War on Cancer marked the moment when the drive of this new
community of biologists to explore life at ever more fundamental levels
came into contact with demands to renegotiate the biomedical settlement.
Whereas the first generation of molecular biologists could rely on support
from universities and philanthropies to sustain the modest needs of their
research, the only organization with the means to support molecular biology's future expansion was the federal government. Yet molecular biologists identified with neither the aims of traditional biomedical scientists

nor those of mission-oriented research. To their dismay, both allies and critics of the biomedical settlement seemed willing to contemplate major changes to the organization of research. The managerial emphasis of the National Cancer Act realized molecular biologists' worst fears about the potential for the federal government to control the intellectual and professional fate of their discipline. In the debate over approaches to cancer launched by the National Cancer Act, this community proposed a new understanding of cancer and cancer research. Cancer viruses, with their ties to the aims of both public health and molecular biology, were both among the most visible therapeutic targets of the proposed War on Cancer and among the topics where molecular biologists would debate questions of what processes drove the biology of cancer, how to study disease, and who, ultimately, would define the future organization of biomedical research.

Autonomy and Accountability for Biological Research

The biomedical settlement, which presented federal support for peer-reviewed biomedical research as a major contribution to the nation's health, had enjoyed a high level of political support throughout the 1960s, with the exception of comparatively minor conflicts and challenges. The Democratic administrations that held power in Washington for most of the decade encouraged the public to see research sponsored by the National Institutes of Health as a major contribution to science and social welfare. Yet as time passed, the settlement's political allies started to question the returns of this arrangement. Before the escalation of the Vietnam War stalled his domestic agenda, President Lyndon B. Johnson sought to make the translation of biomedical research into concrete therapeutic advances a fitting sequel to his "War on Poverty."[8] The central issue, as Johnson saw it, was to establish the medical relevance of biological research. With $800 million (in 1966 dollars) spent annually on biomedical research, he declared, "We must make sure that no lifesaving discovery is locked away in the laboratory."[9]

This aim was in tension with the views of the scientists advising federal research policy, who continued to support grant-based "pure" or "fundamental" research and embrace the creative aspects of scientific research over its practical returns. In a more politically palatable form, this view insisted that independent research was the most efficient means of producing practical returns.[10] The political manifestation of these values at the NIH—the peer-review system—had shielded scientists from public hopes and expectations even as it supported the rapid growth of biological research on the promise that it would aid human welfare. As Nobel Prize–winning molecular biologist Salvador Luria affirmed from his laboratory at MIT, the "future of the health sciences" rested upon "basic science such as biochemistry and microbiology."[11]

Nonetheless, Johnson's efforts to create targets and timetables for moving biological discoveries from the laboratory to the clinic still affirmed the biomedical settlement. However, longtime critics of the NIH focused on the unspoken implication of this effort: that the peer-review system did *not* effectively move discoveries from the laboratory to the clinic. Representative Lawrence Fountain (D-NC), whose initial attacks on the management of the NIH had spurred the creation of the Special Virus Leukemia Program, amplified, during hearings he convened in 1967, the impression that a gap existed between the pursuit of biological research and the advance of medical care.[12] These hearings captured the growing discontent with the operation of grants at the NIH. Peer review, Fountain charged, left the American public at the mercy of a small group of established, intellectually cautious scientists rather than adventurous clinicians who valued the therapeutic aims of research. A former dean of the Duke University School of Medicine testified that NIH support turned medical schools "into research institutes . . . creating the image that research is superior to medical teaching and patient care."[13] With such an emphasis, the peer-review system would not foster research that placed human health as its top priority.

For all the consternation they had caused within NIH, Fountain's first attacks had failed to gain widespread traction. However, the critique expressed at his 1967 hearings resonated with frustrations regarding peer-reviewed academic research in other science policy circles. As Fountain conducted his hearings, the Defense Department released the results of "Project Hindsight," a review of its weapons research-and-development process since 1945. Contrary to the claims of peer-review advocates that basic research obviously provided practical benefits, the authors of this review concluded that such benefits only appeared with the further management of research. "It is unusual for random, disconnected fragments of scientific knowledge to find application rapidly," they wrote. If the fruits of scientific research were to be seen on a "substantial scale in a time period shorter than twenty years," scientists "must put a bigger fraction of their collective, creative efforts into organizing scientific knowledge expressly for use by society."[14] The study's findings caused consternation within the NIH, which launched its own review of the relationship between academic biomedical research and therapeutic advances.[15]

The inability of the NIH to deliver medical progress was all the more galling in light of the apparent success of other federal efforts to channel science toward national goals. The Manhattan Project cast a long shadow over cancer research, and even before it put a human on the moon in 1969, the National Aeronautics and Space Administration's (NASA's) Apollo Program provided an exemplar of how the government could effectively oversee large-scale, mission-oriented peacetime scientific research. The dramatic success of that program in meeting President John F. Kennedy's commitment to place a person on the moon suggested that the right organizational structure could deliver seemingly impossible results on a short timetable.[16] The outgoing director of the NIH, James Shannon, warned members of the American Association for the Advancement of Science in 1968 that academic biological researchers would need to take a broader view of the relationship between their individual projects and the public that they served. Shannon foresaw that, rather than engaging in a "chaotic competition" for grant funds, biologists would need to adjust themselves to mission-oriented research addressing the concerns of the public, "the people who are the consumers" of biomedical work.[17]

News of breakthroughs did little to alleviate these concerns. In 1967 the Nobel Prize–winning biochemist Arthur Kornberg arrived in Washington from Stanford for what was intended to be a victory tour. His laboratory, with financial support from the NIH, had succeeded in creating synthetic DNA identical to that of an organism in nature, the virus Phi-X-174. President Johnson announced this accomplishment at a press conference as the creation of life in a test tube. However, Kornberg could not make a compelling case to the congressional National Commission on Health Science and Society that his biological research advanced medicine, even under prompting from liberal senators. Abraham Ribicoff (D-CT) pressed Kornberg to affirm that his research was inspired by the hope of advancing human health: "Does a gentleman like you ever undertake . . . soul-searching as to the consequences that may come from breakthroughs and achievements in this field?" After Kornberg's noncommittal answer, Walter Mondale (D-MN) followed with a biting question, "Is it any wonder . . . that you are having financial problems?"[18] Dissatisfaction with biologists' commitment to open-ended inquiry for its own sake was growing among the legislative backers of the biomedical settlement. No number of scientific discoveries would dispel this unease as long as these legislators felt that the NIH's organization of scientific research did not reflect a commitment to human welfare.

Renegotiating the Biomedical Settlement

The plateau in appropriations for the NIH in the late 1960s matched the erosion of the system of alliances that Mary Lasker and her allies had painstakingly built in the 1950s around the biomedical settlement. Importantly, Lasker and the American Cancer Society had functioned as the major advocates for cancer research without any grassroots support. Reflecting the views of its scientific and medical membership, the society pushed for increased funding for academic research as long as it was funneled through the peer-review process. In 1968, however, the foundations of this system evaporated. Lasker's two crucial congressional allies, longtime committee chairs Senator Lister Hill (D-AL) and Representative John Fogarty (D-RI), left Congress. James Shannon, the director of the NIH since 1955, retired. Kenneth Endicott, the director of the National Cancer Institute since 1960, also retired. Even worse for Lasker, the election of Richard Nixon ushered in a Republican administration ill disposed to allow a major Democratic donor such as herself the level of access to the inner circles of the executive branch that she had enjoyed during the Kennedy and Johnson administrations.[19]

The impact of this political shift on support for biomedical research was amplified by slowing economic growth and the escalating demands placed on the federal budget by the War on Poverty and the Vietnam War. These developments heightened the difficulty of passing increases in domestic discretionary spending for agencies such as the NIH. Colonel Luke Quinn, Lasker's personal lobbyist in Washington, warned her that without the legislative shield provided by Hill and Fogarty, the ACS needed to take a far more active role in protecting the budget of the National Cancer Institute. Without bringing public pressure to bear, Quinn warned, "we are going to find ourselves standing still or sliding backward in the total effort to combat cancer. This is what has been going on for the past three years, and it will continue to go on unless we can get the ACS to respond to the realities of the situation."[20]

With her position in Washington undermined, Lasker developed new strategies to advance cancer research and ensure the future role of the federal government in biomedical research writ large. She was particularly drawn to the ideas of a University of Missouri pharmacologist named Solomon Garb, the author of *Cure for Cancer: A National Goal* (1968). Garb suggested that popular demands should outweigh expert opinion in shaping research policy. "In a democracy," he maintained, "the fundamental decisions that affect the lives and well-being of the people are supposed to be based on the wishes of those people. . . . We have not made a determined effort to see that a fair share of our national resources is mobilized for a national cancer research program." The problem, as Garb presented it, was that bureaucracy prevented the NCI from acting decisively like a mission-oriented institute such as NASA. Contrary to academic scientists' common assertion that spending more money would not produce breakthroughs, Garb emphasized that an expanded effort would be able to "buy brains" for cancer research.[21]

Whereas Lasker had been satisfied to channel cancer research funds through the peer-review process, Garb attacked that system for relying on committee work that created a "diffusion of responsibility." This structure accounted for "much of the inefficiency" of anticancer efforts. Garb suggested that a national cancer program draw upon a figure such as General Leslie Groves, the former leader of the Manhattan Project, Admiral Hyman Rickover, overseer of the navy's nuclear reactor program, or James Webb, the former administrator of NASA, to provide strong, accountable, centralized leadership.[22] Garb's arguments offered Lasker a blueprint for what a revitalized NCI could hope to accomplish. She distributed copies of his book to her circle of allies and advisers. In early 1969, she funded the creation of a new group, the Citizens Committee for the Conquest of Cancer, under Garb's direction. The committee's populist approach, complemented by Lasker's Washington-based brokering, began to reshape the discussion about the budget and mission of the NCI.[23]

In seeking to broaden public support for federally directed cancer research, Lasker and Garb's Citizens Committee represented a new understanding of the relationship between the public and the federal government's role in biomedical research. Until the late 1960s, public pressure for federal spending on biomedicine was not especially strong, reflecting the origins of the biomedical settlement among the nation's political and professional elite. Most years, budget hearings for the NCI featured doctors and scientists brought in by the ACS, who spoke on behalf of their

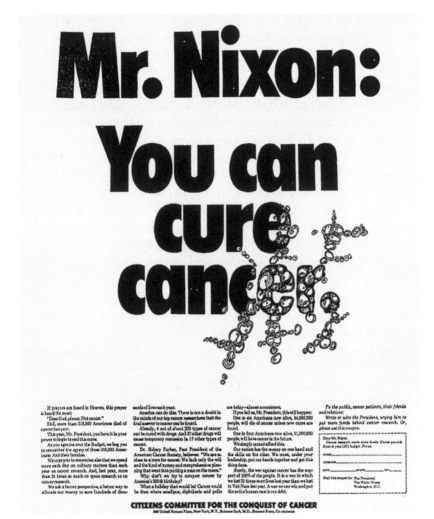

FIGURE 7.1. In December of 1969, the Citizens Committee for the Conquest of Cancer printed this full-page ad in the *Washington Post*, with reprints in other newspapers. The Citizens Committee sought to make cancer research into a national campaign, completing a set of changes started under Mary Lasker in the 1950s.

presumed national constituencies. Lasker and Garb's lobbying, however, presented the government as the standard-bearer of anticancer efforts on behalf of its citizens. That approach echoed the broader rise of civil rights and other mass social movements that focused activism on the federal government rather than on intermediary actors.[24] Seeking a more "grass roots" face for their efforts to increase funds for cancer research, Quinn started to organize and coach groups of activists such as the Candle Lighters, an organization of parents of children with cancer, to lobby Congress directly.[25]

The Citizens Committee's first full-page advertisement in the *Washington Post* at the end of 1969, "Mr. Nixon: You can cure cancer," placed the onus of action on the government emphatically. The advertisement contained the endorsement of Sidney Farber—leading childhood chemotherapy researcher, longtime Lasker ally, and 1970 ACS president—who declared, "We are so close to a cure for cancer. We lack only the will and the kind of money and comprehensive planning that went into putting a man on the moon." The advertisement prompted a sharp spike in correspondence about cancer to the White House. Passing along a letter from a concerned parent, a White House press officer peevishly urged his subordinate to be prompt, mentioning, "Several months ago the Press Office got a lot of flak when the writer, a cancer victim, died before reply was made. Victim this time is a small child."[26] Lasker continued to apply pressure in Congress, obtaining support for a resolution stating, "It is the sense of Congress that the conquest of cancer is a national crusade to be accomplished by 1976," in time for the nation's bicentennial. Thanks to Lasker's efforts, in March of 1970, the Senate voted by a wide margin to appoint the Panel of Consultants for the Conquest of Cancer to realize this goal.[27]

The Panel of Consultants offered a new venue for shaping public discussion about the relationship between biological research and medicine. Its membership was seeded with those likely to endorse a turn toward mission-oriented research. Benno Schmidt, a close Nixon ally chosen as the panel's chair, was the head of J.H. Whitney & Company, an early venture capital firm. Thanks to Schmidt's leadership and reputation as a "gatekeeper" for questions of innovation, J.H. Whitney was poised in the early 1970s to move aggressively into investment in the chemical and computing industries, both settings where goal-oriented research routinely occurred.[28] Under Schmidt's guidance, the twenty-six members of the panel, thirteen doctors and scientists and thirteen laypeople, met several times, always behind closed doors. Schmidt delegated particular technical questions to the scientists and doctors on the panel, reserving the prerogative to frame the overall tone of the final report for himself in close consultation with his administrative staff, Colonel Quinn, and other political allies.[29]

Lasker and Garb intended the Panel of Consultants to produce a strong endorsement for an aggressive and ambitious biomedical research effort against cancer. The general staff of the panel came from the Atomic Energy Commission, an agency with deep experience managing scientific projects. The panel's consultations with the staff of the NCI were most intense with those, like Carl Baker, who were the greatest advocates of planning.[30] With report writing overseen by an "aggressive systems management expert" named Robert Sweek, *Science* reported, the panel was prepared to recommend that biomedical research should be given "the organization and discipline necessary to achieve spectacular results in a reasonable time."[31]

Molecular Biologists Confront Cancer

Throughout the spring of 1970, the Panel of Consultants solicited opinions from a wide range of biologists and physicians about the best means of approaching cancer. This was one of the first times when molecular biologists waded into the controversial questions of how to define biomedical research and to set what its aims should be. This engagement reflected their efforts to translate the methods developed to study bacteria into tools for exploring the behavior of animal cells. After the Second World War, one particular group of molecular biologists, the so-called phage school, studied the genetics of *E. coli* bacteria in single-cell-tall petri-dish "lawns" and the viruses, or phages, that preyed upon them. This system offered a way to study the behavior of viruses and genes quantitatively without having to address the complex chemical basis of that behavior. The phage school, which included numerous future Nobel Prize winners, played a critical role in defining the research agenda of molecular biology.[32]

Methods pioneered in phage genetics did not easily transfer to animal cells. Animal cells were more vulnerable to infection and required more nutrients to stay alive in cultures, so their maintenance was far more expensive and challenging than growing bacteria. After considerable trial and error, Renato Dulbecco, a young member of one of the phage school's important outposts at Caltech, determined how to grow similar lawns of animal cells. Dulbecco's methods suggested that the techniques of the phage school could extend to exploring the behavior of animal viruses, especially the polio virus.[33] In 1958 postdoctoral fellow Harry Rubin and graduate student Howard Temin at Caltech, who were mentioned briefly in chapter 6, used the lawns of chicken cells that Dulbecco developed to study the behavior of Rous Sarcoma Virus, the first known tumor virus, *in vitro*. Whereas both phage and animal viruses killed infected cells, leaving holes in lawns of living cells, cancer viruses "transformed" infected cells and caused them to grow rapidly, forming lumps. Rubin and Temin's innovation was to count these points of rapid increase instead of holes.[34]

Tumor virology researchers now had a system that allowed them to study the genetics of transformation in terms similar to those used in bacteriophage study. Molecular biologists greeted this new experimental system with great interest. The process of cancerous transformation in animal cells seemed to resemble the behavior of certain mutant phage viruses. Upon infecting bacteria, normal phage viruses started to reproduce and were released when the bacterial walls burst. The mutant phage viruses did not appear to harm infected bacteria. Several generations later, however, their progeny would burst open and release phage viruses. The pattern of deaths in these progeny appeared to follow other inherited traits in bacteria, suggesting that the viral infection behaved like a genetic trait. French phage researchers argued that this behavior was evidence that the phage virus had inserted a "prophage," a genetic version of itself, into the bacterial cell.[35]

Analogously, the transformation of healthy cells by cancer viruses suggested that these viruses were inserting a gene into cells responsible for cancer. It offered an approach to the cancer problem at its most fundamental level. A founder of the phage school, MIT biologist Salvador Luria, enthusiastically greeted this new system as the avenue of "most rapid progress" for the analysis of "individual genetic elements in the transformation of normal cells into tumor cells."[36] James Watson, who shared the Nobel Prize in 1962 for his contributions to elucidating the structure of DNA, wrote in his landmark textbook *The Molecular Biology of the Gene* (1965) that the study of tumor viruses provided a tool for seeking the specific mutation that gave rise to cancer among the potentially millions of genes in the mammalian cell in a "straightforward, rational manner."[37] However, optimism about tumor viruses coincided with growing awareness of the expense of this future research and apprehension regarding the stability of future federal support.

As the War on Cancer took shape, the molecular biology community engaged in a two-pronged campaign targeting the federal government and other researchers in the biomedical sciences. On the one hand, they sought to convince the government of the worth of their work, and on the other hand, they aimed to position molecular biology as a fundamental basis for medical research as a whole.[38] In 1969 Paul Berg, who had just started work on the potentially oncogenic Simian Virus 40, wrote to a colleague, "I think that it's important that agencies concerned with the biosciences recognize that the explosive innovations of molecular biology and biochemistry don't bring us to a stopping point." He concluded, "If the cutback in funding is allowed to stand," both the "present momentum" and "the present generation of bright young people" would "be lost." Writing to Senator Charles Mathias (R-MD), the RNA-DNA hybridization specialist Sol Spiegelman argued, "The time has come to make a concerted effort to apply the information garnered from the study of *E. coli* to situations of more direct relevance to the health and welfare of man."[39] In 1970 prominent scientists such as Berg, Watson, and Charles Yanofsky started lobbying the National Science Foundation to establish a program for the "Molecular Biology of the Human Cell." The use of animal tumor viruses to explore the genetics of cellular regulation and transformation figured prominently in their proposal.[40] If support for molecular biology decreased, Watson worried, a new generation of researchers might be lost to "a safe occupation, like clinical medicine."[41]

As Watson's dismissive comment suggests, oncologists and other physicians received the ambitious claims of these molecular biologists with skepticism rooted in the long-running debate over the respective roles of the laboratory and the clinic in medicine. A member of New York Medical College inveighed against cancer research by biologists "who would not know human cancer if they had it."[42] Other authorities asserted that the interdisciplinary nature of cancer research demanded new organizational forms. Frank Horsfall, the outgoing director of the Sloan-Kettering Institute, wrote that laboratory and clinical research represented "a continuum" directed at the common aim of curing cancer. "More often than not," Horsfall argued, "the history of significant achievements in biomedical research tends to show . . . that laboratory studies of models of human diseases . . . provide springboards" for therapeutic discoveries. "More money will not lead to more effective and fruitful cancer research, unless it is accompanied by an excellent 'general staff' to plan and supervise the expenditure of funds," concluded Albert Sabin, the developer of the oral polio vaccine. Despite the role he played in developing methods for studying animal tumor viruses, Harry Rubin dismissed the possibility that virus research would yield major breakthroughs for human cancer. Instead, he urged, the panel should seek to understand cancer through studies of environment and health, which viewed the disease as a "biological process which reflects the lifetime of the individual."[43] While molecular biologists were not the only group of scientific experts lobbying the panel on how to approach cancer, as a new community within biomedicine they had the most to lose or gain from its recommendations.

Passing the National Cancer Act

The Panel of Consultants' report, released in December of 1970, wholeheartedly endorsed the organization of a major anticancer campaign. Of the 200 million Americans alive, its preamble stated, 50 million could expect to be diagnosed with cancer, and 35 million could expect to die of the disease. Against these numbers, current spending on cancer research was "grossly inadequate." For every 125 dollars spent on the Vietnam War, or 19 dollars spent on NASA, just 89 cents were spent on cancer research. While uncertainty about the biological nature of cancer remained, the report summarized, "A national program for the conquest of cancer is now essential if we are to exploit effectively the great advances which are presented" in scientific understanding of cancer. It recommended, rather than continuing to fund cancer research within the bureaucracy of the NIH, that a "systematic attack" on cancer be mounted through the creation of a NASA-like "National Cancer Authority." The proposed office would have wide-ranging powers to accomplish its mission of "conquering cancer." It would be able to issue contracts for research, mass-produce biological materials such as viruses and cells, and take "such actions as may be required for the accomplishment of the mission."[44]

The report further elevated the political profile of cancer research. Although Lasker and her allies had been at pains to present cancer research as a matter of national interest rather than partisan politics, her campaign had advanced largely through the efforts of Democratic legislators. Senator Ralph Yarborough (D-TX) had introduced the resolution calling for the Panel of Consultants, and he now quickly submitted legislation seeking to implement the panel's recommendations. Yarborough left office in January of 1971, and in his place, Edward M. Kennedy, recently reelected as a senator from Massachusetts, emerged as the Democratic Party's new standard-bearer for health policy.[45] Kennedy's visibility and presidential ambitions drew the consternation of the Nixon Administration, which until this time had not displayed interest in Lasker's effort.[46]

Demonstrating his predilection for denying his Democratic opponents political issues rather than adhering to a conservative ideological agenda, an instinct that had already led him to back the creation of the Environmental Protection Agency and federal affirmative action programs, Nixon offered his own cancer bill to implement the panel's recommendations. In doing so, he hoped to deflect criticism for his actions in Vietnam and Cambodia and to deny Kennedy, considered a likely presidential challenger in 1972, a signature issue. Before the Senate could vote on the proposed "National Cancer Authority," Nixon announced in his January 1971 State of the Union Address that he would demand an additional $100 million for the NCI's research efforts, a prelude to the creation of a new "Cure-Cancer Program" with an independent budget and a direct line to the Oval Office.[47] The time had come, Nixon said, "to put our money where our hopes are."[48]

Few could dispute the rhetorical force of the reinvigorated, centralized system for cancer research promised by the new Cancer Authority. Schmidt, in his public remarks, offered a vigorous endorsement of Nixon's proposal to manage biomedical research. He anticipated claims that "the analogy between splitting the atom or the space program" and cancer research "is not valid because we do not have the basic scientific knowledge in cancer that we had in those fields." Echoing Baker's administrative approach to cancer viruses, he retorted, "The valid analogy is not the scientific analogy but the organizational analogy. The cancer program, in order to succeed, needs the same independence in *management, planning, budget presentation, and the assessment of progress that those programs needed.*"[49]

Established biomedical researchers recoiled from this reorganization. In a rare editorial, *Cancer Research*, the flagship journal of the American Association for Cancer Research, asserted that "thoughtful researchers feel that the setting of a date for the cure of cancer, even by implication, would be an act of reckless irresponsibility. . . . A highly structured program . . . will be viewed with suspicion by many who feel that the all-important pioneering exploratory investigations would suffer thereby."[50] Opponents of the new organization, which included the surgeon general, emphasized that "biomedicine is a vastly complex and interdependent" enterprise whose various components were "not easily separated conceptually or programmatically." Privileging cancer research would "distort the flow of personnel into other areas of biomedicine."[51] A representative of the American Society of Biological Chemists feared that spending on cancer would undermine the future of biomedicine because "some individuals will be tempted, particularly if a tight time schedule is imposed on the cancer attack program, to shift the emphasis to short-term gains that may offer only the slightest hope of success and provide little real understanding of the fundamental problems of cancer."[52]

If one only reads the record of hearings in the Senate, critics of the Cancer Authority appear far better represented than its supporters. The witnesses presented a sharp critique of the assumption that biomedical research could be directed toward particular aims. If the horizons of biological knowledge about cancer had been set as the limits of state intervention, this would have posed an insoluble obstacle. However, Lasker and her allies were more than equal to the task of countering this critique. Lasker deepened her cooperation with Garb to mobilize grassroots support for the Cancer Authority. Famously, she prevailed upon her friend Ann Landers, a nationally syndicated advice columnist, to initiate a letter-writing campaign on behalf of the bill. The resulting outpouring of letters, as many as 1 million, inundated the Senate mailroom. With this pressure, the Senate moved rapidly, passing legislation at the end of June 1971 to create the National Cancer Authority by a resounding vote of 79-1.[53] Over the summer, debate shifted to the House of Representatives, where the pace of the legislation's advance slowed and opponents of the Cancer Authority hoped to halt its progress.

Cancer Viruses and the Management of Biomedical Research

Even as heated discussion raged over the merits of the Cancer Authority, it remained unclear what new management techniques would be brought to bear on cancer research. Cancer viruses were at the center of this debate because of their exceptional experimental promise and their preexisting association with the National Cancer Institute's efforts to manage research. If any one individual embodied the ambitions of the vaccination program and the idea of mission-oriented research it represented, it was Huebner, who as the head of the Special Virus Cancer Program's Solid Tumor Virus segment was already a leading member of the NCI's managerial team.

Huebner worried that budget cuts would curtail the NCI's ability to pursue the "exciting findings" in cancer virus research.[54] The success of the basic biomedical sciences until the late 1960s, he explained to a reporter from *Parade* magazine, had been due to the "strong support of a small number of intelligent" legislators, who had not been assisted by "a groundswell of support from the generality of biomedical institutions, scientists, educators, or practitioners." The solution to the problems that biomedical research faced was in its further embrace of big science, even if that ran

FIGURE 7.2. George Todaro, Robert Huebner's collaborator at the SVCP, explains the oncogene theory for the readers of *Medical World News* in the lead-up to the passage of the National Cancer Act in 1971. Photograph by George Tames. Image courtesy of Houston Academy of Medicine–Texas Medical Center, John P. McGovern Historical Research Collections,
used by permission of the estate of George Tames.

against the grain of expert medical and scientific opinion. He accused those
experts of being "traditionalists," with "little confidence in large-scale 'big'
science.'"[55] More management, not less, appeared to be the way forward.

In 1970 and 1971, virus research appeared to provide numerous examples
of the wisdom of Huebner's approach. His proposal of the viral oncogene
theory and the discovery of reverse transcriptase, discussed in chapter 6,
appeared to rebut one of the major critiques of managed research: that
cancer remained too mysterious to study on a timetable. The biological
mysteries of cancer were far more amenable to resolution than biologists
would admit, a point that Huebner and his colleague George Todaro drew
out at a National Academy of Science symposium dedicated to planning
the future of cancer research.[56] Other discoveries in cancer virus research
emerging from the SVCP underlined the necessity and potential of large-
sale research. In the summer of 1971, as the House debated the Cancer
Authority, a team of SVCP-supported researchers at Houston's MD Anderson Hospital and Tumor Institute, led by Leon Dmochowski, a longtime hunter of cancer viruses with the electron microscope, announced

that they had isolated and cultivated an RNA virus from a child's tumor. The *Los Angeles Times* greeted the discovery as a "step towards vaccination."[57] The *Wall Street Journal* noted that while "discovery of human cancer viruses has become almost a 'Holy Grail'" for cancer research, the Texas researchers' discovery might allow the development of a vaccine in the "distant future."[58]

Reports of these advances drew attention to the scale and resources that future work would entail. The *Houston Chronicle* reported that verification of the observations would cost $1 million and "require skimming 7,500 gallons of fluid off growing human cancer cells." Spiegelman argued that if a central facility could supply viruses to overcome the bottleneck of production in individual laboratories, the amount of cancer virus research could be "increased overnight by a factor of 50." Even a molecular biologist such as Berg conceded that there was need for "central facilities which provide resources to many laboratories" and speculated that "other 'big science' facilities could be envisaged."[59]

As the House approached its vote on the National Cancer Authority in the fall of 1971, interest in potential human cancer viruses reached a fever pitch. In a rare move, investigators broke prepublication embargos for their articles at *Nature* and the *Journal of the National Cancer Institute* to announce the discovery of other possible human cancer viruses. The NCI considered establishing a separate journal dedicated to the rapid publication of cancer virus research.[60] This rush of publicity obscured more negative results, such as the rebuttal of Dmochowski's claim by Huebner and other researchers at the NCI. Huebner claimed, correctly, that the MD Anderson group had identified a mouse cancer virus that had contaminated their cultures of human cells.[61]

The climate of anticipation surrounding the discovery of a human cancer virus also resulted in public missteps that revived earlier twentieth-century concerns for professional probity and restraint. Spiegelman felt that his laboratory was drawing close to developing a diagnostic test for the presence of human breast cancer viruses in breast milk.[62] At a press conference at the National Academy of Sciences in October 1971, two of the most prominent scientist advocates for the SVCP, Huebner and Spiegelman, discussed advances in DNA-RNA hybridization for cancer virus research. Huebner himself asked Spiegelman whether he thought "this might be the place for a cancer vaccine." Spiegelman downplayed the idea. However, his subsequent comments on the high incidence of "B-type" viruses in the milk of women with family histories of breast cancer suggested otherwise. In the exchange that followed, Spiegelman was asked whether this meant that mothers with a family history of breast cancer should not nurse their children, which Spiegelman affirmed and then denied (this was a common question extending back to Ludwik Gross's theory of vertical transmission).[63] Fortunately, *Science* reported, the "restraints that govern[ed] the relationship between scientists and the scientific press" prevented Spiegelman's comments from creating a full-blown "scare."[64]

Critics folded the failure of virus research advocates to maintain an appropriate, restrained persona when speaking to the public about cancer into their critique of the theory itself. Shortly before the House voted on the Cancer Authority in December 1971, Nicholas Wade, a reporter for *Science*, used humor to remind his readers how respectable scientists were supposed to deal with the press, recalling criticism of William Gye for announcing his observation of viruses in a newspaper back in the 1920s. In a clear reference to Spiegelman's difficulties at the National Academy of Sciences, Wade published a mock transcript of a press conference from the "Center for Duplicative Results" by "Professor Wangonband," who announced that "five separate research teams under contract to this institute" had isolated the "so-called nobelitis virus" first isolated by "Dr. Medea Courter." At the end of the conference, the source of the "nobelitis" virus was revealed to be the brain of Wagonband himself. The wordplay in Wade's article—portraying a status-seeking researcher following the cancer virus bandwagon and engaging in second-rate research for the purposes of media celebrity—encapsulated many of the attacks that academic biologists leveled at cancer virus research: its emphasis on speed, the sense that its advocates lacked the respect of their academic peers, and the willingness of representatives of the SVCP to foster public hopes for a breakthrough against cancer.[65] However, these academic researchers no longer controlled the machinery of cancer policy. Despite staunch opposition to the National Cancer Act by supporters of the peer-review system, the House passed a version of the bill on December 10, 1971—just as Wade's satire came to press.[66]

A Watershed Moment for the Biomedical Settlement

Nixon signed the National Cancer Act into law on December 23, 1971. The annual budget for the NCI quadrupled in the following two years, and it gained the authority to issue research contracts without the supervision of the NIH. While allies of academic biomedical research had succeeded in keeping the NCI within the NIH, the director of the reinvigorated NCI reported directly to the president, indicating the status of cancer as the preeminent issue for federal biomedical research policy. The newly formed National Cancer Advisory Board, whose members were appointed by the president, replaced the National Advisory Cancer Council, whose members had been determined by the cancer research community. Above the new board, a triumvirate—the President's Cancer Panel— oversaw the creation of the National Cancer Program Plan. Carl Baker, the acting director of the NCI, used the National Cancer Program Plan as an opportunity to realize his long-standing ambition to create systems for the management of biomedical research. Of all the administrators at the NCI, he had remained in closest contact with the Panel of Consultants as it had drawn up plans for the Cancer Authority.[67]

The start of the War on Cancer in 1972 portended a realignment in the biomedical settlement as a whole. While Baker had taken exception to the Panel of Consultants' assertion that the NCI had no coherent organizational plan for research, he was ready to use the authority and resources granted to him to expand efforts to control biomedical research and development. The SVCP, reorganized as the Virus Cancer Program (VCP), provided a template for this larger effort. For Baker and his successors at the NCI, the VCP demonstrated the dynamic possibilities of managing biological research, especially the exciting findings surrounding Huebner's oncogene theory. This research approach rejected the contentions of molecular biologists that cancer was too complex a disease to cure and that the purity of biological research should not be subordinated to therapeutic priorities. Now, Baker had the resources and the administrative structures to make this new vision of biomedical research a reality.

For the molecular biologists whose institutional and intellectual fates were so closely tied to the NCI, however, the VCP presented an existential threat to their identity as creative, independent seekers of knowledge. The shift in their research to animal cells and the proposed changes in the biomedical settlement had awakened molecular biologists to their collective interests in relation to the federal government. But they had been unable to protect these interests in the face of popular enthusiasm for mission-oriented research. The emerging conflict between these two groups showed how apparently separate claims about the biological complexity of cancer, on the one hand, and the appropriate means of managing biomedical research, on the other, were in fact two aspects of a single debate regarding what future system of biomedical research the War on Cancer would endorse. In this debate, the VCP's contracting process became the focal point of a campaign by the molecular biology community to stall and then thwart the further expansion of the government into biomedical research as a whole.

Molecular Biology's Resistance to the War on Cancer

In March 1975, James Watson, molecular biology's ubiquitous ambassador during the War on Cancer, rose to speak at the dedication of the Seeley G. Mudd Building, built to house MIT's new federally funded Center for Cancer Research. The center, one of several throughout the country that the National Cancer Institute sponsored, was to pursue approaches to cancer drawn from molecular biology.[1] One might expect that this moment would provide an opportunity for Watson to celebrate the potential benefits of cancer studies at the molecular level. The NCI already extended similar support to Cold Spring Harbor Laboratory, a center of molecular biology under Watson's leadership. Instead, he characterized most cancer research as "an intellectual graveyard." Molecular biology and the "non-clinical based academic community" deserved more support than cancer research focused on humans. Just as for the War on Poverty, Watson continued, it was time to question whether the objectives of the War on Cancer were reasonable. Although the rewards of "perpetual enthusiasm" for a cure might be "piles of soft money," this effort entailed the sacrifice of the "integrity" of molecular inquiries into the cell. "If we rise and fall as to whether a human cancer virus can be isolated in a vaccine . . . then we are in trouble and the whole community is in deep trouble."[2]

Watson's comments suggest that the national declaration of the War on Cancer was the starting point for major changes in molecular biology and medicine rather than the conclusion. The passage of the National Cancer Act in 1971 ignited conflicts over the organization of biomedical research that smoldered throughout the 1970s. The fate of efforts to manage cancer virus research, in particular, illustrates a phenomenon commonly observed by scholars of American political history but often neglected by historians of science and medicine: the success or frustration of government intervention reflected not only the aims of legislators or administrators but also the will of the populations targeted by new policies. The targets of a state program could use its mechanisms for ends unanticipated by its framers.[3] However, unlike intervention in other spheres of American social and economic life, federal intervention in cancer research stands out in that the populations that were the object of intervention— biomedical researchers and other cancer specialists—were also the groups the federal government looked to as experts on the cancer problem. The politics of cancer research was inseparable from questions about the nature of cancer at a biological level.

As a result of this unique position, molecular biologists developed a collective political interest in shaping the definition of biological research into cancer as a means of defending the future of their discipline.[4] Biologists, clinicians, and cancer research administrators pursued their conflicts not only through legislative advocacy or public testimony, but also through the shaping of political and scientific cultures for biomedical research that conferred (or denied) legitimacy to the government to direct the labors of their professional communities.[5] In particular, appeals to the unpredictability of scientific research and the mystery of cancer were a vital element of seeking to affirm the system of grant-based research administered by the National Institutes of Health, which preserved the status of scientists as independent, creative investigators and, at the same time, dissuaded efforts to plan or manage biomedical research through contracts, which jeopardized this sense of autonomy.

The practice of managing the War on Cancer highlights the degree to which the implementation of state policies hinges on individual expertise and communal knowledge. Specifically, molecular biologists identified ways to use their position within the governance mechanisms of the War on Cancer to limit the extension of the state into their laboratories and their professional lives. Cancer virus studies, with their close links to the expansion of molecular biology and the biomedical management theories of Carl Baker and others at the NCI, were a crucible for these efforts. At first, molecular biologists offered themselves as experts on biomedicine. They presented authoritative opinions on the nature of cancer that might guide the government's effort, arguing specifically that cancer was too complex to cure. Finding, however, that legislators and administrators were unwilling to accept such claims, molecular biologists turned to protest and activism outside the government. If they could not deny public calls that they fulfill the advances in human welfare promised by the biomedical settlement, they would turn back government intervention through guerrilla resistance, denying the NCI the consent that it needed to extend its management efforts to molecular biology. They quickly learned that working the levers of bureaucracy was the most effective means for furthering this goal. This campaign was especially evident in the creation and discussion of "The Zinder Report," a document that helped to stall and then reverse the government's effort to manage cancer research.

The Identity of Molecular Biology

Responding to the public sense that cancer virus research was ripe for acceleration, a view articulated by Robert Huebner and Benno Schmidt, supporters of the autonomy of biology at first sought to show that viruses provided evidence for a more holistic and much less mission-oriented approach by government to biomedical research. James Watson emerged as a visible and incisive spokesperson for this group. Watson had spent the early 1960s fighting to establish molecular biology within the Harvard Department of Biology, and in 1968 he also accepted the directorship of the Cold Spring Harbor Laboratory for Experimental Biology, established in the early twentieth century for the experimental study of heredity—a study it pursued alongside the neighboring Eugenics Record Office. Watson's first act was to focus the laboratory on cancer viruses, DNA tumor viruses in particular, a reorientation that reflected his own intellectual agenda as much as an effort to buttress the laboratory's precarious financial health by attracting the abundant funds designated for cancer research.[6]

Poised between a need for government support and a fear of popular demands for accountability, Watson articulated a framing of the cancer problem that other advocates of basic research also adopted. For Watson, pursuing a biological understanding of cancer might be important, but aiming to gain therapeutic payoffs from this research in the short term was foolhardy. He asserted, "Most people who have been very sincerely working on cancer think that in the short-term the problem is insoluble," and he warned that an emphasis on curing disease created a lopsided pattern of support for biology. The interest in research directed toward cancer cures, Watson claimed, would sustain poor-quality scientific research in one area "even when many other branches of science" were "being cut back to the point that their survival as first rate efforts is coming into question." He argued that studying the molecular mechanisms of carcinogenesis might yield more progress than focusing on cures.[7] This two-part claim, deflating expectations while offering the potential therapeutic payoff of molecular advances in the distant future, aimed to create the space for molecular biology to grow without popular intrusion.

The biological mystery and complexity of cancer often served as a reason to reject research management. Fresh from participating in the large antiwar protests that roiled Cambridge in the summer of 1971, MIT professor David Baltimore, a future Nobel Prize winner for his codiscovery, with Howard Temin, of reverse transcriptase, testified to the House of Representatives, "In general, cancers are still a mystery." To make progress, Baltimore argued, "broad based" research was necessary. Baltimore reminded the representatives that he had not been drawn to his pathbreaking work "because of a direct interest in cancer." His work on reverse transcriptase had arisen out of his interest in the polio virus and had been supported by the National Institute of Allergy and Infectious Diseases. He maintained that many of the successes of cancer virus research could have been achieved through grants and peer review rather than "being decided administratively."[8] Baltimore omitted from his testimony his own involvement with the Special Virus Cancer Program, which had furnished him with large quantities of mouse mammary tumor virus from its stockpiles so that he could isolate reverse transcriptase after his initial observations in a different virus.[9] Indeed, this omission was suggestive of the way molecular biologists minimized certain details of their field's growth to emphasize their individual independence rather than the contributions of infrastructure.

Shortly before the House of Representatives voted on the National Cancer Act in December 1971, *Science* published a long report on the SVCP, which speculated that the program would provide a template for future efforts to manage cancer research. The anonymous sources quoted in the article lambasted the management of cancer virus research at the NCI. Research had "no intellectual base," and "several biologists" were critical of the "moonshot-style approach of the program. . . . The emphasis on finding a human cancer virus is regarded as a political goal." Control over money exercised by the program was galling. The principal

officers of the SVCP held "too much power," lacking the "checks and bal-
ances" that governed other research programs. Outside scientists had an
after-the-fact role in the approval of contracts. This focus would not improve understanding of the "fundamental aspects of cancer and cell biology." Many biologists," the article concluded, "believe a longer-term view
is necessary."[10] Although Watson and other molecular biologists might
argue that the idea of managing cancer research was premature, they were
unable to halt the passage of the National Cancer Act.

Managing Viruses during the War on Cancer

Following the passage of the National Cancer Act, Carl Baker and his
staff at the NCI were ready for the managerial demands of the War on
Cancer. Cancer viruses provided a template for imagining what the management of cancer research might look like. In Baker's view, the NCI had
"pioneered" the modification of management techniques and systems
analysis to biomedical research programs. Through the design of the
SVCP, the NCI developed a cadre of personnel able to show adaptabil-
ity and speed in cancer research. Baker envisioned a system of management very different from peer review: "*Detailed* decision making and review must be decentralized, while *major* decision making and allocation of
blocks of resources remain centralized. For these reasons and to insure the
best use of our cancer R&D [research and development] managers, much
of the work will need to be carried out through a system of lead contractors
who will manage segments of the programs, including major subcontract-
ing efforts."[11] Research planning could be a heady experience. Louis
Carrese, Baker's lieutenant, focused on the creative flow of the process.
Groups participated in sessions that went on "for three to four weeks, un-
interrupted, 10–14 hours a day." It was best if the team was "released entirely from their regular duties" as they mapped new research pathways.[12]
Even before the passage of the act, the NCI gathered hundreds of cancer
researchers in northern Virginia to identify leads for immediate pursuit.[13]
Some of these leads provided the basis for the expanded version of the
SVCP, the Virus Cancer Program (VCP).

To accommodate the rapidly growing need for research space, the
VCP also established laboratories in the facilities of several private contractors near the NCI headquarters in Bethesda. By 1976 VCP project
officers worked at Meloy Laboratories, Microbiological Associates, Flow

Laboratories, and others in the Maryland and Virginia area.[14] Contract work presaged the redistribution of resources at the NCI. Whereas universities were the leading recipients of grants, commercial contractors were among the top recipients of contracts from the NCI.[15] While the VCP oversaw contracts that went to academic researchers, its mixture of commercial and academic contractors became a prominent feature of cancer virus research.[16] The NCI advertised contracts in the *Commerce Business Daily*, a publication for the burgeoning industry of government contractors, rather than in academic publications such as *Science*, heightening the impression that academic research might be displaced by commercial providers.[17] These individual groups were brought together by a new system of meetings established by the VCP in Hershey, Pennsylvania— suggesting that a new community of researchers might supplant academic molecular biologists in cancer virology.[18] A prominent example of this new approach was the conversion of the Fort Detrick biological warfare laboratory into the Frederick Cancer Research Facility.[19]

The use of private contractors accountable to the executive branch to "hollow out" government bureaucracies, as the VCP did, appealed to Nixon's desire to wrest control over domestic policy from established interest groups within the federal government.[20] After his reelection in 1972, Nixon turned his attention to scientific research. The staff of the Department of Health, Education, and Welfare (HEW), which contained the NIH, was one of his first targets.[21] He requested the resignation of two thousand HEW employees. He contemplated appointing a new White House science adviser to subsume both the Office of Science and Technology and the National Science Foundation. In April 1973, the Office of Management and Budget proposed that the system of peer review at the NIH might be entirely removed and replaced by a contracting system.[22]

For concerned academic biologists, Nixon's changes portended a structural shift in how the federal government supported biomedical research. The White House's proposed 1974 budget for the NIH failed to fund most activities at a rate sufficient to keep pace with inflation, but it raised funding for contracts at a rate three times as fast as that of traditional research grants. For one speaker at the Federation of American Societies for Experimental Biology's annual meeting, these trends suggested the beginnings of a reorientation of federal support from universities to private organizations. Contracts were turning the area around Bethesda's highway route 70S into a "biological route 128," a pejorative reference to the dense concentration of engineering defense contractors outside of Boston. The use of research-and-development contracts, this researcher continued, might succeed in "shifting the allegiances" of scientists who depended on federal funds and undermining their "commitment to patient, systematic, and often frustrated discovery-oriented basic research." Yet stemming this tide seemed beyond the power of the scientific community. When the Federation, a self-fashioned voice of biomedical researchers, called for national protest, it elicited just 503 letters.[23]

Molecular Biology's Antiwar Moment

In 1973, the future of the American biomedical research community appeared to hang in the balance. While that community had benefited from the biomedical settlement, it lacked the capacity to define its political contours. Biomedical researchers struggled to translate their fears into the language of politics, ironically resorting to the specter of socialist centralization to resist the actions of a president whose political reputation had been built on anticommunism. A biochemistry professor from Case Western Reserve University warned Mary Lasker that Nixon's planned consolidation of power represented a dark turn toward a "European pattern" of centralization that would extinguish the innovation associated with the "democratic diffusion of support" in America.[24] Another biochemist bemoaned that public demands for "mission-oriented research" overwhelmed the organizational capacity of the biomedical community. Government efforts to impose such a structure would result in "extremes of excess control."[25] The White House was not sympathetic, maintaining that an "excess of qualified manpower" obviated the need for a broad base of biomedical research. It was time for these researchers to start their "transition away from Federal dependency."[26] However, an economic recession in the early 1970s undermined other sources of support for academic research and magnified the implications of the Nixon Administration's changes in the NIH budget.[27]

Yet academic researchers were too few in number and too concentrated in particular parts of the country to launch a broad-based campaign of opposition, as physicians had done against President Truman's plan for national health insurance in the late 1940s.[28] The desultory response to the call of the Federation of American Societies for Experimental Biology for protest underscored this point. To shape federal policy, this community would need to find new means of influence. The debates ignited by the

Nixon Administration's actions gave new urgency to the reauthorization of the National Cancer Act in 1974. Even as the War on Cancer amplified the administration's command and control ethos, it created new avenues for molecular biologists to resist centralization. Molecular biologists grasped that their protests—lodged from outside the planning machinery of the War on Cancer—had thus far not achieved their aims. However, these protests generated a few new opportunities for resistance. Watson was invited to join the National Cancer Advisory Board in 1972. From this post, he was able to give force to the rising tide of molecular biologists' criticism directed at the NCI.[29]

Molecular biologists assumed a leading role in shaping the tone and focus of the academic biomedical community's response to the NCI. Watson conceded that biomedical research had to reckon with new organizational realities. The future resembled the state of physics: "large collective teams" that drew on budgets of millions of dollars and appreciated that "biology now runs fast." Echoing anticommunist critiques, Watson cautioned that these new organizations should not be expected to achieve a "great leap forward."[30] Charging that Nixon's health policy evoked "the image of the political commissar for the first time in American history," he urged "university scientists" and "intelligent men" to band together to advance and shape "high powered cancer research" focused on cellular and molecular biology.[31] Watson disparaged officials in the administration as "political hacks . . . block[ing] the import of new scientific blood into the National Cancer Program."[32] Pivoting to critics of the Vietnam War, Watson rejected the government's wish for "instant victory."[33] What united the pastiche of political metaphors in Watson's statements was hostility to the idea that the work of individual molecular biologists might be subordinated to administrative supervision by the NCI.

For both molecular biologists and the NCI, virus research provided a central stage for defining the legitimacy of the War on Cancer. Baker's sharp personality and blunt confidence in his own opinions had not proved to be popular with other members of the anticancer alliance, and his tenure as NCI director was brief.[34] He was replaced in 1972 by Frank Rauscher, a mouse virologist and an alumnus of the Special Virus Leukemia Program, the predecessor of the VCP. Rauscher wearily noted that most of the criticism he received about the National Cancer Plan concerned virus research and that he would need to allow the National Cancer Advisory Board to appoint another review committee. "Get all your ducks in a line," Rauscher urged the director of the VCP.[35] While this move appeared to satisfy the board's demand, Rauscher knew that as director he retained ultimate discretion over the committee's findings, a fact that was particularly important since he expected harsh judgment from the committee.[36]

As a member of the National Cancer Advisory Board, Watson could not sit on the review committee, but he put forward Norton Zinder, a highly respected phage researcher who had discovered the phenomenon of bacterial transduction, to take his place. A professor at Rockefeller University, Zinder was well poised to defend molecular biology's interest in fundamental research. He had earned the National Academy of Science Award in Molecular Biology in 1966 and was one of a handful of molecular biologists—including Watson and Joshua Lederberg—who had started to appear as the public face of molecular biology in broader debates over medicine and biological warfare. Zinder also possessed a reputation for being "outspoken," iconoclastic, and "brutally frank"—traits that made him well suited to deliver a strong critique of government management efforts.[37] Zinder was keenly aware of the political importance of his mission. As the leader of the review, he aimed to reduce contracts while ensuring that the money devoted to cancer research would remain undiminished and under the direction of peer review. "Should I succeed," he wrote to himself, "it will be a contribution to science probably bigger than any other I can make."[38]

In the summer of 1973, the deliberations of the "Zinder Committee" provided a revealing window into how the culture of molecular biology became bound up with opposition to the legitimacy of state management. At the opening meeting of the committee, on June 8, Zinder noted that the committee's questions about government competence were all the more sensitive in light of the unfolding Watergate scandal.[39] The committee should counter the administration's assumption that grants "permit indolence," one member proposed.[40] The committee's report would defend the utility of "basic research" while avoiding the sense that it flouted the "accountability" associated with contract research.[41] The committee was especially vexed by Robert Huebner's hybrid status as a scientist-manager. After his role in advocating for the National Cancer Act, Huebner played a major part in the VCP, both as a scientific worker and as a contract administrator, embodying the merger of science and administration that the VCP sought.[42]

On one hand, Huebner appeared to have far more control in determining who would get contracts than any one individual would have held in

FIGURE 8.1. James Watson (*left*) and Norton Zinder in 1974. Both of these men emerged as spokes-scientists for the molecular biology community in the 1970s. Watson served on the National Cancer Advisory Board and Zinder oversaw a report critical of the Virus Cancer Program. Watson in particular was an advocate of "first rate" molecular biology at the expense of clinical approaches to cancer. Image courtesy of Cold Spring Harbor Laboratory.

the usual system of peer review. A biochemist told the committee that while working with Huebner, "We vote but that does not necessarily mean that the vote you are taking is in any way going to influence" the final decision.[43] On the other hand, the experience of working on a contract did not seem as constricting as critics feared. Robert Gallo, who later became one of the first to identify the human immunodeficiency virus after working to find human leukemia viruses, maintained, "The contract does not state very clearly what [one] is supposed to do. . . . I think it can work with established persons, just like the grant mechanism." Wallace Rowe, an adviser to the VCP, sharply criticized the effort to separate administration from research: "This is the whole function of Huebner's group. You are saying get rid of Huebner? You would if you put this into effect. What you are saying is that someone with a scientific career should not also be in a position of administering large amounts of money."[44]

The committee's report, presented by Zinder in March 1974, highlighted the threat posed by contract research to the autonomy of molecular biology. The VCP "overshadowed" other areas of research. It had the ability to shape its chosen areas of cancer virus studies without recourse to peer review, spending $42 million on 131 contracts, while the rest of the NIH spent just $58 million on more than 2,000 peer-reviewed grants in cancer virology. Sensing that it would be unwise to challenge the therapeutic ambitions of mission-oriented research directly, Zinder sought to co-opt and repurpose the very ideas and language that had been used to advance contracts at the expense of peer-reviewed research in the 1960s. Rather than providing a unified approach to cancer research, the report maintained, contracts had splintered into "disparate elements." Administrator-researchers were consumed by conflicts of interest. The VCP itself was "overblown," "redundant," and had "high-cost relative to yield"—echoing the charges leveled at peer review a decade earlier.[45] Peer review, in contrast, provided responsive control from expert scientists rather than uninformed VCP officers.[46] Zinder endorsed gutting the contracting apparatus of the VCP and diverting its budget to grants.[47]

Later accounts claim the Zinder Committee's report as a decisive blow against the VCP, but its immediate impact was modest. Rauscher's bureaucratic acumen forestalled the implementation of its recommendations. Although it was presented to the National Cancer Advisory Board, the NCI never officially published the document. The subsequent analysis of the subcommittee of the National Advisory Cancer Board charged with recommending changes in the VCP based on the report did not endorse reducing the number of contracts *or* shifting the VCP's focus away from human cancer viruses.[48] One of the three members of this subcommittee, Howard Skipper, a chemotherapy expert, leveled biting criticism at the Zinder Committee, charging that it "did not spend any time looking (in depth) at the *scientific and/or pragmatic* progress, or lack of progress of the VCP. . . . This is the critical issue, not who they like or don't like, or who has responsibility (power) and who *wants* that responsibility (power)."[49] While funds for contracts in cancer virus research peaked in 1974, they did not decline until several years later.[50] In public, Rauscher could ignore the report's condemnation of contracts and argue that health research was of a different character than biological research and thus in need of different administrative mechanisms.[51]

Molecular Biology versus Big Biology

Molecular biologists lacked sufficient power within the management of the War on Cancer to alter its course. However, they proved extremely savvy in using the controversy surrounding the Zinder Report as part of a guerrilla campaign to slow the momentum toward contract-based biomedical research. Their campaign focused on articulating the identity and ethos of molecular biology in opposition to the legitimacy of the government's authority to direct basic research. Watson, identified as one of the VCP's most "vociferous" opponents, highlighted the idea that the criticism of the VCP was about the distribution of resources: "There are a lot of virologists who share the same goals. . . . The ones in the VCP were very rich. The others, who are just as good, were very poor."[52]

Other critics made it clear that the demographic and intellectual future of molecular biology was at issue. The rise of cancer research came at the expense of other fields. Howard Temin, the codiscoverer of reverse transcriptase, pointedly observed that "the talent and money spent on cancer research are lost to other pursuits." The Stanford biochemist Arthur Kornberg, who had failed to persuade skeptical senators of the social worth of his research in 1967, repeated the accusation that work in cancer was of low intellectual quality and that contracts "narrowed the focus" of research and reduced the opportunities for "serendipitous discovery which have dominated the modern history of medical research." He continued that the "availability of money, jobs, and facilities in the comparatively affluent . . . cancer project seduces students, faculties, and institutions. It may distort their perspective." Kornberg vividly envisioned a dystopian future in which "first rate" researchers in bacterial genetics and virology "are denied NIH support and university appointment while second-rate colleagues working on cancer viruses get both."[53]

The Zinder Committee's work produced further evidence used to criticize the administration of the VCP and the War on Cancer during the renewal of the National Cancer Act, evidence that again raised the accusation that managers were enriching themselves at the expense of good science. Frustrating Rauscher's effort to contain the distribution of the committee's report, a member of the committee passed a copy to a reporter at *Science* before it was even presented to the National Cancer Advisory Board. Coverage in *Science* amplified the concern that contract research practices created a "closed shop." At a time when the expansion of contract research appeared likely, the article concluded, "the report must not be accepted and then shelved."[54] While maintaining that "as research goes" there was no difference between contracts and grants, Zinder argued that the "specified target" of a contract did not fit a basic research effort.[55] Speaking to the *Medical World News*, Zinder commented that there was nothing wrong with the government deciding to focus on cancer, but that "having done so, it should let the scientists decide how they should go about it—at least as long as we need fundamental research to get us off the ground."[56]

The discussion of misadministration signaled that critics of the NCI's management effort had succeeded in finding a place to push back that struck at both the VCP's effective operation and its standing within the scientific community of experts it sought to enroll. The key personnel of the VCP's administrative effort were contract specialists, who provided the "bridge between the scientists who administer research and the administrative mechanism by which the information is implemented through the awarding and fiscal monitoring of a research contract." By combining "scientific comprehension with their administrative knowledge" these specialists were essential to the operations of the program.[57] Close monitoring was essential for these "fast moving" research programs."[58] The NCI had long struggled with the difficulty of recruiting these hybrid scientist-managers. "Many of the staff members who have the technical competence for program management are motivated, by the attitudes of their professional community, to apply this competence in individual research rather than to 'waste' it in management affairs," one review of management priorities at the NIH had concluded in the 1960s.[59]

This situation is perplexing until it is viewed from the perspective of the academic molecular biologists whom the NCI sought to enlist. When Zinder presented his report, he emphasized the importance of this group's consent: "High productivity in research of broad scope was not possible under the control of a small group of individuals. The result has been tension and apathy in the scientific community."[60] These scientists resented the administrators' power and the way it shaped their professional identity. Even if that power was used lightly, the cultural taint of control remained. "Discretion with respect to what to do about new ideas and directions developed by an investigator should lie" with scientists rather than managers, "who appear to view basic science projects and scientist[s] in the same light as house painting and house painters," one virologist tartly wrote.[61]

The debate fostered by Zinder, Watson, and others succeeded in deepening the cultural divide between molecular biologists and the administrative machinery of the NCI. A survey in *Cancer Research* revealed these sharp differences. Of those receiving grants administered through the peer-review process, 76 percent supported a major shift away from contracts, but only 28 percent of those receiving contracts did.[62] Rauscher confessed that, despite abundant funding, the NCI struggled to recruit contract specialists and other administrative staff.[63] The former Atomic Energy Commission program manager Robert Sweek, having had his interest in biomedical research planning piqued by his work for the Panel of Consultants for the Conquest of Cancer in 1971, went on to pursue a doctorate in public administration from American University. His dissertation examined the implementation of the National Cancer Plan. Sweek concluded that the War on Cancer's administrative problems stemmed from the biomedical community's resistance to the training of "technically qualified management talent for large biomedical programs." Based on his survey of biological and physical scientists, Sweek suggested that "a fundamental difference in philosophy" separated the two communities. Physical scientists were much more "program-oriented," while the biological sciences resisted management in favor of individually autonomous "basic" research.[64] He concluded, "The secret of success in administering large scale cancer research seems to have eluded everyone," but it would be more accurate to say that molecular biologists managed to elude large-scale administration.[65]

In refusing to engage with the managerial apparatus imagined in Washington, academic biologists succeeded in denying the NCI access to the tacit knowledge of their field and diminished its capacity to manage or direct their research process.[66] An internal NIH review captured what this lack of knowledge produced, commenting that "NCI contracts are very vague and indefinite about reporting requirements on the part of the contractor."[67] The result was, ironically, that those working on contracts often felt a greater sense of autonomy than they had experienced working under grants. One researcher reported that during his contract to study Epstein-Barr Virus, "I never had the feeling that the VCP has attempted to direct our research. On the contrary, we felt positively stimulated by the support . . . and possibilities for interaction with other investigators."[68]

Unlike engineering and the physical sciences, molecular biology emerged from its brush with national mobilization holding fast to its small-scale and libertarian ethos. In their fight, molecular biologists recognized their collective interests as a scientific group and discovered that melding political values of freedom and biological claims of complexity produced an ideal means for hedging the expansion of the state even as they remained dependent upon it. When their first efforts at grassroots protest or expert critique fell short, they found that controlling the flow of tacit knowledge about their field checked the ambitions of the VCP. Their dim view of federal government soon extended beyond those working directly on virology. The biochemist Arthur Pardee, best known for his work on messenger RNA, wrote to Benno Schmidt criticizing the NCI for its failure to support "innovative, fundamental, cancer research" through grants to individual scientists. "Original research," Pardee explained, "is exploration rather than engineering. . . . I am not at all persuaded that the best expenditure of American taxpayers' dollars is for a small number of second-rate problems that employ numerous third-rate investigators and use supply contract materials made by an army of technicians." Yet Pardee's letter also contained a startling admission for one so confident in his indictment: "I suspect that the Zinder Report attacked this problem, though I have not seen this report (Could you have a copy sent to me?)"[69]

The perception that basic or fundamental research was marked by freedom and serendipity, articulated in the course of resistance to the War on Cancer, reemerged as a powerful principle in renegotiating the biomedical settlement. The president of the Memorial Sloan Kettering Cancer Center concluded at a conference on the fifth anniversary of the National Cancer Act that "really good ideas, the sudden intuitive perceptions of connections between seemingly unrelated bits of information, the sudden overwhelming revelations that make a scientist worry seriously about what would happen to the fate of the world if he were hit by a truck on his way to work, occur in individual minds, and they cannot be programmed or planned."[70] A 1976 congressional panel on biomedical research argued that basic and applied research were divided primarily by their degree of uncertainty and certainty. The methods of managing these kinds of research were "radically different." "Tight organization schedules" should not be applied to problems of high uncertainty, the panel concluded.[71] The complexity of cancer and the freedom of scientists were linked. Watson concluded his chapter on cancer viruses in the 1976 edition of his textbook *Molecular Biology of the Gene* with the warning, "No matter how cleverly we think, no matter how great our capacity to spend vast sums of money for crash programs, the inherent complexity of eukaryotic cells still exceeds our intuitive powers. So all too frequently we must be satisfied with modest objectives."[72]

A Watershed Moment for the Biomedical Settlement

The War on Cancer, like the War on Poverty, marked a watershed for the reach of the government into American society. In the late 1970s, confidence in the government's ability to direct biomedical research eroded nearly as quickly as it had risen in the late 1960s. The VCP mirrored this trajectory. Having failed to identify a human cancer virus, it was abruptly closed in 1978. Some of the reasons for the VCP's frustration were more contingent than the resistance it faced suggested. The resignation of President Nixon in 1974 after the Watergate scandal removed an ardent advocate of centralization from the White House. The most powerful members of the War on Cancer bureaucracy, Rauscher and Schmidt, lacked the administrative acumen (or tenacity) to fully implement and defend the most ambitious managerial goals of the program as they had been framed at its inception. Huebner, the VCP's most energetic and articulate manager, succumbed to Alzheimer's disease as the administrative burdens of his position grew.[73] These many developments blunted momentum toward the management of biomedical research by the late 1970s.

The cultural shift instigated by the War on Cancer, coupled with the controversy over the NIH's regulation of recombinant DNA techniques in the mid-1970s, succeeded in fostering an enduring suspicion of government intrusion in the molecular biology community.[74] Many scientists came to live a double life, publicly claiming that their work might further the quest for a cancer cure while privately admitting that such claims were only a means of attracting funding for the intellectually exciting research they aimed to carry out in biology.[75] The molecular biologists who emerged as prominent authorities on cancer sought to manage this contradiction by embracing the idea of cancer as a complex puzzle. In that view, the returns of the VCP or other efforts to direct biomedical research would always be meager.

Suspicion of management cast a long shadow, even when scientific developments suggested that hope for a vaccine to prevent cancer was not forgone. Just as the VCP ended, virologists and epidemiologists working in other quarters were on the verge of demonstrating a link between infection by the Hepatitis B virus and liver cancer.[76] Yet in offering advice to the NCI as it prepared to establish a task force on Hepatitis B virus in the early 1980s, a molecular virologist repeated Zinder's criticism, lambasting the use of contracts as a device that attracted researchers interested in cancer not "for its inherent intellectual challenges" but in the "advertised funds."[77] The author, Harold Varmus, had in fact been one of those biologists drawn, wittingly or not, into cancer virus research by the ready funding offered by the NCI. The activities of his laboratory at the University of California, San Francisco, considered in chapter 9, illustrate how apparently serendipitous discoveries in the laboratory could possess powerful antecedents in state planning.

CHAPTER NINE

The West Coast Retrovirus Rush and the Discovery of Oncogenes

In the summer of 1970, while national debate over strategy for the War on Cancer mounted, Howard Temin rose to give a lecture on retroviruses as a part of a summer course on animal virology at the Cold Spring Harbor Laboratory. Temin had devoted the previous decade to a lonely effort: studying the reproduction of Rous Sarcoma Virus, an RNA tumor virus, or "retrovirus." If each of the fifteen people in the room understood retroviruses at the end of the course, Temin joked, he would have succeeded in doubling the number of scientists in the world who did.[1] By 1974, however, 450 scientists packed into the Cold Spring Harbor Laboratory's tumor virus meeting, and retroviruses dominated the discussion.[2] What had changed in the interval? In 1970 Temin and MIT virologist David Baltimore identified an enzyme—reverse transcriptase—that allowed retroviruses to change the DNA of cells. As discussed in chapter 6, this discovery energized the ongoing efforts of the National Cancer Institute to understand retroviruses as a critical element of demonstrating the viral origins of human cancers. The National Cancer Institute placed retroviruses and the molecular oncogene theory at the center of its efforts during the War on Cancer, creating an "instantaneously overcrowded field."[3]

In 1976 two of the scientists drawn into this crowded field of retrovirus research, J. Michael Bishop and Harold Varmus, at the University of California, San Francisco, provided evidence that a viral oncogene, *src*, was in fact present in many normal cells. Bishop and Varmus's demonstration of the existence of cancer-causing genes, or "oncogenes," within normal cells marked a decisive step in the ongoing effort to molecularize studies of cellular growth and development as a whole.[4] Awarding Bishop and

Varmus the 1989 Nobel Prize in Medicine, the Nobel Assembly praised
their work for showing that "cancer was not contagious." Their finding,
which indicated that "changes in genetic material constitute the basis for
the development of cancer," provided molecular biologists with a pivot
point from external to internal causes of cancer.[5] In accepting his prize,
Varmus quoted Baltimore's own Nobel acceptance speech from 1976 in
giving thanks that the "virologist . . . can see into his chosen pet down to
the details of all its molecules."[6]

However, the ability to "see into the molecules" was not a feature of
viruses themselves but a consequence of the material and social infrastructures that the NCI had built around retroviruses during the War on
Cancer. The accomplishments of Bishop and Varmus's laboratory in San
Francisco during the early 1970s illustrates a paradox familiar from the
history of physics. Seeing with smaller and smaller resolution—to the
level of the molecule—required larger and larger forms of scientific organization. The infrastructure that the leadership of the War on Cancer put
in place exerted a profound influence on the migration of the molecular
approach to life from bacteria into higher organisms. I explore other aspects of this infrastructure in chapter 10, but in tracing the genealogy of
Bishop and Varmus's identification of *src*, I will provide a concrete example of how social and material infrastructures played a generative, rather
than merely accessory, role in the production of knowledge about cancer
at the molecular level.

The core concept that helps connect the infrastructure of the War on
Cancer to the practices of Bishop and Varmus's laboratory is that of an *experimental system*. Experimental systems, as philosopher of biology Hans-
Jörg Rheinberger describes them, are the "working units" of science,
an arrangement of interpretive theories, technologies, tools, and organisms designed to ask questions rather than provide answers. Because experimental systems are organized around questions, they are by nature
unstable—the research object, or "epistemic thing," at the center of the
system is not yet fully defined. It is the object that the experimenters aim
to learn more about, and to ask questions about the object, they need
to use the methods and tools of the system to bring the epistemic thing
into existence in the laboratory. There are no single experiments that are
decisive moments of "proof": each single movement in the system occurs
in the context of previous experiments. Instead of promising replicability,
experimental systems often thrive on uncertainty and ambiguity—these
are the moments when new kinds of research objects emerge. They are

machines, in the words of the molecular biologist François Jacob, "for generating the future."[7]

This chapter follows the emergence of the theory of cellular oncogenes from its origins in the Special Virus Cancer Program to the experimental practices of Bishop and Varmus in San Francisco. The efforts of their laboratory to construct an experimental system for the study of oncogenes provide an ideal lens through which to explore how the infrastructure of the War on Cancer shaped the creation of new knowledge in molecular biology at the level of the benchtop. Their pursuit of *src* drew upon a range of social, financial, and material resources mustered under the aegis of Robert Huebner's interest in using retroviruses to reveal a link between viruses and human cancer. Few specialists endorsed this theory, but as a leader of the NCI's SVCP, Huebner was able to use the very feature of the War on Cancer that molecular biologists resented—the capacity of administrators to award contracts for research independent of peer review—to foster a wide range of inquiries into retroviruses. Those inquiries provided the background against which Bishop and Varmus started their search for *src*, with unanticipated results. Although Huebner did not intend to reveal cellular oncogenes when he placed the resources of the SVCP behind retrovirus studies, these results illustrate that the futures generated by experimental systems are closely bound up with their past material and social environment.

The Social Infrastructure of RSV Research on the West Coast

Bishop and Varmus were two of the new researchers drawn into the "overcrowded" field of retrovirus research by the SVCP. Though neither had first chosen to pursue cancer virus research, when they arrived in San Francisco they drew on a dense community of retrovirus researchers to make rapid progress in their study of Rous Sarcoma Virus (RSV). The ease with which the two outsiders became fluent in the study of RSV illustrates the importance of the social infrastructure the SVCP helped foster around cancer virology. In 1969 neither Bishop nor Varmus was an obvious candidate for retrovirus research. Both had pursued training in clinical medicine, not biology. The Vietnam War, not the War on Cancer, moved them to pursue biomedical research. During the 1960s, the National Institutes of Health, worried that doctors from elite medical schools were shunning biological research in favor of clinical work,

FIGURE 9.1. J. Michael Bishop, about 1984. Courtesy of General Motors Cancer Research Foundation and the National Cancer Institute.

used the Selective Service System to recruit male doctors for training as physician-researchers at its Bethesda campus. Through an arrangement with the surgeon general and the US Public Health Service, whose Commissioned Corps members could count their work toward military service, the NIH had the ability to offer up to seven hundred doctors each year the chance to pursue research rather than face combat.[8] When Bishop graduated from Harvard Medical School in 1962, he jumped at the chance to avoid the "clutches of the US Army" by studying the biochemistry of the poliovirus at the NIH.[9] His training there included courses on many topics in the biological sciences as well as laboratory research projects—a PhD in miniature.[10]

In 1968 Bishop accepted a job offer from the Microbiology and Biochemistry Department of the University of California, San Francisco (UCSF). Bishop joined an institution deeply reliant on federal support for biomedical research. Founded in the late nineteenth century as a medical school affiliated with the University of California, Berkeley, UCSF attempted to connect training hospitals in San Francisco with biology and biochemistry laboratories across the bay in Berkeley. After the Second World War, UCSF gained its independence and a campus in San Francisco. Yet this independence was costly. Most of the biology and biochemistry faculty, along with their resources, remained at the more prestigious Berkeley campus. Bishop joined the many young professors that UCSF hired to develop its program of laboratory biomedical research. Like other young professors there, he found little institutional support. The new UCSF lacked an endowment to support biological studies that were not directly tied to clinical ends. As a result, the growth of UCSF and the work of its laboratories hinged on external funding to a greater degree than its established, and wealthier, peers.[11]

The start of Bishop's academic career coincided with a period of stagnating federal funds for biomedical research in the late 1960s, discussed in chapter 7, threatening his efforts to continue research on RNA viruses. Although Bishop had a grant to continue his studies of polio, he gravitated toward the stability and growth promised by funding for cancer research, forming a partnership with Warren Levinson, an RSV researcher at UCSF. Bishop's biochemical experience, they hoped, could usefully be applied to studying the replication of RSV and its transformation of infected cells. In Bishop's mind, this was a move from the study of the biochemistry of one RNA virus to another, not a move toward cancer research per se.[12]

Bishop and Levinson approached the SVCP in 1971, requesting support for the study of reverse transcriptase in avian tumor viruses. Given the hundreds of thousands of dollars it spent to support other research contracts, the SVCP was happy to award a contract worth $77,000 annually to sustain Bishop's efforts to purify reverse transcriptase from RSV and detect virus-specific RNA and DNA in infected cells.[13] This amount was modest compared to the program's other expenditures, but for Bishop and Levinson it was a bounty. The funds also arrived quickly. Bishop recalled the extraordinary difference between contract and grant applications: "There was money to be had . . . [and] we petitioned for it and got it. . . . It was *one* month, not six, twelve months . . . then rewrite and rewrite [the grant]."[14] The significance of the contract to their study of RSV was immense. From 1971 to 1972, for example, 85 percent of their external funding to study RSV came from the SVCP.[15] The emphasis of their research, however, was different from that of Temin's work on RSV. In one of their first funding applications, Bishop and Levinson took pains to distance themselves from Temin's controversial theory, writing that their approach took it as "implicit" that the transformation of cells by RNA tumor viruses "is accomplished without the 'integration' of viral genome (or genetic information) into the host genome."[16]

Varmus's path was similarly influenced by the Vietnam War. "Fervently opposed" to the conflict, Varmus also elected to serve at the NIH rather than face the military draft after graduating from the Columbia University College of Physicians and Surgeons in 1968.[17] At the NIH, he initially studied bacterial gene regulation but soon became interested in the possibility of using RNA tumor viruses to understand the genetics of cancer. He first attempted to pursue this project as a postdoctoral fellow at the prestigious Salk Institute under Renato Dulbecco, but Dulbecco, whose focus was on DNA tumor viruses, did not sponsor him. Visiting the San Francisco Bay Area, Varmus first went to Berkeley, where Temin's former collaborator Harry Rubin directed him to UCSF. Varmus visited, "not knowing anything about the group."[18] Despite that ignorance, he was reassured by learning that so many of his potential collaborators had previous experience at the NIH. Moreover, San Francisco's location only hours from the hiking and fishing opportunities of the Sierra Mountains appealed to Varmus, who was an avid outdoorsman.[19]

The NCI's investment in retroviruses not only brought Bishop and Varmus together but also sustained a broader community that aided their studies of RSV. Bishop and Varmus were a node of Robert Huebner's

FIGURE 9.2. Harold Varmus in 1981. Image © 1981 by Janet Fries.

Pacific Tumor Virus Group, one of several regional subgroups established by the SVCP to promote communication among the numerous retrovirus researchers it sponsored after the discovery of reverse transcriptase.[20] Varmus recognized their community of RSV researchers as the "West Coast Tumor Virus Cooperative." Bishop, Levinson, and Varmus at UCSF—along with Peter Vogt at the University of Southern California and Robin Weiss, a former postdoctoral researcher of Vogt's, at the University of Washington—formed the core of the group, and they were later joined by the biochemist Peter Duesberg at the University of California, Berkeley.[21] In 1971, with the aid of the West Coast Tumor Virus Cooperative, Bishop, Levinson, and Varmus set out to expand their laboratory into a center for the use of molecular techniques to hunt for the *src* gene in cells. Whereas Bishop had originally focused on the biochemistry of RSV's reproduction, Bishop and Varmus now sought to detect the presence of viral genes in the genetic material of infected cells.

In shifting their focus from virology to genetics, Bishop and Varmus drew on the work of this vibrant community of retrovirus researchers. The cooperative united skills not usually drawn together. A letter from Weiss to the San Francisco members suggested defensiveness when traditional virologists encountered molecular biology: "I hope you will welcome this as a genuine, practical and intellectual collaboration. . . . You will soon appreciate that I can't tell Watson from Crick . . . but then again you can't tell . . . RSV_α from $_\omega$, can you? Virologists and biochemists of the West Coast: Unite! Spool out the DNA of those embryos that are coming home to roost in a San Francisco Revco [freezer]."[22] While Weiss's study of the mutants of RSV did not equip him for the terminology of molecular biology, the cooperative provided a framework for the two disciplines to meet.

As they embarked on their search for *src*, Bishop and Varmus benefited from the assurance that there was in fact a single gene to find. Unlike molecular biologists, traditional virologists still treated genes as abstract entities for describing inheritance rather than physical fragments of DNA or RNA. After Temin and Rubin's creation of an experimental system for studying the reproduction of RSV, other virologists had attempted to map its genes by tracking mutations over generations, the same strategy that served to map the genes of fruit flies and many other species.[23] Vogt embarked on a long project to generate and catalog mutants of RSV, and the SVCP awarded him $250,000 (1971 dollars) annually to provide these mutants to other laboratories.[24]

Vogt sent some of the mutants to Levinson, a former student, and Levinson shared them with Bishop and Varmus. Bishop and Varmus, in turn, shared these mutants with a postdoctoral researcher, G. Stephen Martin, who was working in Berkeley's Zoology Department under Harry Rubin's supervision. The mutant strains of RSV that Martin received from Los Angeles by way of San Francisco were of a very particular kind. These strains were temperature sensitive. Cultured at 41°C, the average internal temperature of a chicken, this strain of RSV would infect cells and cause them to produce RSV without being transformed. The same RSV strain grown in cells maintained at a lower temperature, 36°C, would infect and transform the cells. To Martin, this finding suggested that within the genome of RSV, the gene responsible for the cancerous transformation of cells after infection was distinct from the genes required for reproduction of the virus.[25]

Another member of the cooperative working in parallel with Martin, Peter Duesberg, soon demonstrated that the existence of a transforming gene posited by Martin's research corresponded to a physical difference in the composition of the genomes of the two RSV types. While he later became notorious for his denial of the link between HIV (another retrovirus) and AIDS, in the early 1970s Duesberg enjoyed a reputation as an expert in the biochemistry of RNA viruses, "one of the best people working on the[ir] molecular biology."[26] He received $150,000 annually from the SVCP to apply the techniques that he had developed to explore the chemical composition of RNA virus genomes to Martin's mutant strains of RSV.[27] Duesberg broke the RNA of the mutant and normal strains into fragments. He then ran these fragments through a gel electrophoresis apparatus, which used an electric charge to sort fragments of RNA by size. The results from this apparatus provided the first evidence that the genomes of the mutant and normal RSV strains differed physically: the mutant was missing a fragment of RNA found in the normal strains of RSV. That fact in turn provided the first suggestion that a tumor virus's transforming power could reside in a single, identifiable, physical gene.[28]

Duesberg's studies raised the possibility that it would be feasible to isolate the RNA fragment corresponding to this transforming gene. Together, Martin and Duesberg used radioactively labeled RNA and electrophoresis to document the presence of an RNA gene fragment in normal RSV but not in temperature-sensitive mutants. They also ruled out the possibility that the additional RNA in a normal RSV genome bore no relationship to the transforming effect of the virus.[29] The transforming

RSV gene became known as *src*, an abbreviation of "sarcoma." Greeting their result, *Nature* predicted, "It should not be long before that part of the Rous sarcoma virus genome is, in useful amounts, separated from the part carrying genes not involved in the transformation of [cells]."[30]

Showing that RSV infection actually placed a copy of *src* into a cellular genome was a far more difficult task. The quantity of DNA in a cell's genome was many, many times larger than the amount of RNA in RSV. The sheer number of fragments overwhelmed Duesberg's electrophoresis apparatus. Bishop and Varmus thought they had a different solution to this challenge. Although neither was well versed in RSV genetics, both were familiar with hybridization reactions between DNA and RNA. Varmus, in particular, had spent much of his time at the NIH using hybridization to study the regulation of *E. coli* genes.[31] Bishop, who had a lingering sense of having been "scooped" by the discovery of reverse transcriptase, pursued the implications of this new method with particular zeal.[32]

Bishop and Varmus aimed to adapt hybridization reactions to the task of hunting for the *src* gene. For molecular biologists, DNA-DNA hybridization was well understood: single DNA strands with complementary sequences of base pairs would bond together to form a stable double helix.[33] While RNA usually existed in more unstable single strands, it could also pair with complementary DNA sequences if the DNA helix was "unwound" under the right temperature or chemical conditions. Hypothetically, RNA *src* genes could hybridize with fragments of uncoiled DNA from infected chicken cells. If successful, such reactions would demonstrate that DNA analogs of the *src* RNA existed in infected cells. However, RNA-DNA hybridization was far more complex than DNA-DNA hybridization; the DNA strands were more willing to fold and pair with themselves than with RNA. Even if conditions could be changed to allow DNA and RNA to hybridize, it was doubtful that the final results could indicate with sufficient precision whether an exact DNA copy of *src* existed in the infected cell.[34] For Bishop and Varmus, reverse transcriptase suggested a different approach to this problem. Starting with RNA corresponding to *src*, reverse transcriptase could be used to create a DNA copy that incorporated radioactively labeled base pairs. This radioactively tagged *src* DNA molecule, which was more stable than *src* RNA, could then be used as the basis of hybridization tests that paralleled those developed at the Salk Institute to study a DNA tumor virus, SV40.[35]

That process, though appealing in its simplicity, required substantial resources. The SVCP was happy to provide these resources, given its hope

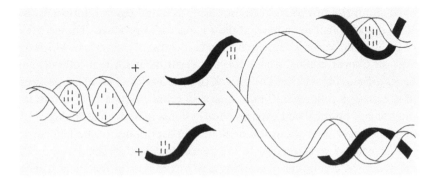

FIGURE 9.3. A condensed illustration of a hypothetical hybridization reaction between an un-known DNA segment (white) and known fragments for comparison (black). The winding and unwinding of DNA helixes as conditions of temperature or chemistry varied provided a very accurate way of comparing sequences of different fragments. The stronger a set of bonds be-tween single strands of DNA, the higher the number of complementary bases. If a fragment of DNA or RNA that a researcher wanted to study (black) could be marked in advance—for example, through incorporating radioactive isotopes into the base pairs—then this marked segment could serve as a probe for similar sequences of base pairs among fragments of DNA (white) as the original DNA helix was unspooled and broken up. For more on the mechanics of hybridization reactions, see note 33 in this chapter. Illustration by Steven Parton.

that one of its contractors would use reverse transcriptase along with DNA hybridization to create "an extremely sensitive device for detecting cryptic virus states" associated with retrovirus infection and human cancer.[36] While the discovery of reverse transcriptase in 1970 had established the theoretical possibility that retroviruses could change the DNA of a cellular genome, by 1972 no such changes had been identified. Huebner and the other leaders of the SVCP's Solid Tumor Virus Segment still hoped to show that retrovirus genes were not only inserted into cells, but became a part of the cell's genetic endowment for generations before the appearance of cancer. Elucidating the character of *src* at a genetic and biochemical level would provide a test for the oncogene hypothesis—proof whether or not a viral oncogene played a role in cancerous transformation.[37]

Practicing Molecular Biology in the Middle of the War on Cancer

With the stability that SVCP sponsorship provided their work, the UCSF group started to increase the scale of their reverse transcriptase studies in preparation for the creation of an *src* probe. Their laboratory would need quantities of reverse transcriptase, RSV, and temperature-sensitive RSV mutants that were beyond their capacity to produce. The progress of Bishop and Varmus's research agenda illustrated the ways in which the War on Cancer created a new form of scientific work that fell between benchtop and big science. The growth of their laboratory and the progress of the probe's refinement in the next few years eloquently illustrate the interplay of political and experimental systems in molecular biology—of growing larger in order to see smaller. The experience of the laboratory also provides a counterpoint to the concerns voiced by Zinder and others about the conditions of working under contract for the Virus Cancer Program.

In late 1971 Bishop wrote to Huebner, "Our principal problem is logistics." Though retroviruses and reverse transcriptase enzymes were becoming commercially available from contractors affiliated with the SVCP, they were not pure enough for the group's sensitive hybridization experiments. However, the DNA hybridization reactions that the group sought to carry out were inefficient, requiring "far more virus than we can presently prepare." Even as its hope of providing standard reagents fell short, the SVCP's support freed Bishop and Varmus from the labor of cultivation, allowing their laboratory to hire technicians and buy equipment to increase their production of RSV "ten-fold." Bishop confidently asserted in one of his first reports, "Whatever the inconvenience and labor may be . . . we intend to continue to prepare our own biological reagents because this provides considerably [sic] economy, assures quality control, and offers greater flexibility in experimental design."[38]

A year later, Bishop stated, "We still need more virus and larger quantities of infected cells of every type. We will be unable to further expand our local production until the University provides us with new space. . . . For the present we must rely on a commercial source of RSV [University Laboratories] to supplement locally produced virus stock, and upon various collaborators to supply us with materials not presently produced in our laboratory."[39] This work was labor intensive. Members of the laboratory needed to cultivate two or three hundred petri dishes of virus to provide the precursors for one experiment.[40] However, grants for RSV research from the NCI and the American Cancer Society provided "no resources for the purification" of the compounds that were an important part of the laboratory's experimental work—this labor was supported by the SVCP and later by its expanded version, the VCP.[41]

As the laboratory grew in size, its fluency with difficult hybridization reactions increased. Bishop and Varmus first used these new tests to argue that viral infections were more widespread in "normal" cells than traditional virologists assumed. In particular, they used hybridization reactions to show that, contrary to the results produced by antibody tests, "clean" chicken embryos showed evidence of RSV infection. Nonetheless, this kind of test was too imprecise to establish whether the transforming genes from RSV were also present in the uninfected cells. Theoretically, the radioactive signature of the synthetic viral DNA (vDNA) probe could be followed to corresponding cellular DNA (cDNA) in a cell during a hybridization reaction. In practice, however, the results were difficult to interpret because of the way the laboratory created its radioactively labeled vDNA.[42] As Bishop and Varmus continued to adjust their techniques, they sometimes found that their estimates of the amount of vDNA copied into a cellular genome needed to be revised sharply downward once they developed more accurate DNA-based assays, sometimes by a factor of ten or more.[43]

Bishop and Varmus had no way to know how many vDNA copies of *src* existed during a synthesis, or whether the entire viral genome participated in their reactions. The length of time for which experiments were allowed to run for "full" synthesis of vDNA from RSV RNA, the appropriate ratio of excess cDNA to use, and the difficulty of gauging the extent of hybridization all complicated the interpretation of the laboratory's experimental results.[44] Reviewers of the laboratory's publications often raised these objections, and the only reply Bishop or Varmus often supplied was to assert the depth of their laboratory's experience with these hybridization reactions.[45] The only definitive means that Bishop and Varmus foresaw of resolving these challenges lay within longer-running and consequently more expensive hybridization experiments.[46]

With support from the VCP, the laboratory's members became the recognized specialists in the use of hybridization and radioactively labeled probes to detect retrovirus genes in cells. The chairs of the VCP lauded the Bishop-Varmus laboratory as "one of the best in the Program" for its mastery of different techniques for the hybridization of nucleic acid. The chairs subsequently voted to increase the laboratory's annual contract from $124,000 to $200,000.[47] At a meeting of the Pacific Tumor Virus Group in Berkeley, otherwise dominated by excitement over the detection of possible retroviruses in human placentas, Bishop announced that his laboratory had developed a "highly sensitive method" for demonstrating equivalence between fractions of the viral genome and DNA sequences in the genome of normal cells.[48] The laboratory expanded its studies to include mouse retroviruses, and in a 1974 contract summary, Bishop made a tongue-in-cheek reference to the laboratory's notoriety for hybridization techniques, writing that "cultured cells had been examined with a newly developed assay based on (what else?) molecular hybridization."[49]

As these studies continued, "small scale" production of reverse transcriptase and single-stranded vDNA probes were "no longer practical," Bishop wrote. Pursuing the "very promising new procedure" of using an excess of vDNA tripled the laboratory's already substantial rate of radioactive isotope purchases. From 1972 until 1976, the laboratory was entirely dependent on its VCP contract to support the purification of RSV and reverse transcriptase. The VCP also supported the purchase of high-speed centrifuges and other equipment unavailable from UCSF. These resources drove the further expansion of the laboratory. Assigning technical staff full-time to the process of synthesizing the probe compelled further production to both meet experimental needs and justify the expense of technicians and equipment devoted to virus culture and enzyme purification.[50]

In their pursuit of *src*, Bishop, Varmus, Levinson, and a fourth colleague, Leon Levintow, grew into a closely knit interdisciplinary research unit at UCSF. The work of this unit, Bishop explained, constituted an "intellectual whole, directed and funded in an interlocking manner." The intermingling of funds and resources created a self-sufficient laboratory complex "without assistance from the university." Even adjusting for the fact that many of Bishop's comments regarding the importance of contracts came in his pursuit of further funding, the transformation wrought on the UCSF laboratory by SVCP support was profound. The laboratory included the four professors, fifteen postdoctoral fellows, numerous graduate students, and ten support staff. The work of the laboratory technicians allowed for the large-scale production of essential laboratory materials, including purified RSV and chicken cell cultures. Contracts from the SVCP underwrote the majority of the supplies required by this operation, including the replacement of ultracentrifuge rotors used in the purification of reverse transcriptase (the laboratory went through three a year), radioactive isotopes, and precision heaters required for long-running molecular hybridization experiments. These were not trivial expenses. Radioisotopes for the production of hybridization probes alone cost $24,250 annually, 150 percent of the salary a skilled technician working in the laboratory earned.[51]

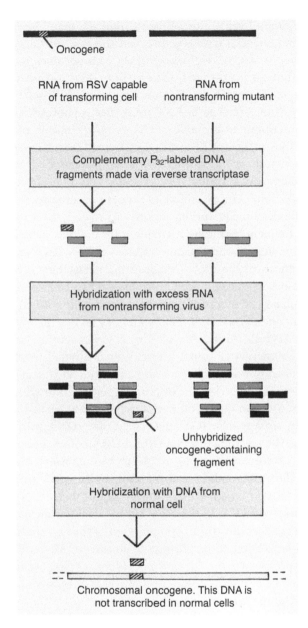

FIGURE 9.4. The hypothetical synthesis of an oncogene probe using mutant RSV, after an explanation in the third edition of James Watson's *Molecular Biology of the Gene* (1976). The diagram emphasizes the conceptual nature of the synthesis, rather than the laborious work of hybridization, electrophoresis, and centrifugation necessary to make this theoretical probe a biochemical reality. Illustration by Steven Parton.

The creation of the *src* probe condensed the many layers of social, material, and financial infrastructure that the War on Cancer provided to researchers at UCSF. Varmus, working with postdoctoral fellow Ramareddy Guntaka, planned to isolate *src* from RSV and use it as the basis for a hybridization probe. Their laboratory would first isolate RNA corresponding to the *src* gene in RSV using electrophoresis, and then use reverse transcriptase to assemble a vDNA copy specifically of *src* from radioactive DNA precursors. The *src*-vDNA could then be used in hybridization reactions to show evidence that a complementary DNA fragment corresponding specifically to the *src* gene was present in cells transformed by RSV infection.[52] Guntaka initially set out to synthesize the *src* probe. However, he was already working on several projects, so the task fell to his friend and frequent lunchtime companion Dominique Stehelin, a visiting scientist from France whose planned studies of RSV RNA had stalled.[53]

The synthesis of the probe required the painstaking use of the full range of resources that had been assembled in San Francisco. Picking up where Guntaka had stopped, Stehelin started with purified RNA from both mutant and normal RSV. The laboratory grew some of the RSV required for this task, but he also used virus stocks sent up from the University of Southern California, where Vogt continued to generate and catalog temperature-sensitive RSV mutants, with VCP support.[54] Drawing on the laboratory's supplies of purified reverse transcriptase and radioisotopes, Stehelin then produced a radioactively labeled vDNA copy of the RNA from regular RSV and mixed that vDNA with the mutant RSV RNA. The resulting reaction hybridized all of the normal RNA except the RNA corresponding to *src*, which he then isolated using centrifuges and chromatography. Next, he used reverse transcriptase and radioactively labeled base pairs to create a radio-labeled DNA *src* probe from that isolated RNA. In the early 1970s, when few means existed for manipulating genes directly, the preparation of the gene probe was itself worthy of publication in the prestigious *Journal of Molecular Biology*.[55] This probe promised to provide proof of Huebner's viral oncogene theory, but Bishop and Varmus soon found that it also suggested a new way of thinking about the origins of cancer.

From Enemy Without to Enemy Within

The emergence of the cellular oncogene theory from this years-long search was an ironic by-product of the infrastructure that the NCI put in place to find proof of Huebner's retroviral oncogene theory. The key discovery of Bishop and Varmus's experimental system was in fact the by-product of a calibration exercise. As the experimental system had been designed, the "epistemic thing" it aimed to identify was the presence of src in a normal chicken cell infected by RSV. In his first efforts, Stehelin had showed with a high degree of precision that src was present in chicken cells infected and transformed by RSV, as Bishop, Varmus, and Huebner had suspected. After this finding, Stehelin went about what should have been a routine experiment, testing the probe against a set of uninfected chicken cells. At the time, researchers active in hybridization studies of RSV assumed that the transforming genes of RSV were not present in normal cells, so the probe should detect nothing in these cell populations.[56] Working alone in the laboratory on a Saturday evening in October 1974, Stehelin was stunned to find that normal, noncancerous chicken cells also contained DNA sequences corresponding to the RSV src gene.[57]

Molecular biologists later heralded the discovery of what became known as cellular src (or c-src, to differentiate it from the src gene found in RSV) as a "revolutionary" event that shifted the causation of cancer from external causes to the study of the genetics of the cell.[58] However, it took another year and a half of work with the src probe to demonstrate why this unexpected finding actually suggested a wholesale reorientation from external to internal causes of cancer. Paradoxically, the reorientation rested on reversing the progress that the UCSF group had made toward creating a specific and precise hybridization probe for src.

From 1970 through 1975, the emphasis of their laboratory's work had been on the comprehensiveness and accuracy of a probe for viral src— ensuring the highest degree of specificity possible between the src probe and the cellular DNA fragments they aimed to identify. Their emphasis on specificity reflected a long-standing approach to immunological tests and other biochemical reactions, which favored precision. Following from this work, the optimal conditions for the probe did not favor hybridization with DNA sequences that were dissimilar from src. As a corollary, the further one pursued the divergence of the src gene from chickens, the less effective the probe became. That finding emerged as a challenge for the laboratory as they sought to locate c-src in other bird species, culminating in the "unpleasant" sacrifice of an emu chick. Although the laboratory found that sequences resembling the src gene were also present in the normal cells of other bird species, they could not identify similar sequences in mammals using their probe.[59] Until the specificity of the hybridization reactions were better understood, it was unclear whether this reflected the conditions under which the laboratory used the probe or the absence of the gene itself.

The concept of evolutionary time and the lack of hybridization provided the ideas from which the UCSF researchers created a new experimental narrative to make sense of the results they had obtained. Huebner's theory had raised the possibility that the oncogenes of retroviruses could persist across generations of normal cells after an initial infection. Indeed, Huebner and Todaro had speculated, oncogenes might be "an essential part of the natural evolutionary inheritance of vertebrate cells."[60] The VCP's leadership speculated that "in an evolutionary context," oncogenes might even have "provided certain advantages at one time in pre-vertebrate periods." The integration of the retroviral oncogenes into cells could have been so successful that the oncogenes were entirely capable of "vertical" transmission from generation to generation rather than by infectious, "horizontal," transmission.[61]

A. C. Wilson, a leader at UC Berkeley in the emerging field of molecular evolution, suggested a new way of interpreting the results of the UCSF researchers' hybridization tests. Previously, Bishop and Varmus had been frustrated by the existence of partial hybridization between their probes and DNA samples. However, the partial hybridization reactions were of interest to molecular evolutionists such as Wilson. Evolutionary biology had a long-standing interest in establishing the degree of relatedness between different species. Traditionally, biologists established these relationships through the collection and comparison of different anatomical specimens. Starting in the early 1960s, some molecular biologists proposed that small changes in the amino acid sequences of common proteins, such as hemoglobin, could serve as a "molecular clock" to mark the divergence of different species.[62] Drawing on this approach, Wilson proposed that immunological tests for protein hybridization could also be used to mark the degree of divergence among species.[63]

Rather than focusing on the specificity of their test, Bishop and Varmus found inspiration in imprecision. Incomplete hybridization was not a problem to be solved, as it might have been in the immunological detection tradition; instead, it was a means of tracing a gene's history. Following Wilson's work, Bishop and Varmus created a new story about their probe and c-src based on their hybridization results. The src probe hybridized best with the c-src gene of normal chicken cells, but not as well with other species. This finding followed from the fact that RSV was adapted to chickens. Bishop and Varmus proposed that the degree of hybridization was in proportion to the evolutionary distance of these other species from the modern chicken. For example, the emu, a "very primitive bird," had a lesser degree of hybridization than a closer evolutionary relative of the chicken, such as the quail. Based on the pattern of decreasing hybridization as evolutionary relatedness decreased, it appeared that the cellular *src* gene had descended from an avian predecessor and that RSV had picked it up from the modern chicken. Therefore, the *src* oncogene was not viral but cellular in origin, and it served "some function," which accounted for its persistence as different species branched from their common ancestor.[64]

However, these results remained limited to bird species. The origins of c-*src* in the cell would be easier to defend if its analogs could be located in other branches of the evolutionary tree. Therefore, the laboratory started to explore means of *reducing* the biochemical specificity of their probe as a means of tracing genetic divergence. The chemical technique of achieving reduced specificity hybridization reactions came from Deborah Spector, a postdoctoral fellow who had trained as a graduate student under Baltimore at MIT and who assumed Stehelin's duties after his departure from the laboratory in 1976. Her experience with RNA hybridization reactions in Baltimore's laboratory gave her insight into how to adjust the conditions for the *src* probe's hybridization reactions. While Spector could not say exactly what the probe hybridized with in the genomes of these cells, her manipulation helped fit the presence of c-*src* into an evolutionary schema that extended beyond birds to other vertebrates and even to sea urchins.[65] Spector's studies continued to draw on the infrastructure of the VCP—because of the decay of the radioactive isotopes used to label the probe, postdoctoral fellows had to prepare new batches of *src* probe roughly twice a month to support the laboratory's ongoing research.[66]

The observations that arose from transferring the *src* probe outside of bird species allowed Bishop and Varmus to argue that the true origins of retroviral oncogenes were in cells without having to resolve the evolutionary question of what function the gene served. A former member of the cooperative, Robin Weiss, prefaced his discussion of Stehelin, Bishop, and Varmus's 1976 paper for readers of *Nature* with a caution that the role of *src* in cell development was unknown. "Of course," Weiss continued, "the isolation of an 'oncoprotein' coded by . . . the 'sarc' gene is a major goal. . . . How such a protein might interact with the host cell is an open question."[67] Baltimore highlighted Bishop and Varmus's results when he accepted his Nobel Prize for the codiscovery of reverse transcriptase in the spring of 1976, but he still maintained that the "isolation of these transforming proteins and elucidation of their mechanism of action . . . is the present challenge of cancer virology." He lingered on the difficulties implied by the detection of viral transforming genes in uninfected cells. "What," Baltimore asked, "is the significance of these genes that look like viruses?"[68]

Writing in *The Molecular Biology of Tumor Viruses* (1982), Bishop and Varmus conceded, "The tools used to trace the lineages of *c*-onc genes [cellular proto oncogenes] are blunt." It remained, however, a "reasonable guess" that cellular oncogenes had been around for millions of years.[69] This insight, however, opened up new questions for cell biology. While postulating a common ancestral gene brought order to the data generated by the hybridization probe, appealing to evolutionary rationality also underscored the importance of showing what function the inactive cellular oncogene served in a normal cell's development or how it might be targeted to interrupt the process of cancerous transformation.[70] This question led molecular biology deeper into questions of cellular development in the ensuing decade.

Systems and Serendipity

Science and other American news outlets greeted news of Bishop and Varmus's Nobel Prize in 1989 with enthusiasm, praising their discovery of c-*src* as a first step toward understanding the biochemistry of cancer development and cheering that these molecular studies were already "paying off clinically" to aid in the diagnosis and treatment of cancer.[71] Stehelin felt differently. Appearing on French television after the announcement of the prize, he argued that the Assembly had erred in giving credit to Bishop and Varmus. He had carried out the critical experiments revealing the oncogene in normal cells, "from A to Z."[72] He was the first author of the groundbreaking *Nature* paper that had appeared in 1976, and Bishop and Varmus readily acknowledged his involvement in the central experiments. Disputes over credit for the Nobel Prize were not new, and Stehelin's complaint echoed that of many "invisible technicians" throughout the centuries who felt that their labor went unrecognized.[73]

Although Stehelin's protest was not unprecedented, the manner in which the scientific press addressed the complaint spoke to the changes that the War on Cancer had set in motion within the molecular biology community. In the 1950s it would have been hard to dissociate an individual researcher from the intellectual work of his or her laboratory, but in 1989 John Maddox, the editor of *Nature*, issued the pointed opinion that while Stehelin might have been the "main pair of hands behind the experiments," he would have failed to carry them out "had he been working at any other laboratory." Moreover, Maddox continued, it was very likely that these experiments would have "emerged from the Bishop/Varmus milieu" even if Stehelin had never traveled to San Francisco.[74]

Although Maddox used the idea of a milieu to describe the research environment at the level of the laboratory, the concept also characterizes the findings of researchers at work within the infrastructure that the administrative approach to cancer viruses created. Following the material and social constituents of Bishop and Varmus's milieu and how it shaped their experimental systems shows how the infrastructure for scientific work extends far beyond the walls of a laboratory to broader legislative debates or administrative decisions. The numerous ways in which this infrastructure touched the patterns of work and thought at the Bishop and Varmus laboratory makes it hard to imagine how the critical experiments could have emerged anywhere else. It provides a counterpoint to the idea, advanced by some molecular biologists, that the serendipity of c-*src*'s discovery was an indictment of the management strategy of the VCP.[75]

This does not mean that all experimental findings follow from having infrastructure in place. Experimental systems often reveal surprising new phenomena. Rather than marking an instance of serendipity, however, those new moments for investigation are generated by the history of the experimental system itself.[76] Understanding the milieu of scientific work draws our attention to not only the ideas but also the infrastructure that facilitated the "empirical roaming around" that produced the discovery of c-*src*.[77] That capacity to roam should be understood as a generative feature that links the infrastructure of big science to the epistemic things produced by particular experimental systems. As molecular biologists united around the idea of the serendipitous turn toward cellular oncogenes in the 1980s, the remnants of the NCI's infrastructure continued to play a vital role in preserving their political fortunes and producing excitement for the future of molecular medicine amid growing dissatisfaction with the War on Cancer.

CHAPTER TEN

Momentum for Molecular Medicine

Even as molecular biologists greeted new developments in retrovirology with excitement, public enthusiasm for the War on Cancer gave
way to frustration. Rates of cancer incidence continued to climb, and the
proportion of persons with cancer cured by chemotherapy, surgery, or radiation remained obstinately low. Raised public expectations for progress
meant that the lack of improvement invited intense disillusionment with
the federal government's campaign against cancer. The outgoing director of the Food and Drug Administration dubbed the War on Cancer "a
medical Vietnam."[1] Nor did advances in molecular biology seem promising. The prominent breast cancer activist Rose Kushner traveled to the
National Cancer Institute in the mid-1970s with "high hopes that a vaccination against breast cancer was imminent . . . or, at least, that a major
breakthrough was expected any day-dawn. . . . Instead, I . . . learned that
cancer may be nature's own solution to the population explosion. An in-
born, suicidal oncogene!"[2] Many biologists were also critical of the biomedical approach to the cancer problem. Accepting his Nobel Prize for
the discovery of reverse transcriptase in December 1975, Howard Temin
pointedly reminded his audience that if they were truly concerned about
rising cancer rates, their efforts would be better spent combating smoking
rather than studying cancer viruses in the laboratory: "Our work has not
yet led to prevention or curing of human cancer."[3]

Most accounts of the War on Cancer have emphasized the hope and
excitement that contributed to the "boom" in cancer research. There
has been less attention to how the biomedical institutions and communities created by this investment weathered the period of disillusionment,
or "bust," that followed.[4] Although cancer specialists and biologists had
been circumspect about the possibility of rapid progress against cancer

from the outset, many feared that popular criticism of cancer research demonstrated that taxpayers were becoming less deferential to the authority of scientists to set their own research priorities.[5] Between 1976 and 1978, the annual budget for the NCI peaked, ushering in a generation of declining or static funding levels. The political consensus on behalf of federally funded biomedical research appeared to be waning. "A whole generation of scientists" had "literally grown up with generous government programs," but they would soon need to live with "cutbacks" in the budget, the chief science adviser of the incoming Reagan Administration warned in 1981.[6]

Indeed, the criticism aimed at the War on Cancer threatened to undermine large parts of the biomedical settlement—pushing federal attention toward preventive research and epidemiology rather than laboratory biology. Molecular biologists, who had found the resources of the War on Cancer an essential part of their migration into exciting new areas of research, faced the prospect that declining funding and competition from other areas of clinical or environmental cancer research would halt the expansion of their field.[7] However, at this moment of peril, a "crescendo" of new discoveries in the field of oncogene research salvaged the political and cultural reputation of the biomedical approach to cancer and opened new frontiers for molecular therapies.[8] Starting with J. Michael Bishop and Harold Varmus's hypothesis of a cellular oncogene in 1976 and ending with the proposal of the first tumor-suppressing gene in 1986, a series of discoveries and conformations helped refashion cancer virology and cancer genetics into endeavors that promised both a fundamental understanding of cancer and a means of producing therapeutic advances.[9]

In this chapter, I lay out how the infrastructure of the Virus Cancer Program and the War on Cancer played a crucial role in fostering the rise of the molecular genetics of cancer and cellular development during the 1980s, a decisive moment in the "molecularization" of biology and medicine as a whole. In addition to its considerable social and technical contributions to the "oncogene paradigm," the infrastructure of the VCP provided advocates of molecularization with a resource that would have been difficult for them to obtain in any other way: a sense that insights from the molecular approach to cancer were arriving at an accelerating rate. The momentum that the VCP's infrastructure lent to molecular studies of cancer provided a necessary political resource as molecular biologists and administrators sought to deflect other approaches to the cancer problem proposed by groups disappointed by the War on Cancer, including environmental activists, labor leaders, and physicians. Oncogenes provided scientists and administrators with both a therapeutically appealing target and a solution to persistent problems that molecular biology had faced in applying its theories to explain the behavior and growth of cells.[10] Although the frustrations of the War on Cancer produced a moment when the biomedical settlement might have been opened to admit a broader range of solutions to the cancer problem, the rapid pace of molecularization drew attention back within the laboratory and deeper into the cell.

The Environmentalist Challenge

Like the Vietnam War and the War on Poverty, the War on Cancer, as its aims became frustrated, left many cynical regarding the government's capacity to protect or advance national health. The outgoing president of the Memorial Sloan Kettering Cancer Center hoped that amid "public disillusionment and discouragement over government and its works," the National Institutes of Health could continue to provide a beacon for federal efforts to advance "human betterment."[11] Yet the NCI's own statistics revealed that the overall death rate from cancer remained stubbornly constant. Critics charged that the leadership of the NCI not only focused on the wrong aspects of the cancer problem but also failed to meet the challenge on terms they had defined for themselves. Although many biomedical experts sought to distance themselves from the promises made by advocates of the War on Cancer, this sense of disappointment shadowed cancer policy discussions in the late 1970s.[12]

Although making cancer policy a matter of democratic concern at the start of the War on Cancer had originally strengthened Lasker's political position, it also created further opportunities for popular movements outside of the biomedical settlement to seek a voice in cancer policy. Early in the 1970s, civil rights groups had drawn attention to racial disparities in cancer care, and the women's movement had worked to compel the NCI to devote more resources to breast cancer screening and treatment.[13] Meanwhile, reforms in the committee structure of Congress and the budget appropriations process allowed more legislators to exercise influence and diminished the power of committee leaders, weakening the system of congressional alliances that had insulated the work of the American Cancer Society and the NCI from public pressure.[14] While these new groups

offered stinging criticism of the NCI, they still sought to further the institute's traditional mission in clinical and biomedical research.

Of the new movements that arose to challenge the NCI, environmentalists offered the most disruptive critique of its operations. Recalling concerns in the early twentieth century that cancer was a "disease of
civilization" itself, environmentalists suggested that neither the laboratory nor the clinic was an appropriate site to grapple with the disease. The
new environmentalist critique, as it emerged in the 1970s, drew on Rachel
Carson's widely read *Silent Spring* (1962), which had increased awareness of the influence of pesticides and other industrial chemicals in the
lives of Americans. "We are living in a sea of carcinogens," Carson wrote.[15]
Environmentalists feared that humans had become "caged canar[ies],
who, through illness, heral[ded] the presence of highly toxic chemicals."[16]
This perspective suggested that neither experimental inquiry nor treatment would alleviate the rising rate of cancer in the United States. Instead, the federal government should take a more active role in regulating carcinogens in the environment. Concern over exposures to harmful
chemicals, including carcinogens, became part of the responsibility of new
agencies such as the Environmental Protection Agency and the Occupational Safety and Health Administration, both of which were established during the Nixon Administration just before the start of the War
on Cancer.[17]

These views were cogently summarized by Samuel Epstein, an environmental and occupational health specialist at the University of Illinois,
in his book *The Politics of Cancer* (1978). Epstein made the claim, based
on his analysis of data from the Occupational Health and Safety Administration, that 20 to 40 percent of cancers were attributable to a handful
of industrial chemicals. Alongside reducing harmful individual behaviors
such as smoking, regulating these chemicals and identifying other chemical carcinogens marked the most important line of research against the
cancer problem. Members of the NCI disputed Epstein's statistics, but
his message resonated with labor groups and their allies in Congress,
who later that year directed the Department of Health, Education, and
Welfare to publish a list of suspected carcinogens and the risk of human
exposure to each compound, with the eventual aim of regulating those
risks.[18] Actions by states such as California suggested what this new kind
of prevention might look like—several years later its voters approved
Proposition 65, which required that warnings be posted about potentially
carcinogenic substances wherever they might appear.[19]

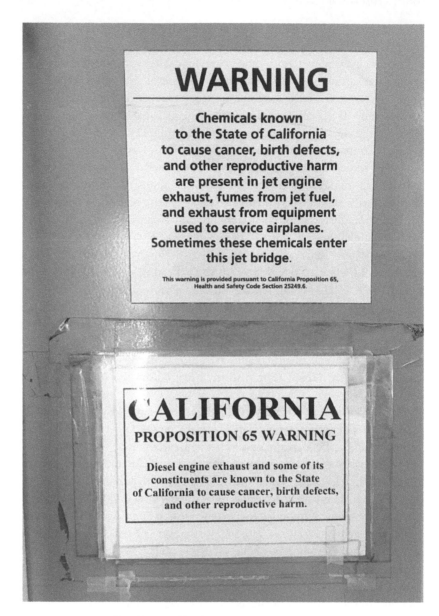

FIGURE 10.1. Proposition 65, passed by the state of California in 1986, required the posting of public notices about potential carcinogens. This requirement reflected the environmental-ist view that the cancer problem was best dealt with as a matter of chemical exposure, not through biomedical research. Photo by author.

Whereas molecular biologists had attacked the NCI's authority to manage research without questioning its biomedical emphasis, Epstein attacked the premise of laboratory-based cancer research itself. Molecular biologists were rudely surprised to find the Zinder Report represented as an indictment of experimental cancer research overall. The report, Epstein argued, did not prove the benefits of peer-reviewed research, but instead showed that biological research sponsored by the NCI was both "professionally self-serving" and "of little merit or relevance" to the aims of the National Cancer Program. Most of the "basic cancer research" sponsored by the NCI, Epstein stressed, appeared "irrelevant to cancer treatment or cancer prevention." This line of research should not be supported by the NCI, Epstein concluded, especially when such a large proportion of cancers might be explained by exposure to carcinogens in the workplace rather than by molecular mechanisms.[20] Two economists argued that screening for potential carcinogens would cost a "tiny fraction" of the funds expended "chasing apparently nonexisting viruses."[21]

For some within the NCI, the focus on environmental carcinogenesis was long overdue. The associate director of the NCI's chemical carcinogenesis program had resigned in 1976, leaving a scathing message regarding the lack of support offered to the screening program for carcinogens: "I cannot accept any longer a situation which . . . deprives the regulatory agencies . . . of data of urgent public health value: it is *people* who are now exposed to toxic agents and who are not protected because the necessary support was not provided in time."[22] Others were less enthusiastic. An NCI staff member groused, "We spent $6 million to see if workers in steel factories at coke-ovens have more lung cancer than expected. 15 were found versus 1.5 expected. Is that worth $6 million[?]" David Baltimore cautioned that "unnecessary hysteria about general environmental exposures" might involve the NCI in "targeted work" that could "taint the mission of the NCI."[23]

Reviewing the progress of the War on Cancer in 1979, Epstein drew on this general sense of disenchantment and concern when he criticized both the management and the emphasis of the National Cancer Program: "We now have abundant evidence that there has been a failure in terms of improvement of overall survival rates. On the other hand, we have abundant evidence on how to prevent cancer, but we are not implementing this for a wide range of reasons, economic, political, and lack of research efforts by the National Cancer Institute." "I would like to see," Epstein continued, "a decrease in emphasis in virology and basic research programs, unless clearly related to prevention programs."[24] Compared to prevention through the regulation of chemical exposures, the promise of a cancer vaccine appeared a distant prospect.

The End of the War on Cancer

In the midst of this brewing controversy, the leadership of the NCI changed dramatically. Since 1960, the directors of the NCI—Kenneth Endicott, Carl Baker, and Frank Rauscher—had all shared a strong presumption in favor of biomedical research. Baker and Rauscher were especially involved with the VCP. However, Rauscher, a Nixon appointee, departed soon after the Democrats regained the White House in 1976. To replace Rauscher, a microbiologist, President Jimmy Carter installed Arthur Upton, from New York University. Upton had trained as a pathologist but had shifted into radiation biology after working at Oak Ridge National Laboratory, a major Atomic Energy Commission laboratory in Tennessee. There, he was involved in debates over the existence of a threshold for harmful radiation exposure. Upton's views were doubly threatening to the VCP—he favored peer review over contracts and environmental carcinogenesis research over laboratory biology.[25]

Upton's emphasis on environmental carcinogens was aligned with the interests of important Democratic party constituencies, such as environmental advocates and labor groups concerned about occupational carcinogenesis.[26] He closed the VCP in 1978, moving its operations into the Division of Cancer Cause and Prevention; the broader name signaled a concern for environmental carcinogenesis.[27] This shift reduced resources for cancer virology. The new division cut some of the VCP's most ambitious efforts to provide materials and resources to working scientists and reduced the NCI's reliance on space rented from contractors in the Washington, DC, area.[28]

As much as Upton's decision to disband the VCP might have pleased the academic molecular biologists who had criticized the encroachment of the contract mechanism, it augured a more general shift in NCI spending from biological to clinical and environmental research. The retrenchment of biological research threatened much more than virus studies. Compared to other branches of the NIH, the NCI's budget sustained a plurality of the inquiries by molecular biologists into higher organisms. Major centers of this research, such as Cold Spring Harbor Laboratory, derived a considerable amount of their annual budget from NCI support, much of it related to cancer viruses. At Cold Spring Harbor in the late 1970s, for example, many of the laboratory's operations were supported by a cancer research center grant from the NCI. The funding for tumor virus research in that grant alone exceeded scientists' grant support from any other part of the NIH. The NCI also provided funding at some level for most active retrovirus and tumor virus researchers at sites such as the University of California, San Francisco, the Salk Institute, the MD Anderson Cancer Center, Sloan-Kettering Institute, Harvard, MIT, and Columbia.[29] This reliance was even higher than formal budgets suggested, since the NCI often redistributed money from its annual appropriation to make up for shortfalls in other branches of the NIH.[30]

Retrenchment of federal funding renewed fears that academic biomedical research faced a crisis of political and professional identity. Speaking on behalf of a group that called itself the "Delegation for Basic Biomedical Research," the president of the Worcester Institute for Experimental Biology wrote to Upton protesting that proposed changes only gave the "appearance" of greater NCI support for "basic biological research" while transferring control over grants into more "clinical hands." Upton's leadership of the NCI risked widening the "philosophical gulf between those who see the NCI as a force to encourage bright and imaginative young scientists to work on their best ideas and those who see it as directing a network of bureaucrat-generated, patient oriented, targeted 'relevant' activities. . . . When knowledge is inadequate, as it certainly is in the case of cancer, we must make the best possible use of the mechanisms for encouraging creative science: the investigator-initiated grant and peer review."[31]

The start of the Reagan Administration might have been expected to revive the fortunes of biomedical research at the NCI. Ronald Reagan was critical of environmental regulation and, by extension, investment in environmental carcinogenesis research. The political benefits of focusing on molecules or viruses rather than on the possible environmental hazards created by business were clear.[32] Indeed, the American Business Cancer Research Foundation, established in 1975 and funded largely by the chemical industry, embraced the idea of fundamental "basic research on the underlying mechanism of carcinogenesis" rather than prevention efforts.[33] However, as helpful as laboratory research into cancer might have been to the conservative movement as a foil to calls for environmental regulation, the Reagan Administration saw supporting biomedical research as a distant priority after its renewed focus on increasing military spending, cutting taxes, and reducing the overall size of the federal government.[34]

The Reagan Administration also indicated that it was prepared to de-emphasize biomedical research in favor of clinical approaches to cancer. Its first appointees to the National Cancer Advisory Board shifted the balance among clinical, lay, and "basic science" members away from scientists. A group of biomedical researchers warned their colleagues that the changing mission of the NCI would have a "direct and deleterious effect on American scientific programs."[35] Watson warned that at Cold Spring Harbor Laboratory, "the increasingly tight federal funding situation" threatened the molecular biology community's ability to work at its rapid "past pace."[36] The president of the National Academy of Sciences foresaw that the "problems of the economy" and the "President's Budget" would do "irreversible damage unless longer term research . . . is protected." The molecular biologist and cancer researcher Arthur Pardee, best known for his work on how messenger RNA aided protein synthesis, worried that the rising cost of biological research, the increasing number of scientists competing for grants, and the rising fraction of indirect costs charged by universities to the NIH were "squeezing" the money left for "actual research." Pardee emphasized: "Again and again basic science has led to products . . . which more than pay for the costs of all research as well as eliminating suffering and grief."[37]

In the face of these darkening political circumstances, molecular biologists had little basis to claim that their work would produce new therapeutic insights into human cancer. The potential for a human cancer vaccine seemed diminished, and the search for cellular oncogenes did not appear to illuminate potential cancer therapies. Although the identification of the cellular *src* oncogene at Bishop and Varmus's laboratory had suggested that cancer was rooted in the cellular genome, these results did not immediately suggest a new way of understanding cancer as a whole. The process by which Bishop and Varmus had identified and isolated *src* depended on particular properties of Rous Sarcoma Virus, a set of properties that not all retroviruses possessed. Temin noted that the strains of RSV used in these studies were not "natural." "RSV is a laboratory creature," he warned, "passaged and preserved by virologists."[38] Despite the confident pronouncement by some molecular biologists that what was true for *E. coli* was true for elephants, a review of the expression of retrovirus genes in *The Molecular Biology of Tumor Viruses* (1973) urged that virologists "should be extremely cautious" as they sought "to extrapolate from one species to another; what holds for mice does not hold for chickens and may well not hold for man."[39]

Recycled Infrastructure and the Acceleration of Oncogene Research

In the early 1980s, centers of molecular biology research across the United States announced discoveries illuminating the molecular mechanisms of carcinogenesis, dramatically reviving the hope that investigations of the fundamental biology of cancer cells would produce medical advances. For many observers the intellectual content of these discoveries was significant enough, but the rapid arrival of the new discoveries and their extrapolation to new biological problems also profoundly contributed to a sense of renewed possibility. These announcements, taken together, were a key moment in the extension of molecular biology into higher organisms, part of a process in which molecular methods became the norm rather than the exception for explanations of cellular development.

The development of the "oncogene paradigm" contains more actors and ideas than this final chapter can cover, so I will focus on two critical moments in its articulation. The first, in 1982, was the identification of the first oncogene associated with a human cancer and its association with a retroviral oncogene, *ras*. This discovery suggested that all cancers might be triggered by a limited set of genes shared across species. The second, in 1983, was the identification of the function of a protein associated with an oncogene—an "oncoprotein." Identifying the function of oncoproteins suggested that it might be possible to interfere with the process of carcinogenesis directly, raising hope for new therapies for cancer.[40] These two moments highlight how the social and material infrastructure of the War on Cancer gave these discoveries momentum, producing a sense of acceleration and a renewed climate of anticipation surrounding further research into the molecular nature of cancer.

The identification of the first human oncogene, and its association with an existing retroviral oncogene, *ras*, gave molecular biologists a means of rapidly transferring to cellular development the knowledge they had assembled with the VCP's assistance, thus reaffirming the idea that laboratory studies would yield therapeutic payoffs. In the 1970s the VCP had sponsored the activities of numerous research groups seeking to understand the genetics of tumor viruses, retroviruses in particular. At first, however, these other researchers did not have a means of following in the footsteps of Bishop and Varmus's discovery of cellular *src*. Bishop and Varmus had depended on a unique collection of temperature-sensitive RSV mutants, whereas other virologists were not so fortunate.

The career of a virologist at the NCI, Edward Scolnick, who eventually isolated the retroviral version of the *ras* oncogene, highlights the challenges that other researchers faced. Like Varmus and Bishop, Scolnick entered the NIH as a way to avoid the Vietnam draft. There, he received training in molecular biology at the National Heart Institute. Huebner and Todaro recruited Scolnick into the staff of the NCI's Laboratory of Viral Leukemia and Lymphoma.[41] Following Todaro's interest in mouse retroviruses, Scolnick collaborated with Peter Duesberg to map the genomes of several mouse and rat retroviruses. His laboratory succeeded in creating DNA hybridization probes for "sarcogenes" in cells a year before the Bishop and Varmus laboratory developed their *src* probe.[42]

As laborious as preparing the *src* probe was in San Francisco, Scolnick's efforts in Virginia were even more difficult. Scolnick's research plan required the time-consuming step of growing and "rescuing" different strains of viruses in mouse cells in an effort to isolate a transforming retroviral gene.[43] Nevertheless, Scolnick was undeterred. Because he was situated in the heart of the NCI at the height of the War on Cancer, resources were never lacking. For years, he oversaw a large group of technicians working in a contract facility that aimed to clone a viral oncogene by growing different variants of the mouse virus in different cell lines. This work was so painstaking and difficult that his laboratory staff threatened to resign. Frustrated himself, Scolnick left to work for a pharmaceutical company, Merck, in 1981.[44]

Recombinant DNA, a new technique that allowed biologists to copy, or "clone," fragments of DNA from many different organisms using *E. coli* bacteria and restriction enzymes, would have accelerated Scolnick's work dramatically, but it was unavailable to his laboratory because of fierce debates in the mid-1970s about its safety.[45] After the development of the technique in 1972, many worried that DNA from cancer viruses might be spliced into a microorganism capable of infecting humans, reviving fears of contagious cancer. This concern was so deep that the NIH placed severe restrictions on the use of the technique by any institution that received its support—effectively restricting the practice at every major center of academic biology in the United States. Researchers employing recombinant DNA had to work in expensive biohazard laboratories that

few universities could afford to build. Avenues for producing potential oncogenes on a large scale were thus limited to the laborious culture of retroviruses until 1979, when the NIH eased its restrictions. It was not until the
early 1980s that NIH restrictions relaxed enough to make the use of recombinant DNA to clone genes commonplace.[46]

Given these regulations, molecular biologists sought ways of identifying
oncogenes that did not depend upon the right combination of viral mutants. This attempt presented a challenge, however. Molecular biologists
were accustomed to studying the behavior of bacterial genes by substituting new genes into new populations. Given that bacteria routinely swapped
fragments of DNA, this was not a difficult step. Eukaryotic cells, such as
those in animals, were a different case. Tumor viruses had inspired enthusiasm precisely because they offered the only means by which genes could
be inserted into animal cells. By the late 1970s, however, a new method—
transfection—allowed molecular biologists to chemically treat eukaryotic
cells so that they would accept DNA fragments.[47] Several researchers started
to develop tests for oncogenic DNA that used this technique to directly
place DNA into cultured cells independent of infection. However, these
tests, like the design of hybridization probes, could be used with different
aims in mind. Some saw the new test chiefly as a way of hunting for retroviral
oncogenes, whereas others saw it as a means of looking for oncogenes unas-
sociated with viruses.[48]

A new assistant professor at the MIT Center for Cancer Research,
Robert Weinberg, was one of the molecular biologists who sought to use
transfection as a tool for identifying oncogenes without recourse to viruses.
Under the leadership of the Nobel Prize–winning phage geneticist Salvador
Luria, the center received generous support from the NCI to explore the
molecular biology of cancer. It had been at the dedication of the center
that Watson had chosen to level some of his most forceful criticisms at
the medical aims of the War on Cancer.[49] After studying tumor viruses at
the Salk Institute, Weinberg was not interested in trying to replicate the
manipulations of retroviral mutants that had yielded the *src* probe when
he arrived at MIT in the mid-1970s.[50]

Weinberg and others understood that the combination of transfection
and other new techniques opened up a new way of studying the behavior of genes in cells. The first of those techniques was the use of restriction enzymes isolated from bacteria, which cut DNA chains at specific
points.[51] These enzymes allowed an experimenter to break the DNA of a
cell into fragments, which could be sorted by size using biochemical methods

such as gel electrophoresis. After the sorting, the transfection test, if performed with appropriately sensitive mouse cells, could be used to determine whether any of the gene fragments prompted cancerous growth.[52] In the late 1970s, Weinberg used this test to look for proof of oncogenes in cells transformed by chemical carcinogens. Weinberg suspected that the DNA of the transformed cells contained new oncogenes unrelated to the viral oncogenes that had been studied before.[53]

The transfection test, as these researchers first used it, produced contradictory results. Weinberg found cancer-associated gene fragments in mutated cells, and other investigators found similar fragments in normal cells.[54] Later work resolved this paradox by suggesting that there were two kinds of oncogenes—tumor-promoting genes and tumor-suppressing genes—but this development took several years to emerge.[55] Even if the results of the transfection test could be reconciled, the identity of the DNA fragments remained unknown: Weinberg and others did not know whether each fragment was unique or whether a limited set of fragments existed. Laboratories using the test lacked the resources to produce fragments on a scale large enough for further comparison.[56] Such problems became more acute as Weinberg and others turned their attention to identifying oncogenes in human cancers.[57]

Molecular biologists using the transfection technique might have labored for several years without agreeing on the number or character of the oncogenes they discovered, but the social and material infrastructure of retrovirus studies allowed them to quickly reach agreement on the powerful theory that cancer emerged from a limited number of oncogenes. By the late 1970s, retrovirologists had cataloged fourteen oncogenes in addition to *src* from viruses infecting chickens, rats, mice, cats, and primates. So many strains of retroviruses emerged that the Retrovirus Study Group of the International Committee on Taxonomy of Viruses formed a special committee to standardize nomenclature, to avoid the problem of the "uses of identical names for genes of unrelated sequence and function." The International Committee's membership—which included Weinberg and Scolnick—presaged the union of retrovirology and cell biology: many of its members were VCP-sponsored retrovirologists.[58] The library of retroviral oncogene probes assembled during the 1970s significantly accelerated the identification of cellular oncogenes using the transfection assay.

In 1982 Weinberg used the transfection technique to isolate a carcinogenic gene from a human bladder cancer tissue specimen. To classify this potential oncogene, Weinberg's laboratory was able to summon an extensive number of DNA hybridization probes initially developed by Scolnick and other researchers affiliated with the vestiges of the VCP. Although Scolnick had already moved to Merck when Weinberg contacted him about his work with *ras*, he was still able to send Weinberg probes he had developed while at the NCI.[59] With probes from Scolnick and from Bishop and George Vande Woude, a virologist at the NCI, Weinberg was able to quickly establish that the gene he had isolated from a human bladder cancer cell line was identical to the v-*ras* oncogene Scolnick had isolated from a mouse sarcoma virus. *Ras* was the first human oncogene identified.[60]

The infrastructure of the NCI soon revealed other potential human oncogenes. The creation of libraries of oncogene clones required recombinant DNA, and the NCI was one of the few places where the manufacture of retroviral oncogene probes had continued on a large scale during the years that this technology was restricted.[61] Vande Woude, working at the Frederick Cancer Research Facility, the group of biological warfare research facilities converted for cancer research by President Nixon, identified a similarity between human DNA and the v-*mos* oncogene of another murine sarcoma virus.[62] Another set of researchers announced that they had identified several human oncogenes simply by testing tumors already collected for other purposes with probes for retroviral oncogenes developed in another laboratory at the NCI.[63] *Nature* acknowledged the lasting legacy of these retroviral oncogenes for the investigation of cellular oncogenes, commissioning a series of reports tracing the discovery of cellular oncogenes in relation to their retroviral cousins.[64] As Weinberg recalled, the retrovirus research that "had seemed to be nothing more than a stamp-collecting . . . mindless cataloging" activity to critics in molecular biology now served as a vital resource for cellular oncogene studies.[65]

Thanks to these collections, the first human oncogene emerged not as an isolated finding but as the first of a rapid volley of publications. *Nature* relaxed its normal process of peer review to keep its readers appraised of these announcements.[66] The speed of the discoveries fostered the idea that a limited number of oncogenes were active across all species. In one of the numerous publications that sprang up to disseminate work in the rapidly emerging field, the editors of *Oncogenes* wrote: "Once it was understood that such unique oncogene(s) could be harbored by retroviruses, there was a movement *en masse* back to the laboratory freezer to recover many 'treasures' that had been stocked away in prior years (dating back, in several instances, more than half a century)." The success of the oncogene endeavor

FIGURE 10.2. This cartoon, created for a meeting at Cold Spring Harbor Laboratory in 1983, illustrates the conceptual realignment posited by the cellular oncogene and some of the community that participated in it. Retroviruses were not the agents of carcinogenesis; they were the unwitting by-products of contact with cellular oncogenes. The cartoon also testifies to the importance of retrovirus research to the study of cellular oncogenes. The retrovirus researchers, represented by the animals whose retroviruses they studied, are (*left to right*) Harold Varmus, J. Michael Bishop, Charles Sherr (NCI), David Baltimore (MIT), Edward Scolnick (NCI and Merck), Robert Weinberg (MIT), Stuart Aaronson (NCI), Inder Verma (Salk Institute), and George Vande Woude (NCI–Frederick). Ed Scolnick and Robert Weinberg are holding their own tails in reference to Weinberg's having erroneously announced the discovery of a new cellular oncogene that was in fact the same as the *ras* oncogene isolated by Scolnick. Illustration by Jamie Simon, reproduced by courtesy of Jamie Simon.

was due to the "faithful investigators" who had remained "much more convinced than many of their colleagues that these treasures would prove to be of extreme importance to the cancer research community."[67]

Oncoproteins and the Therapeutic Promise of Molecular Medicine

Despite the excitement created by using the transfection assay to identify human oncogenes, advocates of a molecular approach to cancer still lacked a sense of the biochemical processes through which these genes gave rise to cancer. As Weinberg mused in a 1983 article explaining on-cogenes in *Scientific American*, "the finding of a precisely defined mo-lecular alteration in one such gene" left "the most difficult problems" still to be resolved: "What is the function of an oncogene?"[68] Oncoproteins provided the biochemical link between a given gene and the physiological processes that produced cancer. While the idea of gene therapy was only a distant prospect, proteins could be targeted by immunological methods, suggesting an immediate avenue for treating cancer. Given these difficul-ties, great excitement greeted two teams of researchers in 1983 when they announced that amino acid sequencing of *vsis*, a retroviral oncoprotein, showed that it matched a growth-promoting protein isolated from blood platelets. Both of their papers revealed a close connection to the work of Stuart Aaronson, a former virus researcher at the NCI.

The realm of protein chemistry was apparently far from the world of viruses, but the study of *vsis* also drew on the accumulated efforts of the VCP. The NCI had taken an intense interest in oncoproteins since the mid-1960s as a corollary to its study of viruses. In the early 1960s, Hueb-ner and others focused on the use of cancer virus protein products, or an-tigens, as a means of detecting hidden infections by tumor viruses such as Simian Virus 40 (SV-40). Before it was possible to manipulate viral RNA using reverse transcriptase, using antibodies to search for virally associ-ated proteins was one of the best techniques for establishing that a cell had been infected by a retrovirus. Huebner's faith that "latent" infections by type-C RNA tumor viruses were an important mechanism of carcinogen-esis magnified the importance of this task. Moreover, the NCI invested heavily in the prospect of blocking proteins associated with cancer viruses as a hypothetical pathway for preventing retroviral oncogene infection from causing cancer.[69]

Aaronson's career illustrates how the War on Cancer accelerated the identification of oncoproteins. Like Varmus, Bishop, and Scolnick, Aar-onson arrived at the NCI in 1967 as a physician seeking to fulfill the terms of the Vietnam draft through biomedical research. Initially, he worked with George Todaro on SV-40, a DNA-based tumor virus, in a contract facility run by Meloy Laboratories, where he sought to develop a test to gauge the susceptibility of human cell cultures to infection with SV-40. This work made use of the immunological tests for the "T-antigen," a viral protein product associated with SV-40.[70] Aaronson gained a permanent post at the NCI in the Viral Carcinogenesis Branch in 1971. He therefore had a ringside seat for the NCI's expansion, serving on the Virus Cancer

Program Coordinating Committee and acting as a supervising project officer for several VCP contractors focused on tasks such as collecting leukemia and lymphoma tissue for screening at the University of Michigan, creating standard cell cultures at Stanford University, and growing retroviruses on a large scale at Electro-Nucleonics.[71]

Promoted to director of the Molecular Biology and Viral Genetics Section of the NCI Laboratory of RNA Tumor Viruses, Aaronson continued research along the lines suggested by Huebner and Todaro's oncogene theory. If cancer could arise from oncogenes hidden in human cells, then it would be useful to have a means of detecting them. By analogy with SV-40's T-antigen, Aaronson speculated that latent infections might display a distinct immunological signature. Toward that end, he supervised the wide-ranging search for proteins that might be immunologically associated with the presence of retroviruses in the tissues of primate species. If those efforts succeeded, Aaronson and his colleagues envisioned preventing cancer by targeting the proteins of viral oncogenes, thus resolving the challenge of how to prevent cancer caused by latent viral infections.[72]

With the support of the VCP, the number of identified retroviral proteins briskly increased. As with many other scientific fields, a clear sign of the rate of expansion of oncoprotein research was the need for the different members of the community to meet and attempt to establish shared conventions for naming and classifying retroviral proteins. Without such coordination, researchers working in far-flung laboratories were able to discuss their work only with difficulty.[73] National coordination of this type was precisely the kind of social infrastructure that the VCP sought to promote.[74] In the late 1970s it sponsored further meetings with the aim of standardizing names and distributing rare antibodies and proteins among contractors and academic laboratories. At the NCI, Aaronson sat at the center of a distribution network of immunological probes for the identification of retroviral proteins and the provision of oncoproteins themselves to laboratories that lacked the capacity to produce these proteins in sufficient quantities for their own research work.[75] Despite the demise of the VCP, Aaronson continued his study of oncoproteins as director of the Laboratory of Cellular and Molecular Biology, overseeing a large team investigating the biochemical and genetic basis of transformation by RNA and DNA tumor viruses, particularly the protein products of these different viral oncogenes.[76]

The stockpiles of purified retroviral proteins accumulated by the NCI proved vitally important when it came to the study of the oncoproteins associated with cellular oncogenes. Such studies relied on possessing a large enough quantity of oncoproteins for biochemical analysis, which was often far more than an individual laboratory could easily produce. Theoretically, recombinant DNA technology offered an easier avenue for producing oncoproteins, but in practice Aaronson and others continued to rely on purified proteins, since the process of using host organisms to assemble recombinant oncoproteins left unanswered too many questions about their biological function relative to the original oncoprotein.[77] For this reason, the identification of proteins associated with oncogenes often lagged behind the study of genes themselves. The *vsis* oncoprotein had not been isolated even after Aaronson's group published a sequence for the v-*sis* gene derived from Simian Sarcoma Virus (SSV) in 1982.[78]

Once protein sequencing advanced, the stockpiles of oncoproteins and immunological tests that Aaronson and his laboratory had accumulated produced an acceleration effect for protein research similar to that provided to oncogene research by the NCI's library of oncogene probes. In 1983 the protein chemist Russell Doolittle at the University of California, San Diego and his collaborators announced that they had associated p28, the protein product of v-*sis*, with a protein found in blood platelets that acted as a growth promoter. The Aaronson laboratory had carried out genetic studies of SSV and immunologically purified the transforming protein, allowing Doolittle and other colleagues to carry out their amino acid sequence comparison with other proteins in their database.[79] A British team working at the Imperial Cancer Research Fund, which shortly thereafter confirmed the resemblance, also relied on the proteins isolated by Aaronson's laboratory.[80] The nearly simultaneous announcement of the findings of two different groups, enabled by their common access to the protein itself, lent fresh momentum to the "oncogene race," according to a writer for the *Journal of the American Medical Association*—a view shared by other commentators.[81]

The association of the v-*sis* oncoprotein with platelet growth factor provided an exemplary case for future molecular studies of cancer; a commentator for *Nature* suggested that the further study of oncoproteins might also reveal that they could be related to molecules with other functions in the body, resolving the mystery of the evolutionary persistence of oncogenes.[82] These breakthroughs also suggested avenues for therapy. Seeking to retain its priority over the discovery, the Imperial Cancer Research Fund put out a press release forecasting that the discovery of the link between growth factor and the oncogene opened the possibility of drugs to block the protein and with it "stop the cancerous growth." A tabloid went further, causing consternation among British cancer researchers by promising a "superjab that will cure certain types of cancer" within the year.[83] Presentations of these discoveries in the United States were no less optimistic. Frank Rauscher, the former director of the NCI and then president of the ACS, told *Newsweek* that these insights would allow chemotherapy to operate with "a rifle rather than a shotgun."[84]

Oncogenes and Oncopolitics

The rapid development of the molecular genetics of cancer in the early 1980s reflected the recycling of the VCP's social and material resources, producing the acceleration of research that Baker and others had hoped for, but with a very different target. Watson captured the sense of excitement that these rapid announcements heralded for molecular biology: "The pace of our research has thus changed . . . from that of an impatient snail to that of an almost uncontrolled tornado."[85] This acceleration provided advocates of molecular biology with a critical resource just as growing dissatisfaction with the dividends of molecular biology threatened to derail federal support for the further molecularization of research. The deputy director of the NCI explained to the House of Representatives that with the discovery of human oncogenes, cancer research had passed through a "biological revolution" that already promised to bring new technologies to the "bedside of the cancer patient." The NCI expected "payoffs in the near future."[86]

The emerging narrative embraced by the NCI and molecular biologists shifted focus from viruses to genes as the key to understanding cancer. Whatever criticism it faced from molecular biologists in the mid-1970s over its mission to produce a cancer vaccine, the VCP was a willing partner in this redefinition. In 1977 its administrators greeted the implications of oncogene research with the claim "Scientists have completely redefined the word 'virus' and no longer think of a tumor virus *only* as a particle that enters a host cell and causes disease, a particle that can be inactivated and turned into a protective vaccine." Therefore, "cancer causation no longer conforms to old concepts of infection and disease. . . . Finding an infectious human cancer virus is no longer necessary to study the role of viral genes in human oncogenesis."[87]

Academic cancer virologists also embraced this shift as a means of redeeming studies of cancer in the laboratory. In 1981 Bishop spent a year as a traveling lecturer on behalf of the ACS, discussing his research into viruses and oncogenes. He summarized the theme of these lectures for the readership of the *Scientific American*, noting that the discovery of oncogenes "heralded a major realignment of experimental cancer research from the search for mechanisms outside the cell responsible for carcinogenesis to mechanisms within." This transition, Bishop wrote, ironically arose from the study of viruses as possible external agents of cancer: "Tumor virology has survived its failure to find abundant viral agents of human cancer. The issue now is . . . how much can be learned from tumor virology about the mechanisms by which human tumors arrived." Bishop explained that his study of RSV showed how "the study of viruses far removed from human concerns has brought to light powerful tools for the study of human disease."[88]

Furthering the idea of a smooth transition from viruses to genes, the 1982 Lasker Prizes honored five scientists deeply involved in retrovirology and cancer genetics. The Albert and Mary Lasker Foundation had established the prizes in 1945 to recognize "major advances in understanding, diagnosis, treatment, cure, and prevention of human disease," and they also served as a potent resource for its lobbying efforts, spotlighting medically promising biological discoveries. At the ceremony, Bishop and Varmus were honored for their work designing the c-*src* probe. Another awardee was Ray Erickson, who had isolated the protein produced by the *src* DNA sequence. The prize lauded these researchers for their "revolutionary discoveries . . . which have provided the long-sought link between viruses and the development of cancer."[89]

The rapid appearance of new findings served to elide molecular biology researchers' continuing doubts about the oncogene theory. The path between oncogenes and cellular development was paved with numerous uncertainties and open scientific questions. In a bluntly titled review for *Nature*, "Oncogenes: We Still Don't Understand Cancer," Temin cautioned, "In spite of the dramatic findings of activation of protooncogenes, transforming DNA, and viruses . . . cancer remains what the classical oncologists have always believed it to be—the final product of a multi-step process."[90] Duesberg pointed out that oncogene scientists could not show by what mechanism individual oncogenes caused cancer. Oncogenes, at least in their inactive form, were in fact common in most normal cells, raising numerous questions regarding how and why a given "proto-oncogene" gave rise to cancer.[91] Harry Rubin observed in a letter to *Science* that "the current rush" to oncogene theories was at best "premature," resting on "flaws in experimental design" and "risky assumptions." "We have," Rubin concluded, "confused advances in molecular biology . . . with deepened understanding of the nature of malignancy."[92]

Scientific objections did not deter the new leadership of the NCI from embracing oncogene research. These doubts certainly mattered less than the sense that they might refocus the NCI on biological research using the excitement generated by oncogenes. Interest in environmental carcinogenesis during the Carter Administration had challenged this focus, but the director of the NCI under President Reagan, Vincent DeVita, was keen to return to biomedical research.[93] He had previously served as director of the Treatment Division, which, like the VCP, used contracts on a routine basis. DeVita was well acquainted with the legislative intricacies of the NCI and the National Cancer Program. He had helped each of the previous two directors, Rauscher and then Upton, prepare for their annual testimony to Congress, and often had joined them. In the late seventies, Lasker had worked closely with him to lobby for increasing the budget of the NCI's chemotherapy programs.[94] However, upon becoming director under President Carter, DeVita confronted the erosion of the NCI's budget and its independence. Budgets for the NCI's grant programs had increased just enough to keep pace with inflation, but that increase had required cutting support for contracts by nearly one-quarter—chemotherapy and virology were especially hard hit. DeVita worried that further cuts to contract research would erode the independence of the NCI.[95]

Although he had overseen these cuts for the sake of supporting research grants to universities and medical schools in the first two years of his tenure as director, DeVita had done so under duress. His aim remained preserving at least a "silhouette" of the more ambitious contract-driven National Cancer Program. DeVita did not share molecular biologists' suspicion of contracts; in fact, like Baker, he appreciated that they gave the NCI leadership a direct say in how money was spent.[96] These instincts formed the basis of his reorganization of the NCI in 1982. DeVita discarded the NCI's environmental carcinogenesis testing program in favor of renewing the institute's commitment to biomedical research. DeVita insisted that the needs of clinical "applied research" would be better served by contracts.[97]

Oncogene research, which promised a unified explanation for cellular development and rapid progress toward cancer therapy, helped heal the political rift between the NCI and molecular biology. Supporting research into the molecular mechanisms of cancer, DeVita maintained, was "important if only for its implications in developmental biology. It needs no other reason for support or excitement. . . . We are clearly optimists, for which no apologies are offered. The danger of overpromise, it seems . . . is exceeded by the risk of failure to pursue and apply one of the most exciting areas of research that brings molecular biology to the crowded bedside of the cancer patient. A good dose of optimism seems about right to make a little room."[98] In this new era of "molecular biology," DeVita wrote, the tempo of the transition between "discovery and application is shortening, and molecular biology is moving closer to the bedside."[99] The president of the ACS invoked the possibility that "accelerated funding in this area of cancer research might speed up the delivery of useful clinical strategies. . . . The Special Cancer Virus Program of the National Cancer Institute is a prime example of a successful 'accelerated funding program.'"[100] Even Zinder conceded that the pace of breakthroughs might at some point justify another "special program" along the lines of the VCP.[101]

DeVita also invoked the intellectual and medical potential of oncogene research when he redefined the mission of the Frederick Cancer Research Facility, a prominent icon of the federal government's war on cancer. He and his deputies contemplated the investigation of the Human T-Cell Lymphoma retrovirus recently discovered by Robert Gallo, or using monoclonal antibodies to target cancer, but they settled on oncogenes as the most promising field of "basic scientific importance," which might also produce "eventual functional spin-offs." The NCI's experience of constructing a library of oncogenes that few other centers possessed provided a useful potential template. The new molecular oncogene approach would allow the facility to merge its efforts in chemical carcinogenesis, immunology, and the development of tumors together in a bid for "national prominence."[102] The "code word of 'oncogenes,'" the NCI decided, provided useful cover for maintaining a broad research program in cellular and molecular biology.[103]

The rehabilitation of the VCP's legacy after the attacks leveled upon it by Watson, Zinder, and others in the molecular biology community during the 1970s and its closure in 1978 were essential elements of DeVita's overall effort to shift the NCI's efforts away from environmental cancer studies and toward molecular studies. "The success of oncogene research," DeVita argued, "can be directly traced to support of virus

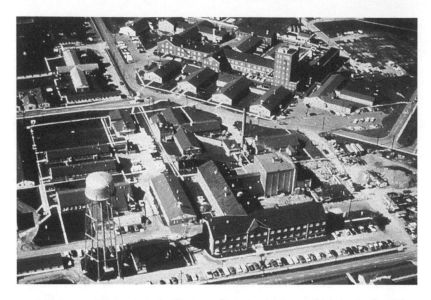

FIGURE 10.3. Frederick Cancer Research Facility, about 1985. Fort Detrick, in Frederick, Maryland, was converted from the center of American biological warfare research efforts into a facility for cancer research after President Nixon signed the Biological Warfare Convention in 1969. Image courtesy of the National Cancer Institute.

cancer research which began in 1964 and expanded significantly in the early 1970s."[104] This forceful conclusion, which underlined the NCI's capacity to anticipate areas for further biological advances, emerged out of a delicate coordination with Watson, one of the War on Cancer's most prominent scientist-critics. Like many other molecular biologists, Watson found the prospect of federal investment shifting from molecular biology to clinical or environmental research more threatening than the prospect of government intrusion.

In the summer of 1982, just as word of the identification of the human *ras* oncogene appeared, Watson placed DeVita at the center of ceremonies for the dedication of a new building for oncogene research at Cold Spring Harbor Laboratory. He timed the ceremonies to coincide with the laboratory's annual conference on RNA tumor viruses, which hosted more than four hundred scientists, creating a clear sense of continuity between cancer virus research and molecular genetics. In his remarks, later published as an essay titled "The Governance of Science," in *Cancer Research*, DeVita argued that the assistance retroviral oncogenes rendered to the identification of cellular oncogenes in humans was strong evidence of how well the VCP had used "the public purse"—a line he borrowed from earlier remarks by Watson.[105]

This emphasis on the links between retroviruses and oncogenes reset the antagonistic relationship between the NCI and molecular biologists. Underlining the intellectual dividends of virus research allowed the NCI to focus on funding cutting-edge biology while also holding out for the promise of therapeutic breakthroughs. Watson argued that the number of scientists at work on cancer virology "in no way reflected the feeling that the cancer problem in any sense was solvable over the short term." Rather, "it reflected . . . a highly intelligent assessment of where the next major advances in molecular genetics would lie."[106] Watson praised De-Vita's remarks, stating that "younger retrovirologists" were "solidly on [DeVita's] side," although many would not say so openly for fear of offending the " 'have nots' " outside of oncogene research who "must remain that way" if the NCI wished to test the "brightest ideas" of the field.[107]

In turn, the NCI's investment in oncogenes allowed this community to grow. During the 1980s and 1990s, the Frederick Cancer Research Facility in Maryland joined the Cold Spring Harbor Laboratory to become one of the hubs of the new oncogene community. More than one hundred post-doctoral fellows trained there in molecular biology, and its annual "Oncogene Meeting" drew around five hundred molecular biologists to Frederick, marking the meeting as an important site for defining the molecular biology community in the late 1980s and 1990s.[108] Addressing Congress for the first time in 1989, DeVita's successor claimed that the NCI had "made considerable progress" in developing "basic science" and converting that progress into prevention and treatment—a set of developments arising from its heavy investment in molecular approaches.[109]

Redefining Cancer as a Molecular Problem

Drawn by the productivity of oncogene research and its ability to illuminate cell biology, molecular biologists flocked to cancer research. A conference convener enthused in 1985: "The 'advances' in human cancer control in the past 30 years have been minor variations of old themes involving little change in our underlying concepts of causes, approaches to treatment or, in my own field, of methods and goals of diagnosis." In this light oncogene research was "especially striking and unusual because it is concerned equally with basic molecular biology, animal models, and clinical cancer. What have been separate worlds for decades now are seen to merge."[110] Weinberg predicted that retroviruses had helped reveal the small number of genes that regulated the "molecular basis of . . . transformation." "The next chapters of the saga," Weinberg predicted, would be the domain of cell physiology and biochemistry, which would finally provide a detailed description of the "wiring diagram of the cell." The tumor virologist Renato Dulbecco endorsed, in *Science*, early proposals to sequence the human genome, suggesting that it might serve as a "turning point" in cancer research.[111]

The events described in this chapter do not exhaust the historical or political developments associated with the rise of the oncogene theory, but they do give a sense of how the previous infrastructure of the VCP helped to produce dramatic changes in the sense of momentum and political possibility attending molecular approaches to cancer in the early 1980s. Contrary to the story of serendipity presented by Watson and DeVita, the expansion of oncogene research did not come at the expense of the study of viral carcinogenesis. Instead, the VCP's infrastructure provided a set of resources that facilitated the emergence of molecular genetics even as interest in cancer viruses remained strong.[112] The discoveries associated with the "oncogene paradigm" might have eventually come about owing to new technologies for manipulating genes unrelated to viruses, but the infrastructure of viral oncogene research allowed those findings to emerge, gain acceptance, and attract further investment far more quickly than they otherwise might have.[113] In turn, the rapid tempo of the discoveries provided important political and cultural momentum for keeping the focus of cancer research squarely on the therapeutic potential of molecular biology, crowding other possible solutions to the cancer problem out to the periphery.

Afterlife, Memory, and Failure in Biomedical Research

This book originated in my effort to explore the history of molecular biology by tracing the development of cancer virus studies in the laboratory. But following cancer viruses led me to a far more complex story, revealing a rich set of interactions among policymaking, American culture, and biomedical science. The multiple nature of cancer viruses as cultural, administrative, and scientific objects, which I had initially found so confounding, was in fact the source of their broad-ranging influence. Although apprehension of cancer as a contagious disease was widespread in the early twentieth century, drawing the federal government into the effort to develop a human cancer vaccine required both new tools in virology and new political alliances. The biomedical settlement's promise that government would confront disease at the bench rather than the bedside brought unprecedented resources to the emerging molecular biology community. However, the same premise also imposed new obligations and expectations on biologists, obligations that scientists themselves were often unwilling to accept. Cancer viruses served as essential tools for the growth of molecular biology as well as prominent objects for defining the role that the government would play in realizing the therapeutic potential of molecular medicine. The infrastructure built to investigate cancer viruses in the 1960s and 1970s, and the debates it inspired, shaped the material and political possibilities of molecular medicine in the 1980s and long afterward.

The mutability and plasticity of cancer viruses as their status developed from possible agents of infection, to targets of vaccination, to administrative objects, to molecular tools, and returned to vaccination, illustrate that any effort to assess their history must attend carefully to the problems that the study of cancer viruses was expected to solve in each era. In his 1959 book *Virus Hunters*, the science journalist Greer Williams, flush with the success of polio vaccination, envisioned a future in which advances in virology would soon "ward off that tragic second childhood of growth, cancer."[1] Today, vaccines against human papillomavirus and hepatitis B virus offer protection against a significant number of cancers. At the end of the 1980s, however, such hopes appeared to be rooted in speculative fiction rather than journalism. The fading chances of a human cancer vaccine were typical of a War on Cancer that was "at first over-hyped and [later] underfunded."[2] One historian of cancer's causes pointedly excluded cancer virology from his survey on the grounds that it was one of the areas, along with molecular genetics, that had received "undue" attention by the scientific community at the expense of research into the roots of cancer as an environmental problem.[3]

The sense of dissatisfaction continued for a generation even among those more favorably disposed toward biomedical resolutions to the cancer problem. The National Research Council's *Large-Scale Biomedical Science* (2003) recycled the Zinder Report's charges, writing that the Virus Cancer Program had been "inaccessible to the virology community" and represented a suspect expansion of the work of a few "program scientists." It was a "significant failure of directed research since it did not lead directly to the identification of any viruses that cause human cancer." The program was redeemed only by its "indirect, beneficial effects on the scientific community."[4] *The Emperor of All Maladies* (2011), a Pulitzer Prize–winning history of cancer, memorialized cancer virus research with the comment that it "siphoned" hundreds of millions of dollars away from other deserving endeavors.[5]

My aim is not to redeem or condemn the investment that the United States made in cancer vaccine research, but to think about how the cycles of fear, mobilization, and frustration that attended this effort were not anomalies but typical of the tensions and hopes that the biomedical settlement created for biology in American society. Questions of success or failure cannot be reckoned with absent an understanding of how the problems of disease are defined by the communities of specialists called upon to provide solutions.[6]

The Afterlife of Infrastructure

Like the flurry of construction that might accompany a gold rush, the structures associated with particular moments of mobilization in biomedical research have a long afterlife. The infrastructure that the War on Cancer established to pursue cancer as a problem of viral infection did not evaporate with that war's political exhaustion. The personnel, resources, and institutions created by the Virus Cancer Program were readily adapted to other diseases. In the early 1980s, reports of a rare cancer known as Kaposi's Sarcoma, along with a rare form of pneumonia, started to appear in unprecedented numbers among men who had sex with men. These were the first signs of a nationwide epidemic of Acquired Immune Deficiency Syndrome (AIDS), whose capacity to weaken the immune system allowed a host of opportunistic infections, including cancer-causing viruses, to enter the human body.[7] Like cancer in the early twentieth century, AIDS inspired shame, fear, and confusion.

Although activists such as the members of the AIDS Coalition to Unleash Power (ACT UP) later introduced new concepts of citizen science to the biomedical settlement as they urged the federal government to respond to AIDS, the National Institutes of Health's initial response rested upon the infrastructure of the War on Cancer.[8] Other federal organizations, such as the Centers for Disease Control, possessed the epidemiological capacity to track AIDS, but only the NIH possessed the capacity to carry out the research that could identify the disease's cause and potentially produce a preventive vaccine or a cure. However, the NIH had largely emphasized research on the problems posed by chronic diseases.[9] The exception to this rule was the investment the National Cancer Institute had made in cancer viruses. In particular, the reverse transcriptase boom left behind a large community of retrovirus researchers with preexisting connections and access to abundant resources. The association of AIDS with rare cancers initially suggested that the National Cancer Institute might take the lead in responding to this new potential virus.[10]

The capacity for retrovirus studies played a key role in the association of AIDS with a virus, Human Immunodeficiency Virus (HIV). Both of the two claimants to the discovery of HIV, Robert Gallo and Luc Montagnier, had previously studied the role of retroviruses in leukemia. In the United States, virologist Gallo had spent most of the 1970s working with the Virus Cancer Program from the Laboratory of Tumor Cell Biology in facilities run by defense contractor Litton Bionetics.[11] Gallo expanded his work overnight by hiring postdocs and acquiring laboratory space.[12] He devoted his attention to one of the original aims of the NCI's mobilization against cancer: identifying a human leukemia virus, presumed to be a retrovirus of the kind that Huebner had predicted.[13] While researchers at academic institutions doubted the existence of human retroviruses and balked at the expense of a search, at the NCI Gallo had the ability to keep searching. Following Huebner's ambitions, and heartened by the discovery of retroviruses in nonhuman primates, Gallo's laboratory worked to culture and observe tissues from different cases of leukemia and lymphomas for the detection of virus particles. In 1979, in a case of T-cell lymphoma, Gallo's laboratory succeeded. A second observation of a "Human T-cell Lymphoma Virus" soon followed. Though these two types of lymphoma were rare, the results inspired confidence that there were other strains of retroviruses responsible for human cancers.[14]

This previous work left Gallo's laboratory rich in experience with retroviruses and the cells present in the immune system—two areas that lent themselves to the exploration of AIDS. In 1983 the laboratory announced it had identified a third Human T-Cell Lymphoma Virus in the blood of a patient with AIDS.[15] Gallo followed this announcement in 1984 with the claim that most patients with AIDS tested positive for antibodies to this virus—suggesting that Human T-Cell Lymphoma Virus III was the causative agent of AIDS.[16] Montagnier, meanwhile, had been pursuing similar studies at the Pasteur Institute in Paris, a center of studies of infectious disease. His work left him well positioned to study the human tumor leukemia viruses announced by Gallo, and, in seeking to identify more of these viruses, he identified the "lymphadenopathy associated virus" (LAV) from a patient in Paris. Encountering difficulty in growing the virus, he exchanged samples with Gallo at a meeting at Cold Spring Harbor in the fall of 1983.[17] The subsequent question that preoccupied American and French researchers—with immense consequences for the patenting rights to different diagnostic tests for AIDS—was whether Gallo had performed his experiments using LAV or his own virus. As the controversy raged, virologists adopted the more neutral "HIV" to designate the retrovirus responsible for AIDS, in the process erasing its associations with the earlier generation of leukemia virus studies.[18]

When it came to the association of HIV with AIDS, the presence of this community created a potent acceleration and convergence effect. The very things that molecular biologists had derided as uninteresting—the capacity to mass-produce viruses and to perform immunological testing for the detection of viruses—now worked to promote the study and identification of HIV. At the Frederick Cancer Research Facility, the NCI mass-produced HIV to encourage development of diagnostic tests for the virus. The rapid development of those tests, which appeared nine months after Gallo's announcement, would have been difficult to achieve without this preexisting capacity.[19] With the diagnostic test in hand, researchers shifted their attention to the development of a vaccine.[20]

Despite challenges to the claim that HIV was the causative agent of AIDS, the biomedical response to AIDS rapidly converged on retroviruses. Reflecting on the response to AIDS from 1983 to 1986, one sociologist noted that "interest in etiology trailed off, while interest in the virus itself exploded," driven by the inquiries of a "small group" of scientists.[21] This reflected the role of the retrovirology community in the study of AIDS: their primary interest was in the virus, not the broader question of etiology. The participants in the HIV vaccine effort included many familiar retrovirus researchers: Gallo, Montagnier, Baltimore, and Varmus, as well as other less well-known figures who became important, such as William Haseltine, who had started their careers working on retroviruses.[22] The migration of this community also explains how the retrovirus boom produced some of the most persistent dissenters from the association between HIV and AIDS, such as Peter Duesberg.[23]

The way this small group became a vanguard for the development of an HIV vaccine exemplifies how biomedical research infrastructure can shape new fields even as the reasons for its creation fade. Indeed, this example echoes the influence that polio researchers had on promoting cancer vaccine research in the 1950s, just as their own funding dissipated. Now, cancer virus researchers endorsed the development of an HIV vaccine as a promising solution to the problem of AIDS.[24] However, these retrovirus researchers also limited the actions that the government might undertake. At the urging of David Baltimore and Howard Temin, who recalled the "stench" of the Virus Cancer Program, the National Academy of Sciences' Institute of Medicine refrained from endorsing a "Manhattan Project" to develop an AIDS vaccine.[25] While activist groups continued to call for more aggressive government intervention in research and development for an AIDS vaccine, efforts moved further into the hands of private biotechnology and pharmaceutical companies rather than the National Institutes of Health.[26]

Memories of State Failure and the Promise of Markets

The terms of the biomedical settlement underwent significant change in the last quarter of the twentieth century, as did the role of the government in American society. Starting in the late 1970s, and accelerating under the "Reagan Revolution" of the 1980s, many policymakers embraced the market over state planning as a mechanism to further public welfare.[27] Central to that development was a new confidence that the translation of biomedical research into therapies was the domain of business, a sector that had been absent from the original biomedical settlement. Many individual changes fostered this shift, including technical advances, patent law rulings, and new investment practices, but uniting these developments was a new understanding of the relationship among market, government, and society in America.[28] Whereas Lyndon Johnson and the National Cancer Institute had offered a version of the biomedical settlement that favored government action, in the 1980s the settlement expanded to include representatives of business, especially the new biotechnology industry. "The private sector, not government," now occupied the "front lines of the cancer war," enthused *Business Week*.[29]

The biomedical settlement was not dismantled by the turn away from state planning, but it took on additional forms and meanings. The most visible icons of the new era were biotechnology and pharmaceutical firms, whose efforts in cancer research gained momentum in the 1980s by drawing upon the community of molecular cancer researchers created by the War on Cancer. In 1983, flush with the therapeutic promise of the first round of oncogene discoveries, venture capitalists hired an entire laboratory of researchers away from the NCI. "It was clear that if you wanted to remain in the research field, you had to leave," said John Stephenson, formerly of the NCI's Laboratory of Viral Carcinogenesis. Oncogene research was moving "from the pure research stage to commercialization."[30]

The government had found it difficult to recruit molecular biologists to serve as program managers in the War on Cancer, but few similar challenges appeared for the biotechnology industry.[31] The movement of molecular biologists into biotechnology reflected a lingering distrust of the federal government's support for academic research as well as enthusiasm over a new form of scientific life seen as more free of intrusion than working on federal contracts or grants.[32] Historians have looked at the excitement of the new industry as creating a "pull" toward biotechnology; however, in the wake of the molecular biology community's negative reaction to the War on Cancer, we may see an antigovernment "push" into that industry as well. Indeed, some figures who had been strident critics of the Virus Cancer Program, such as Arthur Kornberg, viewed their biotechnology ventures as a hedge against their own overreliance on federal funding.[33]

The memory of the VCP's "failure" became a touchstone for those aiming to associate molecular biology with the promise of entrepreneurial biotechnology. The story of the failure of the VCP and its redemption by its serendipitous contributions to molecular genetics had first been told by James Watson and Vincent DeVita as a means of promoting oncogene research. In the hands of others, the idea of failure performed important boundary work for molecular biology as it extended into medicine. It became an exemplar not only of the government's limited capacity to intervene in biomedical research, but also of the futility of managing academic biology toward therapeutic ends.[34] The memoirs and histories written by this group stress the misguided "moonshot" nature of cancer virus research, particularly its emphasis on identifying human cancer viruses.[35] Samuel Broder, the director of the NCI after Vincent DeVita, echoed these assertions of government inefficiency from his new post as chief medical officer of the biotechnology firm Celera Genomics. Eliding the ambitions of the VCP and the resistance of many virologists to the planning efforts that went into the Salk vaccine, Broder charged that the federal government could never have developed the polio vaccine through a "centrally directed program," merely a stopgap such as a better "iron lung." Disease cures could only emerge through "independent, investigator driven discovery research."[36] This perspective has carried through to present-day interpretations of the development of vaccines against human papillomavirus, which have often discounted the contributions of the NCI in favor of following the work of pharmaceutical companies and academic scientists.[37]

The political uses of the VCP's history deepened when molecular biologists responded to the Human Genome Project. The Department of Energy, with its experience managing large projects, appeared to be a logical venue for organizing this work, rather than the NIH. However, the advisers of the Human Genome Project, including James Watson, Norton Zinder, and others who had led the resistance to the VCP, sought to deflect the move toward a large-scale managed program. Instead, they lobbied for an organization within the NIH, where smaller laboratories and peer review would organize the mapping process.[38] Many of the criticisms of the Department of Energy's approach were familiar. Many academic biologists charged that it would divert resources from other fields. Much of the labor of sequencing was portrayed as routine and uninspiring. One molecular biologist even joked that the work of sequencing should be assigned as punishment.[39]

In lieu of planning and organization, the repeated invocations of the VCP's failure elevated serendipity as a mechanism of scientific discovery. The Office of Technology Assessment's influential report *Mapping Our Genes : The Genome Projects : How Big? How Fast?* (1988) reflected this debate. Writing in terms familiar from critiques of the VCP by academic scientists, the Office of Technology Assessment noted that the "small group" should "remain the principal means of studying physiology and disease," although there might be a role for some "Big Science" facilities as long as these small groups were not "starved" of resources to support them. The Office of Technology Assessment cautioned that a potential "targeted" genome project should not become an "instead-of" program rather than an "in-addition to" program.[40]

The VCP, in this set of memories, offered an enduring reminder of the dangers of government overreach, while obscuring the question of what, if any, obligations the biomedical settlement imposed on biological research. Even if the Human Genome Project had adopted a centralized framework, its aims still would not have equaled the ambition of the War on Cancer. The project proposed to survey the genome as a resource for further therapeutic exploration by private companies, not to supervise the translation of findings into therapeutic products itself.[41] Only recently have discussions of the need for federal management of biomedical research reemerged alongside new doubts regarding the "payoff" for human health of federally sponsored biological inquiries.[42]

The Borders of the Biomedical Settlement

As the molecular approach to medicine grew stronger in the late twentieth century, the overall political architecture of the biomedical settlement continued to have an important role in framing and constraining the development of vaccines against human cancer. In 1976 Baruch S. Blumberg, a physician trained in immunology, epidemiology, and public health, received a Nobel Prize for his identification of a new strain of hepatitis virus, hepatitis B, in 1972. Even as he accepted his award, further epidemiological research in Taiwan demonstrated that chronic hepatitis B infection was closely associated with liver cancer, the third-most-common cancer in the world. Moreover, as this epidemiological connection emerged, Blumberg and collaborators at the pharmaceutical firm Merck were poised to develop a vaccine against hepatitis B.[43] Amid condemnation of the VCP as a failure, the first cancer vaccine with the promise of preventing human cancer stood on the verge of mass distribution.

In 1978 Blumberg was one of several experts gathered by NCI director Arthur Upton as he considered new pathways for confronting cancer. The hepatitis B results might have been expected to generate considerable excitement regarding the promise of a cancer vaccine, but they did not. As molecular biologists defined the problem of cancer, the prospect of producing a hepatitis B vaccine did not appear to be helpful. Blumberg entered into a pointed exchange with David Baltimore regarding the feasibility of a cancer vaccine.[44] Echoing the interests of the molecular biology community in fundamental research, Baltimore contended that a hepatitis vaccine could not be fully developed without understanding the mechanism through which the virus caused cancer. Blumberg demurred that in public health, virologists had the "recurrent experience of being able to prevent disease without knowing its mechanism."[45]

This exchange over the nature of the cancer problem, staged between a molecular biologist and a traditional immunologist, was familiar from the debates Huebner and Watson had engaged in around the War on Cancer. But Baltimore's rebuttal to Blumberg was not. He responded, "Even if we cured all the liver cancers in East Africa, we would not have responded to the mandate" to "deal with the problems we have here in the United States" instead of studying the "ways of preventing cancer anywhere in the world."[46]

Baltimore's retort, although blunt, is a reminder that the foremost goal of the NCI's hunt for human cancer viruses was to improve the health of American citizens. Yet the cancer problem, and the urgency of particular solutions, remained very different depending on what national context it was viewed from.[47] The biomedical settlement did not accommodate a vision of the NCI's addressing cancer on purely humanitarian grounds in the 1970s, nor did it recognize liver cancer as a pressing threat in the United States. In America, liver cancer was associated with alcohol use—a population attracting far less sympathy than the vulnerable children who had sustained the early search for a leukemia vaccine. Therefore, the federal government did not invest heavily in the production of the hepatitis B vaccine, whose global distribution took several decades as international organizations struggled to muster the resources for its manufacture and distribution.[48] Likewise, reflecting the unequal global distribution of health care coverage that would pay for vaccination, the first versions of the human papillomavirus vaccine, which can protect against oral, anal, penile, cervical, and other cancers, emphasized strains of the virus found in the United States and Europe rather than those common in Africa, Asia, and South America, where the largest number of deaths from these cancers occur.[49]

*   *   *

Throughout the twentieth century, as the Boardmans of New Jersey contemplated the safety of their mattress, as the residents of Niles, Illinois, sought to understand a mysterious leukemia cluster, or as hundreds of thousands of individuals wrote to President Nixon in support of the War on Cancer, countless Americans looked to biological research as a beacon of hope against cancer. However, even as this hope sustained great advances in our biological understanding of the disease, the prospect of easy solutions to the cancer problem, such as a cancer vaccine, receded further from our grasp. This observation is symptomatic of a paradox that emerges from the biological study of chronic diseases: greater knowledge often results in renewed uncertainty about the nature of the problem itself rather than a clear solution. Although many might aim to resolve this paradox with renewed scientific inquiries, the history of cancer viruses suggests that these paradoxes are not faults but features of the biomedical approach to disease. The strategies that doctors, activists, administrators, and scientists adopted for navigating cycles of hope and disappointment are a core part of the history of biomedicine in the United States. As we contemplate to the challenges posed by cancer in the twenty-first century, we would do well to keep in mind both the occasional rewards of failure and the unanticipated pitfalls of success.

Time Line

History of Cancer	History of Medicine and Biology	United States History
	1880s–1900	
1887: William Halstead starts to develop radical mastectomy	1880: Louis Pasteur coins term "virus-vaccine"	1883: US Supreme Court overturns 1875 Civil Rights Act, setting stage for disenfranchisement of African Americans and power of southern states in Congress
1898: Roswell Park Cancer Laboratory established	1882: TB bacteria isolated by Robert Koch	
1899: Term "cancer problem" first used	1887: Pasteur Institute established	
Harvard Cancer Commission created	1890: Cold Spring Harbor Laboratory established	1887: Marine Hospital Service established
	1895: X-rays described	1893–1897: "Panic of 1893," major economic depression
	1898: Radium isolated	1898: Spanish-American War begins
	1900s	
1901: Rockefeller Institute for Medical Research founded	1904: National Tuberculosis Association founded	1902: Public Health & Marine Hospital Service Hygienic Laboratory established
1908: Chicken leukemia virus observed	1909: Salvarsan chemotherapy for syphilis carried out	Biologics Control Act passed
1910: Peyton Rous carries out observations on "non-filterable" particle later known as RSV	1910–1939: Eugenics Record Office in operation	1906: Pure Food and Drug Act passed
	1910s	
1913: American Society for the Control of Cancer created	1916: 1st polio outbreak in New York City	1912: Public Health Service established to address "diseases of man"
1915: *Mortality from Cancer throughout the World* published	1918–1919: Influenza pandemic	1914–1919: First World War
Rous stops working on RSV		

History of Cancer	History of Medicine and Biology	United States History
1920s		
1921: ASCC launches "Cancer Week"	1925: Analytical ultracentrifuge invented	1920s: PHS malaria eradication campaign in southern states
1925: William Gye announces he can see RSV with microscope	1928: Penicillin discovered	1928: President Hoover elected; associational approach to cancer
1926: Johannes Fibinger awarded Nobel Prize for identification of a cancer parasite		1929: Great Depression
1928: Senator Neely calls for conquest of cancer		1930: Ransdell Act, National Institute of Health created
1930s		
1933: Shope papillomavirus observed	1933: 1st "Presidential Birthday Ball" to raise money for polio	1932: Franklin Roosevelt elected
1936: "Milk Factor" identified by John Bittner; later known as Mouse Mammary Tumor Virus	1934: Rockefeller Foundation launches its program "Science of Man"	1935: Social Security Act passed
	1937: First human viral vaccine (yellow fever)	1937: "Court Packing" controversy
1937: National Cancer Institute established	1938: National Foundation for Infantile Paralysis founded	1938: Federal Food Drug and Cosmetic Act passed, expanding Food and Drug Administration
1940s		
1943: Nitrogen mustard chemotherapy for cancer developed	1940: RCA electron microscope invented	1939–1945: Second World War
1944: Mary Lasker refashions ASCC into American Cancer Society	1942–1945: Office of Scientific Research and Development oversees mass production of antibiotics, antimalarial compounds, and vaccines	1942–1945: Manhattan Project
1945: Sloan-Kettering Institute founded	1945: *Science, the Endless Frontier* published	1946: Hill Burton Hospital Construction Act
1946: "Manhattan Project" for cancer called for	"Phage school" started	1948: Rand Corporation founded
Atomic Bomb Casualty Commission formed	1948: Randomized controlled trial of antibiotics	President Harry Truman reelected
1947–1948: Sidney Farber tests first antifolate compounds; Jimmy Fund created	1949: Linus Pauling publishes "Sickle Cell Anemia, a Molecular Disease"	1949: Failure of Truman health insurance

History of Cancer	History of Medicine and Biology	United States History
	1950s	
1951: Ludwik Gross demonstrates murine leukemia virus	1950: Doll-Hill paper on smoking and lung cancer	1952: Hydrogen bomb
1953: Cancer Chemotherapy National Service Center formed	National Science Foundation created	1954: *Lucky Dragon* radiation incident
Radioactive fallout protests	1953: DNA structure found	
1956: NCI establishes Viruses and Cancer Panel after special congressional appropriation	Virus cultures created	
	Atoms for Peace	
	1955: Salk polio vaccine	
1958: Howard Temin and Harry Rubin create transformation assay for RSV	1955–1968: James Shannon is director of NIH	
Burkitt lymphoma discovered		
Polyoma virus observed		
Charlotte Friend develops leukemia vaccine in mice		
1960: Simian Virus 40 identified in polio vaccine		
	1960s	
1961: Niles, IL, leukemia cluster	1961: Operon theory formed	1960: "Missile Gap" debate between Nixon and Kennedy
1962: Simian Virus 40 shown to be oncogenic	1962: *Silent Spring* published	
	mRNA discovered	1961: Robert McNamara becomes secretary of defense
1964: Special Leukemia Virus Program begun	Hybridization studies	
	1964: Surgeon general's *Report on Smoking and Health*	NASA commits to moonshot
Epstein-Barr virus observed		
1968: Special Virus Cancer Program begun	1966: Australia Antigen isolated; later known as hepatitis B virus	1962: FDA powers expanded by Kefauver amendments
1969: Viral oncogene theory promoted		1963: Atmospheric Test Ban Treaty
Ft. Detrick converted to cancer research		1964: Lyndon Johnson elected
		1965: *Unsafe at Any Speed* published
		1966: Great Society
		1967: Vietnam War protests
		1969: EPA/OSHA
		Moon landing

History of Cancer	History of Medicine and Biology	United States History
	1970s	
1970: Reverse transcriptase discovered	1970: TV and radio advertising for cigarettes banned	1970: 1st Earth Day
1971: War on Cancer	1973: Recombinant DNA regulation begins	1972: Richard Nixon reelected
1972: Virus Cancer Program established	1974: Tuskegee Syphilis Study revelations are made	1974: Nixon resigns
1973: Ames Test		1979: Three-Mile Island incident
1974: Zinder Report	1975: Monoclonal antibodies discovered	1980: Bayh-Dole Act passed
1976: Cellular proto-oncogene *src* discovered	Temin and Baltimore receive Nobel Prize	
Blumberg receives Nobel Prize	1978: Love Canal controversy	
1978: VCP closed		
1979: Human T-cell Leukemia Virus I observed		
NCI removes endorsement for radical mastectomy		
	1980s	
1981: Heptavax-B, first hepatitis B vaccine, approved	1983: HIV associated with AIDS	1981: Ronald Reagan Administration
1982: *ras* identified as first human oncogene	1985: Mapping of human genome proposed	1986: Chernobyl & Strontium 90
1983: *vsis* first oncoprotein to have function identified		
1986: Tumor-suppressing genes identified		
Recombinant Hepatitis B vaccine		
1989: Bishop and Varmus receive Nobel Prize		
	1990s–2010	
2006: Gardasil vaccine against human papillomavirus strains 16 and 18 approved	2000: Human genome sequenced	2009: Increased budget for NIH passed as part of the American Recovery and Reinvestment Act

Acknowledgments

Scholarship is a social and a historical enterprise, and this book reflects the accumulated wisdom and assistance of the communities I have had the privilege and good fortune to encounter while pursuing this project over the past decade.

When I started to follow cancer viruses into the worlds of biology, medicine, and society, Daniel Kevles was a peerless mentor and critic. His support has sustained and guided this project. Jennifer Klein, Bruno Strasser, and John Warner all helped me think more broadly about the connections among policy, medicine, and cancer as my interests moved out of the laboratory and into society. I also drew from the generosity of other faculty at Yale, including Jean-Christophe Agnew, Joanna Radin, Bill Rankin, Naomi Rogers, Paul Sabin, William Summers, and Adam Tooze. These conversations, and the questions they left me with, were essential mileposts as I journeyed from the technical realm of virology and genetics to broader historical and cultural questions about biology and medicine.

Throughout the process of writing and rewriting, I benefited from discussions with friends and colleagues both near and far. At Yale, Mary Brazelton, Lisa Furchgott, Jennifer Lambe, David Minto, Joy Rankin, and Tom Reznick bore the brunt of my thinking out loud. Jennifer and David were indefatigable writing companions. I thank Jonny Bunning, Deborah Doroshow, Ted Fertik, Matt Fraser, Andrew Horowitz, David Huyssen, Antoine Landecker, Todd Olszewski, and Gabriel Weinert for their conversations over the years.

At the American Academy of Arts and Sciences Visiting Scholars Program, I had the luxury of a year of writing in the company of scholars from across the humanities, particularly Brent Cebul, Emily Remus,

Claire Seiler, and S. C. Yang, whose comments and insights on chapter drafts helped me fit the history of cancer into the historical experience of the modern United States. Larry Buell, the program director, provided valuable insight in shaping the manuscript.

At MIT I have found an enthusiastic and welcoming group of colleagues in the History, Anthropology, and STS Program, and further afield. Christopher Cappazola, Michael Fisher, Deborah Fitzgerald, Jennifer Light, Heather Paxson, Harriet Ritvo, and Rosalind Williams have been generous with their thoughts and unsparing with their criticism when necessary. I have especially enjoyed talking with my cohort of early career academics at MIT as we have navigated our first books: Dwai Banerjee, William Derringer, Caley Horan, and Amy Moran-Thomas. Emily Richmond Pollock has been a wonderful revision companion.

During this journey, I am grateful to those further afield who aided my research and thinking. I am especially grateful to John-Paul Gaudillière for hosting me for a day in Paris and allowing me to examine research he had carried out at the National Cancer Institute and to David Cantor for allowing me space to work at the National Institutes of Health Office of History. Special thanks to my fellow participants in "100 Years of Viruses and Cancer": Alex Broadbent, Brendan Clarke, Neerja Shankaran, and Doogab Yi. Doogab in particular has been wonderful and generous in welcoming me to the neighborhood of cancer virus studies. Conversations with Natalie Aviles, Luis Campos, Nathaniel Comfort, Angela Creager, Nathan Crowe, Julia Cummiskey, Helen Curry, Dawn Dirgius, Evan Helper-Smith, Ton Van Helvoort, Nicolas Hopwood, S. Lochlann Jain, Daniel Liu, Marissa Mika, Joshua Nall, Sejal Patel, Simon Schaffer, James Strick, and Carsten Timmerman enriched my thinking. I owe particular thanks to those who lent their expertise to help me think through particular chapters: Andrew Hogan, Sheila Jasanoff, and Jacob Steere-Williams. Ashawari Chaudhuri provided invaluable research assistance in completing the final manuscript.

There is a special class of individuals who read the entirety of my manuscript in its various stages of development and offered immensely helpful critical commentary. I thank the participants in the book workshop sponsored by the MIT Program in Science, Technology, and Society in October of 2017—Soraya de Chadarevian, Stefan Helmreich, David Kaiser, Illana Löwy, and James Sparrow—for reading an entire draft manuscript. Eli Anders provided invaluable final commentary to sharpen my prose as well as an invaluable sounding-board throughout the process of writing. Adam Berger bravely served as my "intelligent, uninformed reader."

As I wrote, I also had the pleasure of teaching students at Yale and MIT. Their thoughts and questions, especially in my course "The Long War on Cancer," have pushed me to understand and explain my scholarship in productive ways.

I could not have written this book without extensive access to archives and research libraries, and I could not have found the materials I needed (and some I did not know that I did) without the guidance and support of archivists and librarians, who went above and beyond to help me examine materials that were of interest. I owe many serendipitous finds to their suggestions. At Yale, Toby Appel and Melissa Graffe helped me locate essential materials as I started this project. Michelle Baildon and the interlibrary loan staff at the MIT libraries helped me locate materials that had evaded my initial research. Further afield I give special thanks to Kathleen Brennan (Memorial Sloan Kettering Cancer Center Archives), Stephanie Bricking (University of Cincinnati Harry R. Winkler Center), Javier Garza (MD Anderson Research Medical Library), Charles Greifenstien and Andrew Lippert (APS Library), Lee R. Hiltzik (Rockefeller Archives Center), Polina Ilieva (UCSF Archives), Phillip Montgomery (Texas Medical Center McGovern Historical Library), Barbara Niss (Mt. Sinai Levy Library), Scott Podolsky (Harvard Countway Library Center for the History of Medicine), David Rose (March of Dimes Archives), and Daniel Smaczny (Niles Historical Society and Museum). Richard Mandel, of the Executive Secretariat of the Office of the Director for the National Institutes of Health, provided access to the Office's local digital archive before its accession to the National Archives. My particular thanks to Barbara Harkins of the National Institutes of Health Office of History for her unwavering efforts to locate elements of Carl Baker's papers and other material as I completed the book.

I am appreciative of the individual scientists who spoke with me about their research and helped me clarify conceptual questions about the state of cancer virology as it stood decades ago. I am especially grateful to J. Michael Bishop, Murray Gardner, Ray Gilden, and Harold Varmus for allowing me to quote directly from their interviews. Paul Levine and Frank J. Rauscher III provided images from their personal collections for the book.

I am also very thankful for the opportunity to refine my ideas in front of many audiences over the years. Audiences at Boston University, the University of Cambridge, the Chemical Heritage Foundation, the Harvard Kennedy School of Government, Harvard Medical School, the Johns Hopkins Institute for the History of Medicine, Kings College London,

MIT, Rensselaer Polytechnic Institute, Stanford University, the Stevens Institute of Technology, the University of Chicago, the University of California, Los Angeles, the University of California, Santa Barbara, and the University of Wisconsin–Madison listened to longer versions of my research and provided very useful feedback in both forming and shaping the final book. Setting out on my project, I benefited from feedback at the Three Societies Meeting of 2013, the Joint Atlantic Seminar for the History of Biology, the History of Science Society, the Organization of American Historians, the American Association for the History of Medicine, and the International Society for the History and Philosophy and Social Studies of Biology. The organizers also provided financial assistance for attending these meetings.

The travel and research that allowed me to visit so many archives and meet so many scholars were only possible through the generosity of many institutions. The National Science Foundation's Graduate Research Fellowship Program gave me the flexibility and freedom to begin research. Yale University provided me with further fellowship funding, including travel funds through a graduate fellowship, a John Enders Fellowship for Summer Research, and a University Dissertation Fellowship. The Program in the History of Medicine provided funds for conference travel as well as "seed" money for research trips. Additionally, the American Philosophical Society, the Rockefeller Archives Center, and the Yale Club of Philadelphia all provided fellowships to support long-term research. At MIT I have benefited from support from the SHASS Dean's Travel Fund and especially the Leo Marx Career Development Professorship, which underwrote many of the expenses involved with the preparation of this book.

Finally, my thanks to the University of Chicago Press for their excitement and enthusiasm for this book, for the editorial guidance of Karen Darling, for the editorial support of Evan White and Susannah Engstrom, and for the copyediting of Lois Crum. I am grateful for the comments of two anonymous reviewers, who helped me hone the manuscript.

Any errors or omissions at the end of this process are my responsibility alone.

My intellectual journey toward this book started long before I encountered cancer viruses. My parents instilled a love of learning in me and a curiosity about both the social and natural sciences. Our dinner table was more than adequate preparation for the most spirited seminar exchanges once I arrived in graduate school, and the exchange has continued over countless hikes in the East Bay hills. It's been a joy to grow up with my brother and see my research through the eyes of his alert questions, as well as to benefit from his own expertise in biostatistics as I approached some of the knottier issues in contemporary cancer research. My in-laws, Steven Parton and Melody James, have welcomed me into their home for many weekends of writing and lent me their artistic perspective on the narrative and visual elements of the book. I am especially grateful to Steven for providing illustrations to help elucidate a few of the more complex scientific concepts.

When I was in college struggling to reconcile my interests in history and chemistry, Ronald Suny suggested to me that a wonderful way of uniting my interests in history, chemistry, and grain elevators might be to study the "history of science" for a living. Cathy Gere gave me my first introduction to the history of medicine and suggested that I should spend some time with a group of very clever people in Cambridge, England, after I graduated. In particular, I want to acknowledge the late Alison Winter, who showed me how vast a continent the history of science is and launched me on its exploration. Her scholarly insight and generosity have set a standard that I strive to emulate. I hope that I will be able to aid my future students the same way that she aided me.

Last, but so very far from least, I give my thanks to Caitlin, for her wit, wisdom, and forbearance! In the midst of her own challenging work she has been both an anchor and a light to me as I have navigated the shoals of writing—reading full drafts no fewer than three times. I depend on her incisive mind, and I am so happy that she is my partner and teammate in this endeavor and so many others. I look forward to seeing what we will take on next. I dedicate this book to her.

Notes

Introduction

1. This account of the outbreak is synthesized from the Minutes of the Niles Board of Trustees Meeting, August 13, 1963, and the following sources: "Seek Leukemia Clew in Study of Niles Cases," *Chicago Daily Tribune*, May 19, 1961; Joseph Hearst, "Cancer Fight Aided by Tests Made at Niles," *Chicago Daily Tribune*, March 18, 1962; "Open Forum on Leukemia Will Be Held," *Chicago Tribune*, July 28, 1963; "Niles Hears Panel's Views on Leukemia," *Chicago Tribune*, August 8, 1963; "Niles Board Cites Disease 'Reportable,'" *Chicago Tribune*, October 6, 1963; Pat McGrady and Murray Morgan, "Cancer Is Yielding Up Its Secrets," *Saturday Evening Post* 237, no. 18 (May 9, 1964): 19–25.

2. The visit took place in 1968, and this story never reached a wider public: the assassination of Robert Kennedy preempted its publication in *Life* magazine. Dr. Paul Levine, interview by author, July 25, 2013, National Institutes of Health.

3. Steven O. Schwartz, Irving Greenspan, and Eric R. Brown, "Leukemia Cluster in Niles, Ill.: Immunologic Data on Families of Leukemic Patients and Others," *Journal of the American Medical Association* 186, no. 2 (October 12, 1963): 106–8; Schwartz is quoted in McGrady and Morgan, "Cancer Yielding Up Its Secrets," 21.

4. Schwartz, Greenspan, and Brown, "Leukemia Cluster in Niles, Ill.," 106; Wendon Wallace, "Cancer Found in Clusters," *Baltimore Sun*, August 4, 1961.

5. "Leukemia Tops Public Enemy List, Says Cancer Society," *Lexington Dispatch* (NC), March 19, 1963, 11.

6. Albert Rosenfeld, "A Superplan to Cut Years off the War," *Life*, November 18, 1966, 110–11. Total spending on the National Cancer Institute's Viral Oncology Programs was about $6.65 billion in 2017 dollars. Budgetary data was obtained from Stephen Hazen, memo for the record, August 10, 1983, NCI LION Database (hereafter "LION"), DC8308. For the Human Genome Project, the total cost was about $2.7 billion in 1991 dollars, or about $6.08 billion in 2017 dollars. The source for cost in 1991 dollars was The Human Genome Project Completion:

Frequently Asked Questions, accessed November 16, 2017, https://www.genome
.gov/11006943/human-genome-project-completion-frequently-asked-questions/.
Spending adjusted to 2017 dollars by the author, using the biomedical research-
and-development price index developed by the National Institutes of Health in
its budgetary calculations. Holloway and Reeb, "A Price Index for Biomedical
Research and Development."

7. Yi, "Cancer, Viruses, and Mass Migration."

8. A version of this statement remains on the American Cancer Society website, accessed March 27, 2018, https://www.cancer.org/cancer/cancer-basics/is
-cancer-contagious.html.

9. For an overview of the natural history of cancer, highlighting its diversity as
a disease category early in the twentieth century, see William Roger Williams, *The
Natural History of Cancer* (New York: William Wood, 1908), 375–454. For a recent
discussion of the diversity of cancer as an expanding disease category in the light of
genomic testing, especially as it pertains to the tailoring of "personalized" medicine, see Jocelyn Kaiser, "Personalized Tumor Vaccines Keep Cancer in Check,"
Science, April 12, 2017; Lean and Plutynski, "Evolution of Failure"; Bert Vogelstein,
Nickolas Papadopoulos, Victor E. Velculescu, Shibin Zhou, Luis A. Diaz, and Kenneth W. Kinzler, "Cancer Genome Landscapes," *Science* 339, no. 6127 (March 29,
2013): 1546–58; Lynda Chin, Jannik N. Andersen, and P. Andrew Futreal, "Cancer
Genomics: From Discovery Science to Personalized Medicine," *Nature Medicine* 17,
no. 3 (March 2011): 297–303. On the challenges of identifying infectious agents with
their diseases, see Broadbent, "Disease as a Theoretical Concept."

10. Wailoo et al., *Three Shots at Prevention*.

11. Zur Hausen, "Search for Infectious Causes." The fraction rises to 26 percent in "developing nations." IARC Working Group on the Evaluation of Carcinogenic Risks to Humans, *IARC Monographs on the Evaluation of Carcinogenic
Risks to Humans* 100B (2012): 39. In conjunction with immunosuppressive infections such as HIV, viral infections account for nearly 40 percent of all cancers in
sub-Saharan Africa. Livingston, *Improvising Medicine*, 35.

12. Paula Kiberstis and Eliot Marshall, "Celebrating an Anniversary," *Science*
331, no. 6024 (March 25, 2011): 1539.

13. Here is a summary of the work that I have drawn from to keep myself
aware of developments on the treatment, causation, and causality of cancer, even
if the path of cancer viruses often did not directly cross these issues. Discussions
of the social history of cancer and its relationship to questions of gender, race,
professional authority, and activism around different treatment regimes include
Aronowitz, *Unnatural History*; Gardner, *Early Detection*; Lerner, *Breast Cancer
Wars*; Löwy, *Preventive Strikes*; Timmermann, *History of Lung Cancer*; Wailoo,
How Cancer Crossed Color Line; Valier, *History of Prostate Cancer*. A second set
of accounts have focused on the development, testing, or marketing of therapies

for cancer; see Daemmrich, "BioRisk"; Goodman and Walsh, *Story of Taxol*; Keating and Cambrosio, *Cancer on Trial*; Mukherjee, *Emperor of All Maladies*; Pieters, *Interferon*. Reflecting interest in the BRCA 1&2 cancer genes, a handful of cases have examined the development of clinical genetics and cancer: Cantor, "The Frustrations of Families"; Necochea, "From Cancer Families to HNPCC"; Palladino, "Between Knowledge and Practice"; Parthasarathy, "Architectures of Genetic Medicine." A final set of works emphasizes the environmental causes of cancer as they were understood during the twentieth century, including Brown et al., "'A Lab of Our Own'"; Clark, *Radium Girls*; Hurley, *Environmental Inequalities*; Langston, *Toxic Bodies*; Spears, *Baptized in PCBs*. For a small sampling of the most recent literature on smoking and lung cancer, see Brandt, *Cigarette Century*; Proctor, *Golden Holocaust*; Rego, "Polonium Brief"; Timmermann, "As Depressing as Predictable?" On radiation biology, see Parascandola, "Uncertain Science"; Lindee, *Suffering Made Real*; Proctor, *Cancer Wars*.

14. Sontag, *Illness as Metaphor*, 16; Kellehear, *Social History of Dying*, 87–168; Wailoo, *How Cancer Crossed Color Line*, 28.

15. Woods Hutchinson, *The Cancer Problem: Or, Treason in the Republic of the Body* (New York: Tucker, 1900), 3.

16. Arthur Newsholme, "The Statistics of Cancer," *Practitioner* 62 (1899): 371–84; "Leading Causes of Death," *Public Health Reports* (1896–1970) 67, no. 1 (1952): 90–95. These issues are discussed contemporaneously in Hoffman, *Mortality from Cancer*. Also see Proctor, *Cancer Wars*, 18–22; Wailoo, *How Cancer Crossed Color Line*, 43–47. Concern for rising rates of cancer has often been attributed to the demographic "epidemiological transition," in which improved nutrition and sanitation resulted in the decline of epidemic illness and the rise of chronic disease as a public health problem. Of course, the manner and form of this transition, as well as its causes, are open to sharp qualification. Jones, Podolsky, and Greene, "Burden of Disease"; Parascandola, "Epidemiologic Transition"; Omran, "Epidemiologic Transition"; Timmermann, "Chronic Illness and Disease History"; Weisz and Olszynko-Gryn, "Theory of Epidemiologic Transition"; Harris, "Public Health, Nutrition, and Mortality." It is worth noting that many speculate that in allowing for the control of infectious disease, advances in late nineteenth-century sanitation and microbiology contributed to making certain diseases, such as cancer, more "visible." Wailoo, *Dying in City of Blues*, 23–24.

17. Lewis Stephen Pilcher, *The Cure of Cancer* (Philadelphia: J. B. Lippincott, 1909), 1. The first use I have found of "the cancer problem" is from 1899.

18. Patterson, "Cancer, Cancerphobia, and Culture"; Pinell, *Fight against Cancer*; Proctor, *Nazi War on Cancer*; Pickstone, "Contested Cumulations"; Toon, "'Cancer as the General Population Knows It.'"

19. Patterson, *Dread Disease*, 35; Cantor, "Introduction," 1–3, 5–7.

20. For more on the importance of surgical intervention in reshaping under-standings of the cancer problem, see Aronowitz, "Do Not Delay"; Gaudillière, "Cancer"; Löwy, *Preventive Strikes*; Gardner, *Early Detection*.

21. See chapter 7.

22. Brandt and Gardner, "Golden Age of Medicine?"; Patterson, *Dread Disease*, 171–72. What counted as biomedicine exhibited great variation across different institutional and national communities—some settings privileged the laboratory, whereas others did not. The American version, however, placed great emphasis on laboratory studies of disease. Sturdy, "Political Economy of Scientific Medicine"; Sturdy, "Knowing Cases"; Sturdy, "Looking for Trouble"; Kohler, *From Medical Chemistry to Biochemistry*; Cambrosio and Keating, " 'Going Monoclonal.' "

23. See chapters 1 and 2.

24. I thank Rosalind Williams for the phrasing "American exceptionalism in a medical key." James Sparrow calls attention to the remarkable nature of this expansion and the reckoning after the Second World War. Sparrow, *Warfare State*, 5–12. On the role of the state in twentieth-century American society, a vast field, I have benefited from reading Brinkley, *End of Reform*; Fraser and Gerstle, introduction to *The Rise and Fall of the New Deal Order*; Katznelson, "Was Great Society Lost Opportunity?"; Korstad, *Civil Rights Unionism*; Rodgers, *Atlantic Crossings*; Stein, *Pivotal Decade*.

25. Funigiello, *Chronic Politics*, 6–37; Sledge, *Health Divided*, 2–3; Weisz, *Chronic Disease in Twentieth Century*, 56–76. In fact, the majority of the National Cancer Institute's budget in the first years of its operation went to spon-sor the distribution of radium for radiation treatment, before this program was shut down by opposition from doctors. Cantor, "Radium and National Cancer Institute."

26. There is a vast literature comparing the trajectory of the development of health insurance in the United States with such trajectories in other countries, but very little of it focuses on biomedical research as a component of the social welfare state. This is discussed at greater length in chapters 3 and 4. For an overview of these matters, see Carpenter, "Is Health Politics Different?" For some useful texts on the US national health insurance debate, both in its domestic political con-text and in comparison to other nations, see Chapin, *Ensuring America's Health*; Derickson, *Health Security for All*; Gordon, "Why No National Health Insur-ance?"; Gordon, *Dead on Arrival*; Hollingsworth, Hage, and Hanneman, *State Intervention in Medical Care*; Hacker, "Historical Logic of Health Insurance"; Quadagno, *One Nation, Uninsured*.

27. Bayne-Jones, *Advancement of Medical Research*, 13.

28. The other major areas were the regulatory efforts of the Food and Drug Administration and the funds for hospital research awarded under the Hill-Burton Act of 1946, and later Medicaid/Medicare payments to hospitals. Carpenter,

Reputation and Power; Field, "How Government Created Hospital Industry";
Ludmerer, *Time to Heal*; Tobbell, *Pills, Power, and Policy*.

29. On the idea of a "settlement," see Jasanoff and Metzler, "Borderlands of
Life." Jasanoff calls these settlements responses to constitutional or "bioconsti-
tutional" moments when different groups are involved in defining a new social-
political order. This was the case in the United States in the 1930s and 1940s.
Jasanoff, "Introduction," 3. An extensive literature exists on the changing role
of the federal government in American social and economic life during the early
twentieth century, particularly on the question of whether its mobilization during the New Deal marked a moment of continuity or of rupture with the state of
American politics—the case of biomedical research tends to favor the idea that
these developments were more continuous with prior roles than revolutionary,
even if their impact was dramatic. Amenta, *Bold Relief*; Brinkley, *End of Reform*;
Cowie, *Great Exception*; Kessler-Harris, *In Pursuit of Equity*; Klein, *For All These
Rights*; Rodgers, *Atlantic Crossings*.

30. Rosenberg, "Rationalization and Reality," 403–4, 413; Kimmelman and
Paul, "Mendel in America," 282.

31. On the reasons for favoring the National Institutes of Health, see chapter 4.
On its academic impact, see Geiger, *Research and Relevant Knowledge*, 181–83.
On the expansion of the National Institutes of Health into a grant-giving body, see
Appel, *Shaping Biology*, 30–34; Fox, "Politics of NIH Extramural Program." The
issue of the importance of biomedical research against disease for state-building
projects has been raised in the case of France by Gaudillière, *Inventer la biomédecine*, 369–71.

32. Grant, "National Biomedical Research Agencies"; John W. Gardner, "The
Government, the Universities, and Biomedical Research," *Science*, 153, no. 3744
(1966): 1601–3. A sense of the degree to which the United States surpassed other
nations in biomedical research may be found in the fact that medical and scientific
publications were shocked to report—in 2014—that the United States no longer
outspent the rest of the world combined. Justin Chakma, Gordon H. Sun, Jeffery D.
Steinberg, Stephen M. Sammut, and Reshma Jagsi, "Asia's Ascent—Global
Trends in Biomedical R&D Expenditures," *New England Journal of Medicine* 370,
no. 1 (January 2, 2014): 3–6.

33. Jasanoff, "Ordering Knowledge, Ordering Society."

34. Phillip Pauly made this point in general about biologists in twentieth-
century American society. Pauly, *Biologists and American Life*, 244.

35. Feinstein, "Intellectual Crisis in Clinical Science," 216. On the question of
"pure" versus "applied" research and its relationship to "basic" or "fundamental"
science, see Stokes, *Pasteur's Quadrant*. I am in agreement with those who have
observed that the dividing line between "pure" and "applied" research, as well
as between synonyms such as "fundamental" and "basic" research, is historically

specific rather than transhistorical and universal. Clarke, "Pure Science with Practical Aim"; Gooday, "'Vague and Artificial'"; Bud, "'Applied Science.'"

36. Linus Pauling, Harvey A. Itano, S. J. Singer, and Ibert C. Wells, "Sickle Cell Anemia, a Molecular Disease," *Science* 110, no. 2865 (November 25, 1949): 543–48. On Pauling's work, see Kay, *Molecular Vision of Life*, 256–59. The therapeutic relevance of this discovery was not immediately apparent, despite later memorialization of this moment as a landmark for molecular medicine. Strasser, "Linus Pauling's 'Molecular Diseases.'"

37. Jacques Monod and François Jacob, "General Conclusions: Teleonomic Mechanisms in Cellular Metabolism, Growth, and Differentiation," *Cold Spring Harbor Symposia on Quantitative Biology* 21 (1961): 393.

38. *Basic Issues in Biomedical and Behavioral Research, 1976: Hearings, Day 1, Before Subcomm. on Health of the Committee on Labor and Public Welfare*, 94th Cong. 387 (1976) (report of the president's Biomedical Research Panel, Appendix A, Part I: The Place of Biomedical Science in Medicine and the State of the Science, 1).

39. De Chadarevian, "Whose Turn?"

40. Abir-Am, "Molecular Transformation of Biology"; Landecker, *Culturing Life*, 14–16.

41. De Chadarevian and Kamminga, introduction. De Chadarevian and Kamminga's definition provides a corrective to the assumption that molecularization is an event that emerged out of molecular genetics in the late twentieth century. While this latter moment is undoubtedly an instance of molecularization, it is not necessarily its defining moment. For a framing that emphasizes the late twentieth-century origins of molecuarization, see Rose, *Politics of Life Itself*, 5–6.

42. I have reviewed the historiography of the biomedical sciences at greater length in Scheffler and Strasser, "Biomedical Sciences." This insistence on the biological basis of disease distinguishes biomedicine from "scientific medicine," "experimental medicine," "rational therapeutics," and even "translational medicine." While all of these terms indicate the importance of science to medicine, only biomedicine makes an ontological, as opposed to methodological, claim about the practice of medicine. In this sense, I adopt a narrower definition of "biomedicine" than that of other authors. On these differences, see Crowe, "Cancer, Conflict, and Nuclear Transplantation"; Keating and Cambrosio, "Does Biomedicine Entail Reduction?"; Löwy, "Historiography of Biomedicine"; Marks, *Progress of Experiment*, 1–5; Clarke et al., "Biomedicalization."

43. Gradmann, *Laboratory Disease*, 2.

44. Creager, *Life of a Virus*, 20–42; Helvoort, "History of Virus Research"; Hughes, *Virus*.

45. A. Lwoff, "The Concept of Virus," *Microbiology* 17, no. 2 (1957): 240.

46. Kay, "Conceptual Models and Analytical Tools"; Summers, "How Bacteriophage Came to Be Used."

47. Morange, *History of Molecular Biology*, 219–30; Müller-Wille and Rheinberger, *A Cultural History of Heredity*, 164; Rheinberger, "What Happened to Molecular Biology?"; Yi, "Cancer, Viruses, and Mass Migration."

48. Gaudillière, "Cancer," 496–98. Nicolas Rasmussen has recently noted the special status of molecular biology in these decades. Rasmussen, *Gene Jockeys*, 27.

49. James Watson, "Getting Realistic about Cancer—CSHL Director's Report," January 1975, 4–5, in *CSHL Archives Repository*, Reference JDW/2/4/1/26, accessed April 9, 2013, http://libgallery.cshl.edu/items/show/52313.

50. Following the circulation of particular experimental systems, models, organisms, or materials between different settings offers a means to synthesize local accounts with emerging "big picture" narratives that promise to link the practice of biology in the laboratory with developments in national histories. De Chadarevian, "Microstudies versus Big Picture Accounts?"; de Chadarevian and Strasser, "Molecular Biology in Postwar Europe"; Creager, "Timescapes of Radioactive Tracers"; Krige, *American Hegemony*; Secord, "Knowledge in Transit." This emphasis both concurs with and departs from a prior generation of sociology of scientific knowledge, which influenced studies that emphasized the particular historical circumstances and social work surrounding the replication of exemplary experiments, most famously Robert Boyle's air pump. This more recent generation of circulation studies departs from this earlier cohort of literature in that it is more sensitive to the fact that social factors are in themselves caught up in the coproduction of scientific knowledge. Collins, "The TEA Set"; Schaffer and Shapin, *Leviathan and the Air-Pump*; Shapin, "Here and Everywhere."

51. This follows Hans-Jörg Rheinberger's reading of Louis Althusser's concept of ideology in Rheinberger, *Epistemology of the Concrete*, 46–47.

52. Capshew and Rader, "Big Science"; Collins, *Gravity's Shadow*; Forman, "Behind Quantum Electronics"; Galison, *Image and Logic*; Galison and Hevly, *Big Science*; Kaiser, "Booms, Busts"; Kevles, *Physicists*; Munns, *Single Sky*; Westfall, "Rethinking Big Science."

53. Creager and Landecker, "Technical Matters"; Creager, *Life Atomic*, 9, 62–106; Daemmrich and Shaper, "Gordon Research Conferences"; Edwards et al., "Introduction"; Edwards, "Infrastructure and Modernity"; Gaudillière and Löwy, "General Introduction"; Star and Bowker, "How to Infrastructure"; Williams, "Cultural Origins."

54. Creager, *Life Atomic*. The entanglement of molecular biology with the reconstruction and expansion of biological and biomedical research during the Cold War was even more apparent in continental Europe than in the United States and Britain. De Chadarevian and Strasser, "Molecular Biology in Postwar Europe."

55. Endersby, *Guinea Pig's History of Biology*; Creager, *Life of a Virus*; Kohler, *Lords of the Fly*; Creager, Lunbeck, and Wise, *Science without Laws*; Rader, *Making Mice*.

56. Clarke and Fujimura, *Right Tools for the Job*; Creager, "Paradigms and Exemplars Meet Biomedicine."

57. I address this idea at greater length in chapter 9, but I draw upon Rheinberger, *Toward History of Epistemic Things*.

58. De Chadarevian, *Designs for Life*; Kay, *Molecular Vision of Life*; Strasser, *Fabrique d'une nouvelle science*; Yi, *Recombinant University*.

59. Landecker, *Culturing Life*, 14–16.

60. David Bloor raises this point in his review of *Toward a History of Epistemic Things*. Bloor, "Sociology of Epistemic Things," 303–5. This point has also been raised by science studies scholars, although never with the intent of following the social relations of an individual laboratory as symmetrically as they might study activity within the bounds of a particular scientific dispute. Latour and Woolgar, *Laboratory Life*.

61. Davies, Frow, and Leonelli, "Bigger, Faster, Better?"; Kevles, "Big Science and Big Politics"; Vermeulen, *Supersizing Science*; Hilgartner, *Reordering Life*.

62. Knorr-Cetina, *Epistemic Cultures*, 79–82, quote on 81. Steven Hilgartner has recently reiterated this claim, following Knorr-Cetina. Hilgartner, *Reordering Life*, 53. On the Human Genome Project, see Kevles and Hood, *The Code of Codes*; Lenoir and Hays, "The Manhattan Project for Biomedicine"; Cook-Deegan, *Gene Wars*.

63. Here, I draw upon the idea of "following" objects such as tracers articulated by Angela Creager. Creager, "Timescapes of Radioactive Tracers."

64. Aronowitz, "Do Not Delay"; Good, "Practice of Biomedicine"; Jain, *Malignant*, 46–66.

65. *Report to the President: A National Program to Conquer Heart Disease, Cancer and Stroke* (Washington, DC: US Government Printing Office, 1964), 1:21.

66. Engerman, "Introduction"; Gilman, *Mandarins of the Future*; McCray, *Visioneers*; Beckert, *Imagined Futures*.

67. Pickstone, *Ways of Knowing*.

68. Latour, "On the Partial Existence"; Taussig, Klaus, and Helmreich, "Anthropology of Potentiality in Biomedicine"; Fischer, *Anthropological Futures*, 50–113.

69. On the presentation of cancer as a "riddle," see Löwy, "Cancer," 474–75. The idea of ordering cancer and the process of cancer research to make these activities "legible" to state intervention, and the process by which communities might resist this effort, is an idea I have adapted from Scott, *Seeing like a State*, 2–3.

70. These might be described as a species of *sociotechnical imaginaries* that were not only rhetorical gestures but also political acts. The acts created material conditions that steered the course of research, confusing the standard lines of expertise in science policy. Jasanoff, "Future Imperfect."

71. Patterson, *Dread Disease*, 255–94; Schneyer, Landefeld, and Sandifer, "Biomedical Research and Illness." For more contemporary critique of the biomedical

approach to disease at a conceptual level, see Ankeny and Leonelli, "Model Organisms"; Lynch, "Use with Caution."

72. I take the idea of recalcitrance from Carsten Timmermann, who notes that although historians of science and medicine have justifiably become wary of narratives of progress, most stories of disease, especially cancer, implicitly adopt a progressive framework by virtue of focusing only on therapeutic advances or scientific discoveries. However, it is equally important to think about how medicine absorbs frustration or failure as well as how it addresses success, and indeed to think about what we mean when we use these words. The time frame and settings through which I follow cancer viruses help to construct a different narrative—one in which the question of success or failure is always qualified by the recalcitrance of cancer on many levels. Timmermann, *History of Lung Cancer*, 8–10.

73. These dynamics—the tension between experience and expectation and the effort to define the horizons of progress in reference to the past—are ideas that have been developed by Koselleck, *Futures Past*, 277–83.

74. Here, I draw upon literature on hope, hype, and promissory politics that has usually been articulated in regard to entrepreneurial biotechnology. Brown, "Hope against Hype"; Fortun, "Genomics Scandals"; Fortun, *Promising Genomics*; Fortun, "The Human Genome Project"; Rajan, *Biocapital*, 123.

75. In addition to the literature on anticipation and financial speculation in biotechnology cited above, the idea of boom and bust as a narrative has been offered in the case of the development of settler-colonial societies. Belich, *Replenishing the Earth*.

76. Gaudillière, "Molecularization of Cancer Etiology." Today, other cancer viruses have reemerged as tools for new medical technologies, such as vectors for gene therapy or immunotherapy. Melissa A. Kotterman, Thomas W. Chalberg, and David V. Schaffer, "Viral Vectors for Gene Therapy: Translational and Clinical Outlook," *Annual Review of Biomedical Engineering* 17 (2015): 63–89; V. Boisguérin, J. C. Castle, M. Loewer, J. Diekmann, F. Mueller, C. M. Britten, S. Kreiter, Ö. Türeci, and U. Sahin, "Translation of Genomics-Guided RNA-Based Personalised Cancer Vaccines: Towards the Bedside," *British Journal of Cancer* 111, no. 8 (October 14, 2014): 1469–75.

Chapter One

1. The tumors Rous isolated were sarcomas, or tumors of the muscle and connective tissue. These are distinct from carcinomas, which are tumors of the organs and are more common, or leukemias and lymphomas (cancers of blood cells and lymph, respectively). Leukemia was not classified as a cancer at the beginning of the twentieth century, so even though two researchers in Copenhagen, Vilhelm Ellerman and Olaf Bang, are recognized for observing a leukemia virus in

chickens in 1908, Rous was under the impression in 1910 that he had found the first filterable agent associated with cancer. Although Ellerman and Bang's experiment is technically the first evidence of an animal cancer virus, their results did not have as much impact as Rous's results because their contemporaries did not classify leukemia as a cancer. Becsei Kilborn, "Scientific Discovery and Scientific Reputation"; Helvoort, "Century of Research," 299–301. A sample of Rous's early articles: Peyton Rous, "A Transmissible Avian Neoplasm (Sarcoma of the Common Fowl)," *Journal of Experimental Medicine* 12, no. 5 (1910): 696–705; Peyton Rous, "A Sarcoma of the Fowl Transmissible by an Agent Separable from Tumor Cells," *Journal of Experimental Medicine* 13, no. 4 (April 1911): 397–411; Peyton Rous, "On the Causation by Filterable Agents of Three Distinct Chicken Tumors," *Journal of Experimental Medicine* 19, no. 1 (January 1914): 52–68. For press coverage, see "Clue to Parasite as Cause of Cancer," *New York Times*, February 14, 1912.

2. Richard Boardman to Rockefeller Institute for Medical Research, October 29, 1913, in Rockefeller University Archives, Business Manager Subject Files, RG 210.3, business, box 1, folder 3, Rockefeller Archive Center. Although only Richard Boardman is identified in this letter, the 1910 federal census records a lawyer named Richard Boardman who lived in Union City, New Jersey, next to Jersey City, and further, that he was married to Dorcas Boardman. Since it seems likely this is the lawyer who wrote to Peyton Rous, I have interpolated Dorcas into the dispute rather than referring only to "Mrs. Boardman," as she appears in the letter.

3. Rockefeller Institute for Medical Research to Richard Boardman, October 30, 1913, in Rockefeller University Archives, Business Manager Subject Files, RG 210.3, business series, box 1, folder 3, Rockefeller Archive Center.

4. Ewing, *Neoplastic Diseases*, 114.

5. Kevles, "Pursuing the Unpopular."

6. Peyton Rous, "The Challenge to Man of the Neoplastic Cell," *Science*, 157, no. 3784 (July 7, 1967): 26.

7. I take the idea of a "regime of perceptibility" from Murphy, *Sick Building Syndrome*, 24.

8. To these sources of skepticism one might add the anecdotal observation on the part of doctors that cancer patients were noncontagious. Rous is often portrayed as having been shunned by the scientific community because of the controversial nature of his ideas, but recent scholarship has placed that history in the context of Rous's own self-creation after tumor virology rose in prestige during the 1960s and 1970s. Becsei Kilborn, "Scientific Discovery and Scientific Reputation"; Fisher, "Not beyond Reasonable Doubt."

9. Nancy Tomes makes this point in regard to the rise of "germ" theory in the United States. Tomes, *Gospel of Germs*. Here, I follow a suggestion by Michael Lynch that historians of science should not seek to proceed from a set of prior philosophical commitments but to examine how actors in particular contexts construed the existence of particular objects. Lynch, "Ontography." My interpretation

remains focused on human actors because cancer viruses themselves bridged the world of the real and the hypothetical. How one speaks of the agency of a hypothetical nonhuman entity in the spirit of Actor-Network Theory is a question that history of science and science and technology studies have yet to fully grapple with. Law and Lien, "Slippery."

10. The notion that the rise of microbiology was a "revolutionary" event in the scientific sense has been sharply qualified by recent historians of nineteenth-century medicine. However, the idea has such force in retrospective accounts given by physicians and scientists that I use the term here advisedly. For further discussion of the reception of germ theory and the place of the laboratory, see chapter 2. For a brief overview, see Worboys, "Was There a Bacteriological Revolution?"

11. Gosset, "L'hôpital des cancérés," 17–22; Stolberg, "Metaphors and Images of Cancer," 69. On the eighteenth-century idea that air itself could be a means of transmitting cancer, see the discussion of odor below.

12. Stolberg, "Metaphors and Images of Cancer," 55–62; Skuse, *Constructions of Cancer*, 61–73.

13. Pelling, "Contagion/Germ Theory/Specificity"; Worboys, "Contagion."

14. Rather, *Genesis of Cancer*, 10, 98; Sontag, *Illness as Metaphor*, 16–18.

15. Wolff, *Science of Cancerous Disease*, 433–35.

16. A translation of Rigoni-Stern's paper appears in Scotto and Bailar, "Rigoni-Stern and Medical Statistics," 71–72.

17. Löwy, *Woman's Disease*, 129–30; Nolte, "Carcinoma Uteri and 'Sexual Debauchery,'" 31–36.

18. Hoffman, *Mortality from Cancer*, 205–6.

19. Alfred Haviland, *The Geographical Distribution of Disease in Great Britain* (London: Swan Sonnenschein, 1892), 335. For more on Haviland's medical geography, see Barrett, "Geographical Distribution of Diseases." Also see Agnes Foster-Arnold's forthcoming article in *Social History of Medicine*, "Mapmaking and Mapthinking: Cancer as a Problem of Space in Nineteenth-Century England."

20. L. Nicholls, "Is Cancer Contagious?," *Lancet* 130, no. 3353 (December 3, 1887): 1145.

21. Barnes, *Great Stink of Paris*, 254–59; Valenčius, *Health of the Country*, 109–32; Kiechle, *Smell Detectives*, 170–97.

22. Rather, *Genesis of Cancer*, 108–12; Landecker, *Culturing Life*, 86–91.

23. Coleman, *Biology in the Nineteenth Century*, 22–23; Rather, *Genesis of Cancer*, 108–12.

24. Edward Tibbits, "On the Probable Relationship of Syphilis, Scrofula, Tubercle, Cancer, and Other Allied Morbid Conditions," *Lancet* 108, no. 2783 (December 30, 1876): 922–23.

25. Rutherford Morison, "An Address on Some Points concerning Tubercle, Syphilis and Malignant Disease," *British Medical Journal* 2, no. 2603 (November 19, 1910): 1573–75.

26. Gradmann, *Laboratory Disease*, 2; Steere-Williams, "Performing State Medicine."

27. Tomes, *Gospel of Germs*, xv.

28. Tomes and Warner, "Introduction to Rethinking the Germ Theory." On contention between laboratory and clinic, see chapter 2, as well as Sturdy, "Looking for Trouble."

29. Charles Ryall, "The Technique of Cancer Operations, with Reference to the Danger of Cancer Infection," *British Medical Journal* 2, no. 2492 (October 3, 1908): 1005.

30. "Dr. Burnette's Death of Cancer: Medical Men Take Great Interest in the Case—Contagiousness of the Disease," *New York Times*, September 27, 1895.

31. A. Lapthorn Smith, "Is Cancer Contagious?," *Journal of the American Medical Association* 46, no. 3 (January 20, 1906): 207.

32. Patterson, *Dread Disease*, 23; Kisacky, *Rise of the Modern Hospital*, 152–53.

33. D'Arcy Power, "An Experimental Investigation into the Causation of Cancer," *British Medical Journal* 2, no. 1760 (September 22, 1894): 637–38.

34. D'Arcy Power, "Cancer Houses and Their Victims," *British Medical Journal* 1, no. 1745 (June 9, 1894): 1240. Power's search was of a piece with other bacteriological efforts to locate sources of germs outside the body in environmental media—an early kind of disease ecology. Steere-Williams, "Performing State Medicine."

35. Summarized in Triolo, "Nineteenth Century Foundations," 5–8. I am grateful to Nathaniel Comfort for drawing my attention to this source, and for his own discussion of cancer and germ theory in "Rous's Reception," 3–9.

36. Samuel G. Shattock, "The Morton Lecture on Cancer and Cancerous Diseases," *British Medical Journal* 1, no. 1742 (May 19, 1894): 1065–67.

37. Helvoort, "Dispute over Scientific Credibility," 320.

38. Roswell Park, speech to American Medical Association, as quoted in "Dr. Roswell Park on Cancer: The Disease Is Parasitic in Origin, He Says," *New York Tribune*, July 30, 1901.

39. Williams, *Natural History of Cancer*, 233.

40. "More Cancer Germs," *Literary Digest*, December 7, 1912. I am grateful to Nathaniel Comfort for drawing this reference to my attention.

41. Charles Powell White, *Lectures on the Pathology of Cancer* (Manchester, UK: Manchester University Press, 1908), 70.

42. Herbert Snow, "The So-Called 'Parasitic Protozoa' of Mammary Carcinoma," *Lancet* 142, no. 3663 (November 11, 1893): 1183.

43. "The Contagiousness of Cancer," *Lancet* 160, no. 4136 (December 6, 1902): 1559.

44. Charles Plumley, *The Control of a Scourge; or, How Cancer Is Curable* (New York: Dutton, 1907), 34–35.

45. "The Bacillus of Love," *Baltimore Sun* (1837–1988), July 10, 1902.

46. Discussions of "oncology," derived from the Greek for mass or lump, were seen as preferable to naming cancer itself because of the fear that it inspired. See "Oncology, n.," *OED Online*, accessed May 20, 2017, www.oed.com/view/Entry/236123; Patterson, *Dread Disease*, 30–32.

47. Even according to Halsted's own optimistically skewed statistics—he did not count a patient as dying of cancer if they died from cancer at a site other than the one he had operated upon—nearly two out of every three women who underwent surgery still died from cancer. Aronowitz, *Unnatural History*, 99–104; Lerner, *Breast Cancer Wars*, 18–29.

48. Patterson, *Dread Disease*, 38–41.

49. The legal action included a successful suit by Henry Hoxsey against the president of the American Medical Association, Morris Fishbein, for slander and libel. Hoxsey v. Fishbein et al., US District Court, ND, Dallas Division no.3803, 83 F.Supp. 282 (1949).

50. Samuel Adams, "What Can We Do about Cancer?," *Ladies' Home Journal* 14 (May 1913): 21–22.

51. "Preliminary Report of the Field Service of the American Society for the Control of Cancer," March 1–June 15, 1922, in Commonwealth Fund Archive, RG 18, series I, box 12, folder 123, Rockefeller Archive Center.

52. Joseph Bloodgood, "Control of Cancer," address to Lehigh Valley Medical Society, June 7, 1913, reprinted in *Journal of the American Medical Association* 61, no. 26 (December 1913): 2282–86.

53. "Medical Practitioner and the ASCC," February 15, 1923, in Commonwealth Fund Archive, RG 18, series I, box 12, folder 123, Rockefeller Archive Center.

54. Cantor, "Uncertain Enthusiasm," 42–43.

55. "Cancer and the Public," *Lancet* 2, no. 5111 (August 13, 1921): 347.

56. "The Control of Cancer," *Huron Evening Huronite* (SD), November 18, 1922.

57. "Assails Overeating as Cause of Cancer," *New York Times*, September 23, 1926.

58. American Society for the Control of Cancer, bulletin 5, as reprinted in "The New Idea of Cancer: Many Cases Can Be Cured If Reported Promptly," *Hartford Courant* (1923–1989), December 1, 1927. Indeed, remarkably similar language may be found on the current ACS website offering facts about cancer, accessed May 10, 2017, https://www.cancer.org/cancer/cancer-basics/is-cancer-contagious.html.

59. "The Present State of Scientific Knowledge of Cancer," *Science*, 64, no. 1657 (October 1, 1926): 320.

60. Less than 2 percent of newspaper articles discussing cancer mentioned the ASCC or its "Cancer Week" campaigns in the nine years after its founding. "American Society for the Control of Cancer" appeared in 612 of 48,605 articles mentioning cancer that appeared in the Proquest Historical Newspapers and Chronicling America newspaper databases from 1913 to 1922; "Cancer Week"

appeared in 394. By comparison, "American Cancer Society," the society's successor organization, appeared in 21 percent of the 217,480 articles mentioning cancer
appearing in the Proquest Historical Newspapers Database from 1970 to 1980, the
decade of the War on Cancer (search conducted February 20, 2015).

61. "Mississippi: Hotels, Restaurants, and Boarding Houses: Regulation and
Inspection of. (Reg. St. Bd. of H., Aug. 20, 1912)," *Public Health Reports* (1896–
1970) 28, no. 19 (May 9, 1913): 916; "Spokane, Wash.: Restaurants and Eating
Places. License. Sanitary Regulation. (Ord. C 1548, Nov. 17, 1913)," *Public Health
Reports* (1896–1970) 29, no. 31 (July 31, 1914): 2055–56.

62. Edric Smith (RIMR) to Campbell's Soup Company, September 12, 1914;
and J. Worrence to Smith, September 24, 1914, in Peyton Rous papers, series I,
Chicken Tumor 7 (1913–26), American Philosophical Society Library (hereafter
"Rous papers," with series and folder numbers).

63. Stolley and Lasky, "Fibiger and His Nobel Prize"; Stolt, Klein, and Jansson,
"Analysis of Wrong Nobel Prize."

64. This point has been well documented in the context of other kinds of scientific observation. See, for example, Daston and Galison, *Objectivity*; Schaffer and
Shapin, *Leviathan and the Air-Pump*.

65. "Claims Isolation of Cancer Germ: German Physician Describes Bacillus
as Visible under Microscope," *Hartford Courant* (CT) (1923–1989), February 17,
1926; Rene Bache, "The Germ of Plant Cancer Found," *Salt Lake Tribune*, July 14,
1912, magazine sec.; "Declares a Germ Is Cause of Cancer," *New York Times*, August 21, 1926, shipping and mails sec.; "Photographic Proof of the Cancer Germ,"
Salt Lake Tribune, September 29, 1912, morning ed.

66. W. E. Gye, "The Aetiology of Malignant New Growths," *Lancet* 206, no. 5316
(July 18, 1925): 109–17; J. E. Barnard, "The Microscopical Examination of Filterable
Viruses Associated with Malignant New Growths," *Lancet* 206, no. 5316 (July 18,
1925): 117–23; J. E. Barnard and W. and W. E. Gye, "Note on the Preceding Papers,"
Lancet 206, no. 5316 (July 18, 1925): 123.

67. "Immunity from Cancer: Dr. Gye on His Next Task Looking for a Vaccine,"
Manchester Guardian (1901–1959), July 23, 1925; "The 'Germ' of Cancer: Rejoice
and Beware the New Discovery and Its Possibilities," *Observer (London) (1901–
2003)*, July 19, 1925.

68. Edward B. Krumbhaar to Peyton Rous, August 13, 1925, in Rous papers,
series I, Krumbhaar, Edward B., no. 1.

69. Francis Carter Wood to Krumbhaar, August 29, 1925, quoted in Patterson,
Dread Disease, 99.

70. Editorial, *London Times*, November 11, 1925.

71. "Cancer Yields a Secret," *The New Republic* 43, no. 556 (July 29, 1925): 250–51.

72. "Filterable Germ Forms Seen with New Super-Microscope," *Science News-
Letter* 20, no. 557 (December 12, 1931): 371. This report was carried by the Science
News Service and in dozens of local newspapers.

73. Edward C. Rosenow, "Observations with the Rife Microscope of Filter-Passing Forms of Microorganisms," *Science*, 76, no. 1965 (1932): 192–93.

74. The history of the Rife Microscope as scientific heterodoxy is summarized in Hess, "Technology and Alternative Cancer Therapies," 662–66.

75. Rasmussen, *Picture Control*, 222–56.

76. R. E. Seidel and M. Elizabeth Winter, "The New Microscopes," *Journal of the Franklin Institute* 237, no. 2 (February 1944): 103–30.

77. John J. Bittner, "Some Possible Effects of Nursing on the Mammary Gland Tumor Incidence in Mice," *Science*, 84, no. 2172 (1936): 162; Klein, *Atheist and the Holy City*, 122; Gaudillière, "Cancer entre infection et hérédité," 62–69; Rader, *Making Mice*, 184–86.

78. Eugene H. Kone (public relations consultant for Rockefeller Institute for Medical Research) to director of information, US DHEW, November 1, 1957, in Rockefeller University Archives, Richard E. Shope papers, RG 450Sh77, box 1, folder 13, Rockefeller Archive Center.

79. Rous to Ann Hotchkiss, August 28, 1962, in Rous papers, series I, folder "Life." The original article was Albert Rosenfeld, "Clues to a Deadly Riddle," *Life*, June 22, 1962.

80. A. B. Gilfillan to Rous, n.d., in Rous papers, series I, folder "Life."

81. R. B. Morrison to Rous, December 17, 1962, in Rous papers, series I, folder "Life."

82. Ann Yockey to Rous, September 18, 1962, in Rous papers, series I, folder "Life."

83. Rosamond Campion, *The Invisible Worm* (New York: Macmillan, 1972), 16–26.

84. "Counterattack on Infectious Cancer," *Science News* 104, no. 23 (December 8, 1973): 358.

Chapter Two

1. "Dr. M.P. Neal Tells Women about Cancer," *Columbia Evening Missourian*, November 17, 1922.

2. William Seaman Bainbridge, *The Cancer Problem* (New York: Macmillan, 1914), 144.

3. De Chadarevian and Kamminga, introduction.

4. Worboys, "Was There a Bacteriological Revolution?" For a similarly critical discussion of the idea that the entry of the laboratory into medicine was a "revolutionary" event, see Bynum, *Science and Practice of Medicine*; Cunningham and Williams, *Laboratory Revolution in Medicine*.

5. This point is discussed at greater length below, but for an overview of these changes see Hughes, *Virus*.

6. Gradmann, *Laboratory Disease*, 72–84.

7. Mendelsohn, "'Like All That Lives,'" 3–4.

8. For an intellectual-biographical sketch, see Geison, *Private Science of Louis Pasteur*, 22–50.

9. The ethical and practical ambiguities by which Pasteur prepared and tested his vaccine are discussed in ibid., 177–256.

10. This episode is related in greater detail in Hansen, *Picturing Medical Progress*, 51–76.

11. Cunningham, "Transforming Plague."

12. Worboys, "From Heredity to Infection?," 90–93.

13. Mendelsohn, "Medicine and Bodily Inequality," 21–31.

14. Ottomar Rosenbach, *Physician versus Bacteriologist*, trans. Achilles Rose (New York: Funk & Wagnalls, 1904), ix, xi.

15. Kohler, *From Medical Chemistry to Biochemistry*, 7–8.

16. Helvoort, "Dispute over Scientific Credibility," 320.

17. Oberling, *Riddle of Cancer*, 44–59.

18. Indeed, that Rockefeller should have funded a medical research institute investigating cancer is oddly appropriate, considering that his father had made a start in business selling cancer cures in the nineteenth century. Hollingsworth, "Institutionalizing Excellence in Biomedical Research," 19–21. The role of philanthropy in smoothing the troubled political fortunes of the Rockefeller holdings has also been noted in the context of the history of industrial philanthropy. Brown, *Rockefeller Medicine Men*.

19. Becsei Kilborn, "Going against the Grain," 60–61.

20. Löwy, "Experimental Systems and Clinical Practices," 403–5, 411–17; Comfort, "Rous's Reception," 9–10.

21. Leo Loeb, "General Problems and Tendencies in Cancer Research," *Science*, 43, no. 1105 (March 3, 1916): 293.

22. Peyton Rous, "A Transmissible Avian Neoplasm (Sarcoma of the Common Fowl)," *Journal of Experimental Medicine* 12, no. 5 (1910): 696–705; Peyton Rous, "An Experimental Comparison of Transplanted Tumor and a Transplanted Normal Tissue Capable of Growth," *Journal of Experimental Medicine* 12, no. 3 (May 1, 1910): 344–66.

23. Peyton Rous, "A Sarcoma of the Fowl Transmissible by an Agent Separable from Tumor Cells," *Journal of Experimental Medicine* 13, no. 4 (April 1911): 397–411, quote on 409. On the Berkefeld Filter, see G. Sims Woodhead and G. E. Cartwright Wood, "An Inquiry into the Relative Efficiency of Water Filters in the Prevention of Infective Disease," *British Medical Journal* 2, no. 1767 (November 10, 1894): 1053–59.

24. Rous papers, Series. 1, Chicken Tumor Folders (14 total).

25. "Prisoner Seeks to Be Inoculated with Germ to Prove Whether or Not Cancer Is Contagious," *Cincinnati Enquirer* (1872–1922), December 2, 1914.

26. Peyton Rous, "Report of Dr. Rous," in Report of Director of the Laboratories of the Rockefeller Institute for Medical Research (January 1911), Scientific Reports, Rockefeller University Archives, RG 439, box 1, folder 2, Rockefeller Archives Center.

27. Rogers, *Dirt and Disease*, 22–23; Marcus, *Malignant Growth*, 90–91.

28. Simon Flexner and Paul A. Lewis. "The Nature of the Virus of Epidemic Poliomyelitis," *Journal of the American Medical Association* 53, no. 25 (December 18, 1909): 2095.

29. Creager, *Life of a Virus*, 28–32; Helvoort, "History of Virus Research," 189–202; Hughes, *Virus*, 29–41.

30. Summers, *Félix d'Herelle*, 82–96.

31. "Clue to Parasite as Cause of Cancer," *New York Times*, February 14, 1912. This story was reprinted in the *Los Angeles Times*.

32. "Is Cancer Infectious?," *Cincinnati Enquirer* (1872–1922), February 16, 1912.

33. The most comprehensive survey of responses to Rous's experiments is in Becsei Kilborn, "Going against the Grain," 102–6; Helvoort, "Start of Cancer Research Tradition," 196–98.

34. Amédée Borrel, as quoted in W. E. Gye and W. J. Purdy, *The Cause of Cancer* (London: Cassell, 1931), 22, author's translation from the original French.

35. Helvoort, "Start of Cancer Research Tradition," 192–93.

36. Ewing, "Pathological Aspects of Cancer Research," 79.

37. Ewing, *Neoplastic Diseases*, 123.

38. Becsei Kilborn, "Going against the Grain," 106–9.

39. See Patterson, *Dread Disease*, title of chap. 4.

40. Maud Slye, "Cancer and Heredity," *Annals of Internal Medicine* 1, no. 12 (June 1, 1928): 951–76. On the spread of Mendelian genetics in the United States, see Kevles, "Genetics," 441–55; Rosenberg, "Rationalization and Reality."

41. Holmes, *Reconceiving the Gene*, 5–28.

42. Rader, *Making Mice*, 80–95.

43. Letter to Maud Slye, quoted in Proctor, *Cancer Wars*, 220, quoted by Proctor from McCoy, *Cancer Lady*, 92.

44. Carl V. Weller, "The Inheritance of Retinoblastoma and Its Relationship to Practical Eugenics," *Cancer Research* 1, no. 7 (July 1, 1941): 534.

45. Patterson, *Dread Disease*, 17–28.

46. Proctor, *Cancer Wars*, 30–36.

47. Shaughnessy, "Story of American Cancer Society," 14.

48. Ibid., 87–88.

49. Quotations are taken from the ASCC minutes of the Executive Committee, as quoted in ibid., 131–34.

50. Rivers, "Viruses and Koch's Postulates."

51. Creager and Gaudillière, "Experimental Technologies and Techniques," 207–9; Shope, "Evolutionary Episodes," 262–68.

52. Rivers, "General Aspects of Filterable Viruses," 21–22.

53. Richard E. Shope, "A Transmissible Tumor-Like Condition in Rabbits," *Journal of Experimental Medicine* 56, no. 6 (December 1, 1932): 793–802.

54. James Murphy, a researcher at the Rockefeller Institute of Medical Research, claimed that a longer discussion of tumor viruses was omitted because of the animus of the Yale physician and bacteriologist Stanhope Bayne-Jones. James Murphy to Carl Volgetlin (National Institute of Health), November 17, 1938, in James Murphy papers, box 29, folder "National Institutes of Health: Cancer 1938," no. 4, American Philosophical Society Library (hereafter "Murphy papers," with box, where available, and folder information).

55. Stanhope Bayne-Jones to Murphy, July 17, 1938, in Murphy papers, box 29, folder "National Institutes of Health: Cancer 1938," no. 4.

56. Stanhope Bayne-Jones, Ross G. Harrison, C. C. Little, John Northrop, and James B. Murphy, "Fundamental Cancer Research: Report of a Committee Appointed by the Surgeon General," *Public Health Reports* 53, no. 48 (December 2, 1938): 2130.

57. Creager, *Life of a Virus*, 123–30; Gaudillière, "Rockefeller Strategies." Ironically, however, the power of this influenza vaccine was remarkably limited; see John M. Eyler, "De Kruif's Boast: Vaccine Trials and the Construction of a Virus," *Bulletin of the History of Medicine* 80, no. 3 (September 2006): 409–38.

58. J. W. Beard, "The Fallacy of the Concept of Virus 'Masking': A Review," *Cancer Research* 16 (1956): 279–91; J. W. Beard, "Review: Purified Animal Viruses," *Journal of Immunology* 58, no. 1 (January 1948): 49–108; Creager and Gaudillière, "Experimental Technologies and Techniques," 214–19.

59. Creager, *Life of a Virus*, 147–48, 171–75; Kevles, "Dulbecco and New Animal Virology," 420–21; Landecker, *Culturing Life*, 129–37; Paul, *History of Poliomyelitis*, 316–19, 373–81.

60. Robert J. Huebner, "70 Newly Recognized Viruses in Man," *Public Health Reports* 74, no. 1 (January 1, 1959): 6–12.

61. Robert J. Huebner, "Implications of Recent Viral Studies," *Public Health Reports* 72, no. 5 (May 1, 1957): 380; Engel, "Toward Conquest of Virus," *New York Times*, March 24, 1957.

62. Morris Schaeffer, Roslyn Q. Robinson, and Andrew R. Fodor, "New Viruses and Virus Diseases of Man," *Annual Review of Microbiology* 13 (October 1959): 345.

63. Rasmussen, *Picture Control*, 28–70, 197–221. On the importance of visual representations in the sciences writ large, see Daston and Galison, *Objectivity*, 363–415.

64. R. D. Passey, Leon Dmochowski, William Astbury, and R. Reed, "Electron Microscope Studies of Normal and Malignant Tissues of High- and Low-Breast-Cancer Strains of Mice," *Nature* 160, no. 4069 (October 27, 1947): 565; R. D. Passey Leon Dmochowski, William Astbury, R. Reed, and G. Eaves, "Electron

Microscope Studies of Human Breast Cancer," *Nature* 167, no. 4251 (April 21, 1951): 643–44; "Briton Says Virus May Be Cancer Cause," *Baltimore Sun*, July 29, 1954; "Dr. Leon Dmochowski, Researcher in Virology," *New York Times*, September 28, 1981, obituaries sec.

65. W. Bernhard, "The Detection and Study of Tumor Viruses with the Electron Microscope," *Cancer Research* 20, no. 5, part 1 (June 1, 1960): 712–27; Leon Dmochowski, "Viruses and Tumors," *Science*, 133, no. 345 (February 24, 1961): 551–61; Leon Dmochowski, "Viruses and Tumors in the Light of Electron Microscope Studies: A Review," *Cancer Research* 20, no. 7, part 1 (1960): 977–1015; Leon Dmochowski and Clifford E. Grey, "Studies on Submicroscopic Structure of Leukemias of Known or Suspected Viral Origin: A Review," *Blood* 13, no. 11 (November 1958): 1017–42; Wendell M. Stanley, "Relationships, Established and Prospective, between Viruses and Cancer," *Annals of the New York Academy of Sciences* 71, no. 6 (September 1, 1958): 1100–13.

66. "Scientist Finds Viruses 'at Scene' of Cancer," *Boston Globe*, March 10, 1957.

67. Patrick MacCormack, "Blood Cancer Virus Is Found," *Washington Post and Times Herald* (DC), February 26, 1958.

68. Peyton Rous to Joseph Beard, August 22, 1957, in Rous papers, series I, Beard, Joseph W., no. 4.

69. Several researchers maintained an interest in the Rous Sarcoma Virus and other avian cancer viruses, which were joined by the studies of Richard Shope on papillomaviruses and Francesco Duran-Reynal's work on coinfection and helper viruses. These are summarized in J. W. Beard, "Viruses as a Cause of Cancer," *American Scientist* 46, no. 3 (1958): 226–54; F. Duran-Reynals, "Neoplastic Infection and Cancer," *American Journal of Medicine* 8, no. 4 (April 1950): 490–511.

70. Gross worked with sarcoma 37, a fast-growing mouse tumor that could be transmitted by grafting parts of the tumor from mouse to mouse. "The Search for Viruses as Etiological Agents in Cancer and Leukemia: How the Mouse Leukemia Virus Was Discovered," 3–8, n.d., in Ludwik Gross papers, box 1, folder 12, National Library of Medicine. (hereafter "Ludwik Gross papers," with box and folder information)

71. Ludwik Gross, "Is Cancer a Communicable Disease?," *Cancer Research* 4 (May 1944): 296, 300–301.

72. Ludwik Gross to Dr. A. C. Denny (Wyeth, Inc.), March 8, 1945, in Ludwik Gross papers, box 3, folder 29.

73. From Jacob Furth, a researcher at Cornell Medical School in New York, Gross obtained the "AK" strain, 90 percent of which developed "spontaneous" leukemia. Bessis, "Mouse Leukemia Virus," 291–95; Ludwik Gross, "A Filterable Agent, Recovered from Ak Leukemic Extracts, Causing Salivary Gland Carcinomas in C3H Mice," *Proceedings of the Society for Experimental Biology and Medicine* 83, no. 408 (June 1953): 414–21; Ludwik Gross, " 'Spontaneous' Leukemia

Developing in C3H Mice Following Inoculation, in Infancy, with AK-Leukemic Extracts, or AK-Embryos," *Proceedings of the Society for Experimental Biology and Medicine* 76, no. 381 (January 1951): 27–32.

74. Bessis, "Mouse Leukemia Virus," 296–97.

75. Ludwik Gross, "Mouse Leukemia," *Annals of the New York Academy of Sciences* 54, no. 6 (July 1, 1952): 1184–96.

76. "Filterable Agent Causing Mouse Salivary Gland Carcinoma," *Journal of the American Medical Association* 153, no. 2 (September 12, 1953): 150; "Transmissible Leukemia in Mice," *Journal of the American Medical Association* 148, no. 9 (March 1, 1952): 746.

77. John B. Moloney, "Properties of a Leukemia Virus," in *Symposia: Tumor Viruses*, ed. Joseph W. Beard (Washington, DC: US Department of Health, Education, and Welfare, Public Health Service, 1960), 7–38.

78. B. E. Eddy, S. E. Stewart, R. L. Kirschstein, and R. D. Young, "Induction of Subcutaneous Nodules in Rabbits with the SE Polyoma Virus," *Nature* 183, no. 4663 (1959): 766–67; Sarah E. Stewart and Bernice E. Eddy, "Tumor Induction by SE Polyoma Virus and the Inhibition of Tumors by Specific Neutralizing Antibodies," *American Journal of Public Health* 49 (1959): 1493–96; Bernice E. Eddy and Sarah E. Stewart, "Neoplasms in Hamsters Induced by Mouse Tumor Agent Passed in Tissue Culture," *Journal of the National Cancer Institute* 20, no. 4 (April 1958): 747–56; Bernice E. Eddy and Sarah E. Stewart, "Neoplasms in Mice Inoculated with a Tumor Agent Carried in Tissue Culture," *Journal of the National Cancer Institute* 20, no. 6 (June 1958): 1223–36; Bernice E. Eddy and Sarah E. Stewart, "Induction of Tumors in Rats by Tissue-Culture Preparation of SE Polyoma Virus," *Journal of the National Cancer Institute* 22, no. 1 (June 1959): 161–70. For further discussion of Gross and the work of Sarah Stewart and Bernice Eddy, see Morgan, "Gross, Stewart, and 1950s Discoveries."

79. Charlotte Friend, "Immunological Relationships of a Filterable Agent Causing a Leukemia in Adult Mice I. The Neutralization of Infectivity by Specific Antiserum," *Journal of Experimental Medicine* 109, no. 2 (February 1, 1959): 217–28. For reception of these discoveries and other work carried out by the Memorial Sloan Kettering Cancer Center, see the clippings files maintained on microfilm at the Rockefeller Archives Center, Memorial Press Clippings, vols. 49 and 50 (reel 7).

80. *Labor-Health, Education, and Welfare Appropriations Act for 1957: Hearings, Day 11, Before Comm. on Appropriations*, 84th Cong. 1412 (1956) (statement of Wendell Stanley, director of UC Berkeley Virus Laboratory); Gaudillière, "Molecularization of Cancer Etiology," 150–58.

81. Wendell M. Stanley, "The Virus Etiology of Cancer," in *3rd National Cancer Conference Proceedings* (Philadelphia: J. B. Lippincott, 1957), 47. Stanley went on to reiterate this claim for many other audiences, even though he himself performed almost no cancer virus research in his Berkeley lab. Wendell M. Stanley, "The Virus Etiology of Cancer," *CA: A Cancer Journal for Clinicians* 7, no. 3

(May 1, 1957): 97–100; Stanley, "Relationships between Viruses and Cancer"; Wendell M. Stanley, "Virus-Induced Neoplasia—Outlook for the Future," *Cancer Research* 20, no. 5, part 1 (June 1, 1960): 798–804.

82. Oshinsky, *Polio*, 255–57.

83. Greer Williams, *Virus Hunters* (New York: Knopf, 1959), 4. The title was a self-conscious play on Paul de Kruif's famous *Microbe Hunters* (1926).

84. Ludwik Gross, *Oncogenic Viruses* (New York: Pergamon Press, 1961); Ludwik Gross, *Oncogenic Viruses*, 2nd ed. (Oxford: Pergamon Press, 1970).

85. ACS Brochure on Leukemia, in Mary Lasker papers, box 102, folder "ACS General 1960," Columbia University Rare Book and Manuscript Library (hereafter "Lasker papers," with box and folder information).

86. J. L. Melnick, "Discussion: Electron Microscopy of Tumors of Known and Suspected Viral Etiology," in *Viruses and Tumor Growth* (Houston: University of Texas MD Anderson Hospital and Tumor Institute, 1957), 270–71.

87. Bernhard, "Detection and Study of Tumor Viruses," 723.

88. Wallace P. Rowe, "A Survey of the Tumor Virus Problem from an Epidemiologic Standpoint," *Cancer Research* 25 (1965): 1277.

89. Subcomm. on Government Research, Research in the Service of Man: Biomedical Knowledge, Development, and Use, S. Doc. No. 90-55, at 163 (1966) (Conference Rep.).

Chapter Three

1. "Voluntary Reporting of Cancer Aids Massachusetts Cancer Studies," *Public Health Reports* (1896–1970) 42, no. 38 (September 23, 1927): 2338–39; "Highest Death Rate from Cancer Here," *Boston Daily Globe* (1923–1927), December 16, 1925; "Data on Cancer Sought by State," *Boston Daily Globe* (1923–1927), June 7, 1925.

2. Francis George Curtis and George H. Bigelow, "Cancer Morbidity," *Journal of the American Medical Association* 89, no. 10 (September 3, 1927): 809.

3. George H. Bigelow, "Is the State's Cancer Program State Medicine?," *New England Journal of Medicine* 200, no. 9 (February 28, 1929): 438–39; George Hoyt Bigelow and Herbert Luther Lombard, *Cancer and Other Chronic Diseases in Massachusetts* (Boston: Houghton Mifflin, 1933), 86–88. For further background on the debate over the Massachusetts Cancer Program, see Rosenkrantz, *Public Health and the State*, 162–76.

4. Rosenberg, *Explaining Epidemics*, 294.

5. Harden, *Inventing the NIH*, 9–43; Sledge, "War, Tropical Disease."

6. Hoffman, *Mortality from Cancer*. On the importance of insurance companies and the "insurance experience" in shaping public health approaches to cancer, see Wailoo, *How Cancer Crossed Color Line*, 44–47.

7. Hacker, *Divided Welfare State*, xiii.

8. Hart, "Herbert Hoover's Last Laugh."

9. Morris, *Limits of Voluntarism*, xxxi–xxxiv; Zunz, *Philanthropy in America*, 18–40. On the idea of these organizations as a response to social disorder, see Wiebe, *Search for Order*.

10. Rothman, *Living in Shadow of Death*, 179–92.

11. Feldberg, *Disease and Class*, 36–110; Worboys, *Spreading Germs*. For the standard treatment of tuberculosis as a "social disease," see Dubos and Dubos, *White Plague*.

12. Shryock, *National Tuberculosis Association*, 46–51.

13. "Rich Women Begin War on Cancer," *New York Times*, April 23, 1913.

14. New cancer research efforts included the Collis Huntington fund at General Memorial Hospital in New York City (1902), Roswell Park in Buffalo, NY (1898), the Harvard Cancer Commission (1899), the American Oncologic Hospital of Philadelphia (1904), the St. Louis Skin and Cancer Hospital (1905), and the Institute for Cancer Research at Columbia University (1910). Patterson, *Dread Disease*, 70–71.

15. *American Society for Control of Cancer*, 5–7; Cantor, "Uncertain Enthusiasm," 42–49; Patterson, *Dread Disease*, 68–86; Shaughnessy, "Story of American Cancer Society," 13–15.

16. Shaughnessy, "Story of American Cancer Society," 122–26, quote about quacks by Dr. George Soper, 123.

17. Hoffman, *Wages of Sickness*, 68–91; Derickson, *Health Security for All*, 12–16; Starr, *Social Transformation of American Medicine*, 180–206, 216–18.

18. Gardner, *Early Detection*, 70–92; Morris, *Limits of Voluntarism*, 1–34.

19. Patterson, *Dread Disease*, 97.

20. Weisz, *Chronic Disease in Twentieth Century*, 37–55.

21. Harden, *Inventing the NIH*, 55–56.

22. Matthew Neely, "Cancer: Humanity's Most Deadly Scourge," *Cong. Rec.* S9048–49 (May 18, 1928).

23. Strickland, *Politics, Science, and Dread Disease*, 8–9.

24. As quoted in Harden, *Inventing the NIH*, 140–41.

25. Daniel Sledge considers a contrasting case of how the Public Health Service became involved in the fight against endemic infectious and parasitic diseases in the South, such as hookworm and yellow fever. The federal support that the Public Health Service enjoyed here, Sledge concludes, had a great deal to do with the power of southern Democrats in Congress and the direct economic benefits thought to accrue from controlling these diseases. I thank Professor Sledge for providing me with a copy of his manuscript in advance of publication. Sledge, *Health Divided*.

26. *Hearings on S.1171: An Act to Establish a National Institute of Health, Day 1, Before Comm. on Interstate and Foreign Commerce*, 71st Cong. 9, 15 (1930) (statement of Charles Bloodgood, Johns Hopkins University).

27. Strickland, *Politics, Science, and Dread Disease*, 11–14.

28. *Cancer Research: Joint Hearings on S.2067, H.R. 6100, H.R. 6767, & H.J. Res. 428, Day 1, Before Senate Subcomm. on Commerce and House Subcomm. on Interstate and Foreign Commerce*, 75th Cong. 29 (1937) (statement of Thomas Parran, surgeon general).

29. Indeed, there is good reason to think that dependence on these voluntary associations was a much more important part of the expansion of the federal government during the 1930s than the standard narrative of centralization suggests. I return to this point in chapter 4. Balogh, *Associational State*, 139–71; Hart, "Herbert Hoover's Last Laugh."

30. James Ewing to Senator Royal S. Copeland (D-NY), June 3, 1937, as quoted in *Cancer Research: Joint Hearings*, 75th Cong. 59 (1937) (statement of C. C. Little, president of the American Society for the Control of Cancer).

31. Strickland, *Politics, Science, and Dread Disease*, 14.

32. *Cancer Research: Joint Hearings*, 75th Cong. 53–55 (1937) (statement of C. C. Little, president of the American Society for the Control of Cancer); Rader, *Making Mice*, 137.

33. Patterson, *Dread Disease*, 130.

34. National Cancer Act, Senate Bill 2067, enacted August 5, 1937 (Public Law 244).

35. NACC Meeting Minutes, December 4, 1939, p. 91, in Murphy papers, folder "National Advisory Cancer Council—Eleventh Meeting #1."

36. Erdey, "Armor of Patience," 105–9.

37. Stanhope Bayne-Jones, Ross G. Harrison, C. C. Little, John Northrop, and James B. Murphy, "Fundamental Cancer Research: Report of a Committee Appointed by the Surgeon General," *Public Health Reports* 53, no. 48 (December 2, 1938): 2130.

38. Dupree, "Great Instauration of 1940"; Galison, *Image and Logic*, 239–312; Kevles, *Physicists*, 302–38; Rhodes, *Making of the Atomic Bomb*.

39. Creager, " 'What Blood Told Dr Cohn' "; Gaudillière, "Rockefeller Strategies"; Kay, *Molecular Vision of Life*; Lenoir and Hays, "Manhattan Project for Biomedicine"; Neushul, "Science, Government and Penicillin"; Rasmussen, " 'Small Men,' Big Science."

40. George Gallup, "Public Backs Cancer Research with Tax-Raised Funds," *Washington Post*, June 12, 1946; Gerald Gross, "Could a 'Manhattan Project' Conquer Cancer?," *Washington Post*, August 4, 1946; George Gallup, "Heart Disease Is No. 1 Killer but People Fear Cancer Most," *Washington Post*, August 20, 1947.

41. State chapters of the ACS also feared widespread public disillusionment when such an effort failed to make progress. "Report on American Cancer Society Directors' Meeting," *California Medicine* 66, no. 5 (May 1947): 318–19. Thomas Parran, the surgeon general of the United States, testified that there were too few qualified researchers to effectively utilize such an allocation. *H.R.4502: Hearings*

on H.R. 4502, Day 1, Before the Committee on Foreign Affairs, 79th Cong. 3–9 (statement of Thomas Perran, surgeon general).

42. Thomas Parran, "A National Cancer Policy," *Journal of the National Cancer Institute* 7, no. 5 (April 1, 1947): 301.

43. Reminiscences of Mary Lasker (1962), part 1, pp. 160–161, in the Center for Oral History Archives, Columbia University Rare Book and Manuscript Library; Strickland, *Politics, Science, and Dread Disease*, 41–50.

44. Kevles, *Physicists*; Kohler, *Partners in Science*; Noble, *America by Design*; Zunz, *Philanthropy in America*. The most ambitious expression of this impulse was the mathematician and scientist-administrator Warren Weaver's ambitious and well-funded research program in "molecular biology" at the Rockefeller Foundation in the 1930s. Peripatetic Foundation officials supervised and coordinated researchers in fields from physiology to physical chemistry but still gave deference to individual scientists. Kohler, "The Management of Science." For more on the values of efficiency, see Alexander, *Mantra of Efficiency*; Haber, *Efficiency and Uplift*; Hays, *Conservation and Gospel of Efficiency*.

45. Opponents of this kind of direct intervention on the National Advisory Cancer Council soon managed to redirect the money intended for radium into more modest forms of research and physician training. Cantor, "Radium and National Cancer Institute."

46. Brandt, *No Magic Bullet*, 130–32, 161–63; Löwy, "Immunotherapy of Cancer"; Ward, "American Reception of Salvarsan."

47. Fox, "Politics of NIH Extramural Program," 452, 456.

48. Quinn, "Rethinking Antibiotic Research"; Mochales, "Forty Years of Screening Programmes." For standard histories of penicillin, see Bud, *Penicillin*; Neushul, "Science, Government and Penicillin"; Podolsky, *Antibiotic Era*.

49. Because of security restrictions, the nitrogen mustard findings were not published until 1946. In particular, the explosion of a ship carrying mustard gas in the harbor of Bari during the invasion of Italy resulted in the exposure of military personnel and local fisherman to mustard gas, killing many and calling attention to the effect that nitrogen mustard had on the growth of white blood cells. Smith, *Toxic Exposures*, 95–114; DeVita and Chu, "History of Cancer Chemotherapy"; Mukherjee, *Emperor of All Maladies*, 88–93.

50. On Rhoads's view, see C. P. Rhoads, "Rational Cancer Chemotherapy," *Science*, 119, no. 3081 (January 15, 1954): 77–80.

51. Chandler, *Visible Hand*, 81–120, 145–85.

52. Hounshell and Smith, *Science and Corporate Strategy*; Hughes, *American Genesis*, 53–95; Reich, *Making of American Industrial Research*.

53. For the institutional growth of chemotherapy, see Bud, "Strategy in American Cancer Research," 439–42; DeVita and Chu, "History of Cancer Chemotherapy," 8644; Heller, "Rhoads, Leader in Cancer Research." For an overview of public coverage of these first few chemotherapy agents, see Pat McGrady, "Man

against Cancer: Drugs Can Be Helpful—and Harmful," *Washington Post* (1923–1954), March 31, 1951.

54. M. J. Shear, "Role of the Chemotherapy Research Laboratory in Clinical Cancer Research," *Journal of the National Cancer Institute* 12, no. 3 (December 1, 1951): 574.

55. Alfred Gellhorn, "A Critical Evaluation of the Current Status of Clinical Cancer Chemotherapy," *Cancer Research* 13, no. 3 (March 1, 1953): 205, 214.

56. Frank H. J. Figge, review of *Review of Approaches to Tumor Chemotherapy: A Symposium of Papers and Discussions on Various Aspects of Tumor Chemotherapy, Developed from the Summer Meetings of the Section on Chemistry (C) of the American Association for the Advancement of Science at Gibson Island, Maryland, 1945–1946*, by Forest Ray Moulton, *Quarterly Review of Biology* 24, no. 1 (1949): 70.

57. "Chemotherapy Most Popular Method, Cancer Researchers Say," *Chemical & Engineering News* 29, no. 20 (May 14, 1951): 1949–50.

58. Report of Ad Hoc Research Policy Survey Committee (American Cancer Society, Inc.), December 8, 1955, p. 17, in Rockefeller University Archives, Edward Tatum papers, RG 450-T189, box 14, folder 1, Rockefeller Archive Center.

59. Shaughnessy, "Story of American Cancer Society," 257.

60. "American Cancer Society Grants in Cancer Research," August 1, 1948, in Murphy papers, folder "National Research Council #2."

61. Hansen, *Picturing Medical Progress*, 251–52. For a description of the controversies and arguments about the announcement itself, as well as a crisis when it seemed that impure lots of the vaccine threatened to cause polio outbreaks, see Oshinsky, *Polio*, 199–207; Paul, *History of Poliomyelitis*, 432–36.

62. Ironically, the shift in polio incidence was likely caused by the success of earlier efforts to provide clean water and dispose of human waste. These advances left children's immune systems underexposed to the polio virus, which was carried in fecal matter. Paul, *History of Poliomyelitis*, 148–60; Rogers, *Dirt and Disease*, 30–71. On tuberculosis fund-raising, see Zunz, *Philanthropy in America*, 46–49.

63. Oshinsky, *Polio*, 38–51; Paul, *History of Poliomyelitis*, 300–307.

64. Oshinsky, *Polio*, 52–53.

65. Creager, *Life of a Virus*, 147–48, 171–75; Kevles, "Dulbecco and New Animal Virology," 420–21; Landecker, *Culturing Life*, 129–37; Paul, *History of Poliomyelitis*, 316–19, 373–81.

66. Oshinsky, *Polio*, 81–83; Paul, *History of Poliomyelitis*, 308–16; Rogers, *Dirt and Disease*, 172.

67. Oshinsky, *Polio*, 91–95, quote from David Bodian on 95.

68. John Rowan Wilson, *Margin of Safety* (New York: Doubleday, 1963), 81. For a historical perspective on this same sense of tension, see Mawdsley, *Selling Science*, 53–57.

69. Dr. Hart Edgar van Riper to "NFIP Personnel Making Speeches for MOD [March of Dimes]," December 7, 1950, in folder Speeches and Speaker's Bureau,

Medical Program Records, series 14: Poliomyelitis, carton 15, March of Dimes Archives.

70. As quoted in Oshinsky, *Polio*, 129.

71. Basil O'Connor's comment in marginalia on 1945 NFIP Annual Report is quoted in Creager, *The Life of a Virus*, 153.

72. H. M. Weaver, *The Research Story of Infantile Paralysis* (New York: National Foundation for Infantile Paralysis, 1948), 17, as quoted in Paul, *History of Poliomyelitis*, 412. In a quasi-facetious memo that he sent to O'Connor, Weaver described the beginning of his efforts: "1946: became convinced that prevention . . . was the only promising avenue of approach. 1947: [after a roundtable] recognized for the first time that an appalling few of the NFIP's grantees were really trying to solve the problem of poliomyelitis in man . . . 1948: recognized the need of, and began providing for, more research specifically directed at solving the many problems obviously pertinent to poliomyelitis." Harry Weaver to Basil O'Connor, June 27, 1955, in folder 11 (American Cancer Society), Public Relations Records, carton A, March of Dimes Archives.

73. Creager, *Life of a Virus*, 171–75.

74. Lindenmeyer, *A Right to Childhood*; Meckel, *"Save the Babies"*; Skocpol, *Protecting Soldiers and Mothers*; Connolly, *Saving Sickly Children*, 1–41.

75. *Cancer and Polio Research: Hearings on H.R. 977, H.R. 3257, and H.R. 3464, Day 1, Before House Comm. on Interstate and Foreign Commerce and House Subcomm. on Interstate and Foreign Commerce*, 80th Cong. 39 (1948) (statement of John J. O'Connor on behalf of National Foundation for Infantile Paralysis).

76. Ibid., 41.

77. Morris, *Limits of Voluntarism*, xxxi–xlii.

78. *Cancer and Polio Research: Hearings*, 80th Cong. 15–16, 17 (1948) (statement of Representative Hugh Scott, R-PA).

79. In 1953 the NFIP spent fifteen times as much on polio research as did the National Institutes of Health. Smith, *Patenting the Sun*, 249.

Chapter Four

1. Sparrow, *Warfare State*, 9–12. An extensive literature exists on the changing role of the federal government in American social and economic life during the early twentieth century, especially its mobilization during the New Deal. See Amenta, *Bold Relief*; Brinkley, *End of Reform*; Kessler-Harris, *In Pursuit of Equity*; Klein, *For All These Rights*; Rodgers, *Atlantic Crossings*.

2. For examples, see the essays in Sparrow, Novak, and Sawyer, *Boundaries of the State*.

3. Hacker, *Divided Welfare State*, xiii.

4. My usage of "political culture" follows that of Julian Zelizer in his analysis of social security and tax policymaking in Congress. Zelizer, *Taxing America*, 10–12, 84.

5. Patterson, *Dread Disease*, 172. See also biographical information at the National Library of Medicine's "Profiles in Science" page for Mary Lasker, accessed April 12, 2018, https://profiles.nlm.nih.gov/ps/retrieve/Narrative/TL/p-nid/199.

6. Reminiscences of Mary Lasker (1962), part I, pp. 82–83, in the Oral History Research Office Collection of the Columbia University Libraries (hereafter "Lasker Oral History," with part and page numbers).

7. Lasker Oral History, part I, p. 132.

8. Mary Lasker, interview related in Shaughnessy, "Story of American Cancer Society," 221–22.

9. By 1946 the society had $2.5 million to spend on research. Lasker Oral History, part I, p. 132.

10. Drew, "Health Syndicate"; Patterson, *Dread Disease*, 172–76; Strickland, *Politics, Science, and Dread Disease*, 41–44.

11. Patrick McGrady, science editor, ACS, to ACS-funded scientists, June 1, 1950, in Franceso Duran-Reynals papers [H MS c194], box 3, folder "Committee on Growth," Center for History of Medicine, Francis A. Countway Library of Medicine.

12. Ibid., November 3, 1954, box 3, folder ACS, VD8-VDK, 1947–52.

13. "Committee on Growth of the National Research Council," *Cancer Research* 6, no. 1 (January 1, 1946): 29. However, from the outset the priorities of the scientist-advisers from the National Research Council and the leadership of the ACS over the extent to which clinical research counted as a basic science created ongoing tension. See the correspondence between James Adams, chairman of the ACS Executive Committee, and Louis Weed, chairman of the Medical Sciences Division of the National Research Council, on February 5 and 14, 1947, in Murphy papers, folder "American Cancer Society Grant #2."

14. Robinson, *Noble Conspirator*, 59–63.

15. Bell, *Liberal State on Trial*, 67–84; Derickson, *Health Security for All*, 110–13; Quadagno, *One Nation, Uninsured*, 17–48. On the broader effects of early anticommunism and the ways the 80th Congress helped check the New Deal, see Korstad, *Civil Rights Unionism*, 301–34; Storrs, "Red Scare Politics"; Fraser and Gerstle, introduction.

16. On the contentious process surrounding the National Science Foundation, see Appel, *Shaping Biology*, 9–37; Dennis, "Reconstructing Sociotechnical Order"; Kevles, "National Science Foundation"; Reingold, "Bush's New Deal for Research"; Wang, "Liberals."

17. Mary Lasker to Louis Katz, March 9, 1947, in "The Mary Lasker papers," Profiles in Science, National Library of Medicine, accessed January 12, 2017, https://profiles.nlm.nih.gov/ps/retrieve/ResourceMetadata/TLBBFZ.

18. Lasker Oral History, part I, pp. 82–83, 77–78.

19. Annual Report of the American Cancer Society, August 31, 1947, p. 7, quoted in Shaughnessy, "Story of American Cancer Society," 247.

20. In this effort Lasker joined a long tradition of individuals seeking to make change not through direct democratic processes but through a mastery of policymaking

machinery. Sheingate, "Terrain of the Political Entrepreneur." Lasker's campaign also recalls the point made by historians who have examined the political actions of women earlier in the twentieth century: although they might have been denied access to formal political organizations, women were able to wield influence through the public culture that existed beyond formal politics. Gilmore, *Gender and Jim Crow*, 178; Sklar, *Florence Kelly*, xii–xv.

21. Luke Quinn to Mary Lasker, April 23, 1953, in Lasker papers, box 491, folder "Quinn, Col. Luke (i)."

22. Ibid., April 21, 1954, box 491, folder "Quinn, Col. Luke (ii)."

23. Lasker to John Furcolo, Senate candidate in Massachusetts, October 21, 1954, in Lasker papers, box 491, folder "Quinn, Col. Luke (ii)."

24. See, for example, Representative John Fogarty to Lasker, June 20, 1952, an invitation to a barbecue, in "John Fogarty Papers," Profiles in Science, National Library of Medicine, accessed December 22, 2016, https://profiles.nlm.nih.gov/ps/access/HRBBCK.pdf.

25. Strickland, *Politics, Science, and Dread Disease*, 79.

26. Senator Lister Hill was an especially useful ally because, unlike the House of Representatives, where members of the Appropriations Committee were kept separate from other committees, the Senate allowed Hill to sit on both the Appropriations and the Health, Education and Welfare committees—Hill was thus in the position of approving budget increases that he himself recommended. Ibid., 94–96.

27. Quinn to Lasker, May 17, 1955, in Lasker papers, box 491, folder "Quinn, Col. Luke (iii)."

28. Ibid., January 29, 1954, box 491, folder "Quinn, Col. Luke (ii)."

29. Ibid., April 1, 1955, box 491, folder "Quinn, Col. Luke (iii)."

30. Ibid., May 17, 1955.

31. Strickland, *Politics, Science, and Dread Disease*, 75–108; Swain, "Rise of a Research Empire."

32. Quinn to Lasker, August 23, 1954, in Lasker papers, box 491, folder "Quinn, Col. Luke (ii)."

33. Strickland, *Politics, Science, and Dread Disease*, 154–56.

34. Ibid., 102.

35. Sidney Farber to Lasker, September 4, 1965, in Lasker papers, box 171, folder "Sidney Farber #1."

36. Creager, "Mobilizing Biomedicine," 182–83.

37. Warren Shields to Wendell Stanley, October 5, 1959, in Wendell Stanley papers, carton 13, folder "ACS Research Committee 1959–1965," Bancroft Library, University of California, Berkeley; Harry Weaver, outgoing research director, ACS, memo, "The Role of the American Cancer Society in Research," December 2, 1960, p. 2, Lasker papers, carton 102, folder "ACS Research 1960." For published literature on this realignment, see Graham DuShane, "New Directions in

Research Support," *Science*, 123, no. 3196 (March 30, 1956): 525; Morris, *Limits of Voluntarism*, 35–78; Patterson, *Dread Disease*, 180–84; Zunz, *Philanthropy in America*, 169–200.

38. Aronowitz, *Unnatural History*, 115–43; Wailoo, *How Cancer Crossed Color Line*, 66–91.

39. As described by Krueger, "Death Be Not Proud." For a similar interpretation of disease visibility in the case of sickle-cell anemia and malaria, see Wailoo, *Dying in City of Blues*, 23–25.

40. Krueger, *Hope and Suffering*, 32–39, 52–54. On the image of pediatric leukemia in the United Kingdom during the same era, see Barnes, "Cancer Coverage."

41. Laszlo, *Cure of Childhood Leukemia*, 17–19.

42. Krueger, *Hope and Suffering*, 103–4.

43. W. M. C. Brown and R. Doll, "Leukaemia in Childhood and Young Adult Life," *British Medical Journal* 1, no. 5231 (April 8, 1961): 981–88; Alexander G. Gilliam and William A. Walter, "Trends of Mortality from Leukemia in the United States, 1921–55," *Public Health Reports* 73, no. 9 (1958): 773–84; Alice M. Stewart and D. Hewitt, "The Epidemiology of Human Leukaemia," *British Medical Bulletin* 15, no. 1 (January 1, 1959): 73–77.

44. The Jimmy Fund Building is now known as the Dana-Farber Cancer Center. As revealed years later, "Jimmy" was Einar Gustafason. Krueger, " 'For Jimmy,' " 78–85. There is a more extensive literature on the ethics of the use of "poster children" in the history of polio literature, which Krueger addresses in note 5 of her essay, but for the purposes of my discussion here, it is enough to note the power of these images. On the growth of consumer culture and its relationship to Americanism, see Cohen, *Consumer's Republic*.

45. Mintz, *Huck's Raft*, 273–79. For a broader discussion of how politics, sexuality, and domesticity converged in debates over the nature of the family, which also impacted the value given to children, see May, *Homeward Bound*.

46. Zelizer, *Pricing the Priceless Child*.

47. This is a form of what later anthropologists of biomedicine have called "anticipatory" biomedical politics *avant la lettre*, which often forms around the diseases of particular "vulnerable" populations. In this sense I push the origins of what anthropologists have discussed as "biosociality" back by several decades. Adams, Murphy, and Clarke, "Anticipation."

48. Zubrod et al., "Chemotherapy Program," 351.

49. Frank A. Howard to Alfred Sloan and Laurance Rockefeller, August 31, 1956, in Laurance Rockefeller papers, RG 2, series OMR-MSKCC, box 40, folder 527, Rockefeller Archive Center.

50. *Health Inquiry into Heart Disease, Cancer, Day 2, Hearings Before the Comm. on Departments of Health, Education, Welfare, and Related Agencies Appropriations*, 83rd Cong. 303 (1953) (statement of Dr. R. Keith Cannan of the National Research Council).

51. Rockefeller to Senator H. Alexander Smith (NJ), July 12, telegram, as quoted in Smith to Rockefeller, August 14, 1950, in Laurance Rockefeller papers, RG 2, series OMR-MSKCC, box 26, folder 360, Rockefeller Archive Center.

52. Here, the ACS and its allies emphasized not the need for federal funding leadership, but that federal resources were one of several streams dedicated to chemotherapy. *Departments of Labor and Health, Education, and Welfare Appropriations for 1954: Hearings on Department of Health, Education, and Welfare, Day 5, Before the Subcomm. on Departments of Health, Education, Welfare, and Related Agencies Appropriations*, 83rd Cong. 1381–82 (1953) (exchange among Sidney Farber, Cornelius Rhoads, Senator A. Willis Robertson [D-VA], and Senator Edward J. Thye [R-MN]).

53. *DOL and HEW Appropriations for 1954: Hearings*, 83rd Cong. 1395 (1953) (statement of Dr. Charles S. Cameron, medical and scientific director for the American Cancer Society).

54. *Department of Labor—Federal Security Agency Appropriations for 1952, Day 3, Hearings Before Subcomm. of Committee on Appropriations*, 82nd Cong. 189 (1951) (statement of Dr. Joseph Burchenal, Cornell Medical School).

55. *DOL and HEW Appropriations for 1954: Hearings*, 83rd Cong. 1385 (1953) (statement of Dr. Sidney Farber, Boston Children's Hospital).

56. Zubrod et al., "Chemotherapy Program," 354.

57. Sidney Farber, as quoted in Claude Stanush, "Medicine's Greatest Hunt: For Chemicals to Starve Out Cancer," *Colliers*, November 23, 1956, 27.

58. Minutes, National Advisory Cancer Council, February 16–18, 1954, p. 6 in National Archives and Records Administration II (NARA II), RG 443 A1-27, carton 14, NACC 1954-2-16.

59. *Departments of Labor and Health, Education, and Welfare Appropriations for 1955: Hearings on Department of Health, Education, and Welfare, Day 5, Before the Subcomm. on Departments of Health, Education, Welfare, and Related Agencies Appropriations*, 84th Cong. 1036 (1954) (letter from W. H. Sebrell, director of the National Institutes of Health, in association with testimony of Dr. John Heller, director of the National Cancer Institute).

60. Zubrod et al., "Chemotherapy Program," 354.

61. Gaudillière, "Cancer entre infection et hérédité"; Löwy, *Between Bench and Bedside*, 41–48; Keating and Cambrosio, "Cancer Clinical Trials"; Keating and Cambrosio, "From Screening to Clinical Research"; Krueger, *Hope and Suffering*, 95–99.

62. Kenneth M Endicott, "The Chemotherapy Program," *Journal of the National Cancer Institute* 19, no. 2 (August 1957): 275, 293.

63. Keating and Cambrosio, *Cancer on Trial*, 54–84.

64. Gaudillière, "Circulating Mice and Viruses"; Rader, *Making Mice*, 216–23.

65. Kenneth M. Endicott, introduction to *Perspectives in Leukemia*, ed. William Dameshek and Ray M. Dutcher (New York: Grune & Stratton, 1968), 2.

Endicott was the director of the NCI. Indeed, recent scholarship on the use of animal models for biomedical research in the United Kingdom suggests exactly this point: prewar intellectual controversies were not so much resolved as supplanted by the institutionalization of animals for basic biological testing (as opposed to experiments). Kirk, " 'Standardization through Mechanization' "; Kirk, "Brave New Animal"; Kirk, " 'Wanted—Standard Guinea Pigs.' "

66. NCI historical appropriations data, adjusted to constant 2017 dollars by the Biomedical Research Price Index.

67. Keating and Cambrosio, *Cancer on Trial*, 85–116.

68. Historical data on ACS grants awarded for cancer virology, 1946–1980. Chemotherapy was the other field of a possible breakthrough. *Departments of Labor and Health, Education, and Welfare Appropriations for 1959: Hearings on Department of Health, Education, and Welfare, Day 5, Before the Subcomm. on Departments of Health, Education, Welfare, and Related Agencies Appropriations*, 85th Cong. 390 (1958) (statement of John R. Heller, director of the National Cancer Institute); *Annual Report of Program Activities, National Cancer Institute* (hereafter "NCI Annual Report"), 1958, 1:1–4, in NIH Library.

Chapter Five

1. Albert Rosenfeld, "A Superplan to Cut Years off the War," *Life*, November 18, 1966, 110–11.

2. Carl G. Baker, "Closing Remarks," in *Methodological Approaches to the Study of Leukemias*, ed. Vittorio Defendi (Philadelphia: Wistar Institute Press, 1965), 215–16.

3. The first year's budget for the SVLP was $10 million, and it rose steadily thereafter. The total value (in 1964 dollars) of cancer virology grants handed out by the ACS was $346,918; in 1969 (in 1969 dollars) the total value of grants handed out by the ACS was $990,805. ACS Historical Data on Cancer Virology Grants, 1946–1980, ACS National Cancer Information Center, Atlanta, contacted July 29, 2011.

4. For example, the canonical collection of historical essays on "big science" includes examples from physics, chemistry, and engineering—but not biology and medicine. Galison and Hevly, *Big Science*. A review of a genealogy of the term reflects a similar focus. Capshew and Rader, "Big Science." On the Human Genome Project as a debate about organization, see Francis S. Collins, Michael Morgan, and Aristides Patrinos, "The Human Genome Project: Lessons from Large-Scale Biology," *Science* 300, no. 5617 (April 11, 2003): 286–90; Fortun, "Human Genome Project"; Kevles and Hood, *Code of Codes*; Lenoir and Hays, "Manhattan Project for Biomedicine."

5. Office of the NCI director to Senator Lister Hill, June 12, 1957, in RG 443, UDoD6, carton 93, RES 9-7 National Cancer Institute (NCI) 1937–68, NARA II.

6. Derrida has discussed the "fabulously textual" nature of nuclear war–planning discourses as a means of representing what could not be experienced or described directly. Derrida, "No Apocalypse, Not Now," 23. This point is also amplified by recent work on the textual and material aspects of paperwork in bureaucracy. Gitelman, *Paper Knowledge*; Hull, *Government of Paper*; Riles, *Documents*.

7. Latour, "On the Partial Existence"; Gaudillière, "Genesis and Development."

8. Doogab Yi has also recently examined how the NCI turned to the management of viruses to address concerns for its administrative legitimacy, but with a slightly different emphasis. Yi, "Governing." In examining the emphasis of acceleration in the NCI's management program, my aim is to provide a concrete historical case as a counterpoint to contemporary discussions of anticipation by anthropologists and sociologists of the biological and biomedical sciences, as well as a recent interest in futurity as a practical and affective stance in different strains of modernist thinking. For example, Adams, Murphy, and Clarke, "Anticipation"; Birch, "Introduction"; Milburn, *Nanovision*; Jameson, "Utopia as Method"; Saint-Amour, *Tense Future*, 23–33, 37–43.

9. This idea echoes James Scott's concept of legibility, in that efforts to render nature "legible" for state planning are also efforts to order unruly communities or social practices—in this case the preference of biological researchers for unplanned, "free" inquiry. Scott, *Seeing like a State*, 25–33, 347–49.

10. Minutes, NIH Group for Discussion of Research Administration, April 24, 1957, p. 20, in Carl Baker papers, unsorted, National Institutes of Health Office of History (hereafter "Baker papers," with best identifying information). These minutes reference the H. G. Wells quote as reproduced in Martin Gardner, ed., *Great Essays in Science* (New York: Pocket Books, 1957), 357.

11. Fox, "Politics of NIH Extramural Program," 447–52.

12. Erdey, "Armor of Patience," 24–134.

13. John Heller (director NCI) to John Enders (Children's Cancer Research Foundation), January 15, 1959, in John Enders papers (MS 1478), series I, box 12, folder 246, Yale University Archives and Special Collections (hereafter "Enders papers").

14. "Policies and Procedures of Public Health Support of Research Development and Training," June 15, 1958, attached to minutes of November 11, 1958, NACC meeting, in RG 443 A1-27, carton 15, NARA II.

15. "The Role of Viruses in Relation to Malignancy: Statement by the Virology and Rickettsiology Study Section," September 18, 1958, Appendix A to minutes of November 11, 1958, NACC meeting, in RG 443 A1-27, carton 15, NARA II.

16. "Report of the Planning Committee for the Council," June 1, 1956, Appendix for NACC meeting on June 25, 1956, in RG 443 A1-27, carton 15, NARA II. "For the first time, we as scientists are beginning to hold responsibility for the future in our own hands," wrote one Johns Hopkins biology professor. "There is

a choice before us between free and design research . . . between supporting the man or the experimental design. Let us support the man." Curt P. Richter, "Free Research versus Design Research," *Science*, 118, no. 3056 (July 24, 1953): 91.

17. "Explanation of Statement of Grant Award" (July 1958), in Charlotte Friend papers, box 10, folder NCI Grant, Arthur H. Aufses Jr., MD Archives, Icahn School of Medicine at Mount Sinai (hereafter "Friend papers," with box and folder information).

18. William V. Consolazio and Margaret C. Green, "Federal Support of Research in the Life Sciences," *Science* 124, no. 3221 (September 21, 1956): 522–26.

19. Bayne-Jones, *Advancement of Medical Research*, 4–7.

20. Other members included Hilary Koprowski, John Enders, and Albert Sabin, the developer of the oral polio vaccine. "Highlights of an Inter-Section Study Meeting on the Role of Viruses in Relation to Human Malignancies," September 16, 1958, in Enders papers, series I, box 12, folder 246.

21. "Profile: Research Team," *Cancer News* 10, no. 2 (Spring 1956): 18–19.

22. The National Cancer Advisory Council invited a prominent polio researcher, Hilary Koprowski, to speak with them about the potential properties of a human cancer virus. Minutes of the National Advisory Cancer Council, March 2, 1959, in RG 443, A1-27, carton 15, NARA II.

23. Virology and Rickettsiology Study Section, "The Role of Viruses in Relation to Malignancy," September 18, 1958, included as Appendix A to the Minutes of the National Advisory Cancer Council for November 3, 1958, in RG 443, A1-27, carton 15, NARA II.

24. "Statement of the Current Areas of Interest, Virus and Cancer Panel," June 9, 1960, in Enders papers (MS 1478), series I, box 12, folder 247. At midcentury, many biologists were becoming aware of the need for a system of infrastructure to support the expansion of their research. Creager and Landecker, "Technical Matters."

25. Viruses and Cancer Panel, Minutes of First Meeting (February 11, 1959), in LION, AR005150.

26. Heller (director, NCI) to Enders (Children's Cancer Research Foundation), January 15, 1959, in Enders papers, series I, box 12, folder 246.

27. Viruses and Cancer Panel, Minutes of Second Meeting (May 15, 1959), in LION, AR005150.

28. Viruses and Cancer Panel, Minutes of Eighth Meeting (October 20–21, 1959), pp. 3–4, in LION, AR005150.

29. Viruses and Cancer Panel, Minutes of Third Meeting (July 31, 1959), in LION, AR005150.

30. Indeed, this style of rhetoric was a central feature of many debates over the expansion of the federal government after the Second World War. Zelizer, *Taxing America*.

31. Fox, "Politics of NIH Extramural Program"; Greenberg, *Politics of Pure Science*, 169–77.

32. Nate Haseltine, "Lack of Proper Policing Called Chief Fault of NIH Grant Plans: No Wrong-Doing Found," *Washington Post*, May 1, 1961; Nate Haseltine, "NIH Accused of Improper Accounting," *Washington Post*, April 6, 1962; Joseph Hearst, "Research Administration Blasted by House Group," *Chicago Daily Tribune*, July 2, 1962; Robert C. Toth, "Overspending on U.S. Medical Research Cited," *Los Angeles Times*, July 2, 1962; Robert C. Toth, "Stanford Denies Building Pool with U.S. Grant for Research," *New York Times*, December 21, 1962.

33. *The Administration of Grants by the National Institutes of Health: Hearings, Day 1, Before Subcomm. of Committee on Government Operations*, 87th Cong. 21 (1962) (comments of Lawrence Fountain, chair).

34. *Committee on Appropriations, Departments of Labor, and Health, Education and Welfare, and Related Agencies Appropriation Bill*, 1965, H.R. Rep. No. 88-69, at 23 (1964), comments of Representative John Fogarty (D-RI).

35. "The PHS and the Fountain Committee," March 15, 1963, memo; "Draft List of 'authorities on NIH machinery and procedures, for decision making,'" September 4, 1964, both in Baker papers, Virus and Cancer Programs Backup Memoranda 1963–1964, 1964.

36. NCI Director (Kenneth Endicott) to chief of NCI Virology Research Resources Branch (Harvey Scudder), November 6, 1961, in Baker papers, Virus and Cancer Programs Backup Memoranda 1950–1962, 1961.

37. Minutes of Viruses and Cancer Panel 8th meeting (October 20–21, 1961), pp. 6–7, in LION, AR005150.

38. NCI Office of Information and Publications, *Progress against Cancer:* 1961 (Washington, DC: Department of Health, Education, and Welfare, Public Health Service, 1962), 9.

39. Carl G. Baker and Alton Meister, "Studies on Fatty Acid Oxidation by Normal and Neoplastic Liver," *Journal of the National Cancer Institute* 10, no. 5 (April 1, 1950): 1191–98.

40. Baker did attempt to return to research in the mid-1950s, but his asthma, aggravated by work with lab animals, had not abated. Stevenson, "Interview with Dr. Carl Baker," 1; Frank J. Rauscher, "In Memoriam: Carl G. Baker (1920–2009)," *Cancer Research* 69, no. 11 (June 1, 2009): 4935–36.

41. Carl Baker, responses to Scientist-Administrators at the National Institutes of Health Questionnaire, 1969, in Baker papers, unsorted.

42. Stevenson, "Interview with Dr. Carl Baker," 2.

43. Hollinger, "Free Enterprise and Free Inquiry."

44. The National Science Foundation itself had been proposed by Vannevar Bush to avert a more aggressive plan for a more populist, government-directed agency proposed by the liberal New Deal senator Harley Kilgore. Hollinger's essay "Free Enterprise and Free Inquiry" is an especially useful analysis of the growth of scientific autonomy as an idea and a rhetorical tool in these debates in the context of growing anticommunism. Appel, *Shaping Biology*, 47–54; Dennis, "Reconstructing Sociotechnical Order"; Hollinger, "Defense of Democracy";

DeJong-Lambert and Krementsov, "On Labels and Issues"; Kevles, "National Science Foundation"; Reingold, "Science and Government"; Wang, "Liberals."

45. Kohler, "Management of Science"; Kay, *Molecular Vision of Life.*

46. Discussion materials, NIH Group for Discussion of Research Administration Meeting, January 23, 1957, p. 1, in Baker papers, unsorted.

47. Ibid., pp. 2–3.

48. Ibid., pp. 4–5.

49. Ibid.

50. Ibid., p. 6.

51. Ibid., p. 4.

52. Ticks and check marks left by Baker on his copy of Memo: Book Collection, Division of Personnel (January 30, 1959), in Baker papers, Virus and Cancer Programs, Backup Memoranda 1950–1962, 1959–60.

53. Robert D. Livingston (NIMH), "The Evaluation of Scientific Endeavor Part I," in *Digest of Proceedings* for *Management Institute for Leaders in Scientific and Professional Programs*, May 25–June 3, 1959, pp. 47–48, Baker papers, NCI papers, 1950–64.

54. For example, at a seminar titled "Introductory Systems Analysis" offered at the NIH on November 26–30, 1962. See Baker papers, NCI papers, 1950–1964.

55. Dennis, "Accounting for Research"; Hounshell and Smith, *Science and Corporate Strategy*; Noble, *America by Design*, 110–67; Swann, *Academic Scientists and Pharmaceutical Industry.* The historian of technology David Edgerton, in his introduction to *Industrial Research and Innovation*, distinguishes between industrial *research* and industrial *innovation* in the twentieth century, especially in light of the weight given to economist Joseph Schumpeter's view that large firms are more innovative. Edgerton argues that the Schumpeterian faith in large firms and innovation is misplaced and that we need to know more about how industrial R&D was actually practiced, and Baker's exploration of these themes using systems analysis illuminates this point. Edgerton, introduction, x–xii.

56. For a partial listing of the impact of cybernetics on these fields and others, see, among others, Galison, "Ontology of the Enemy"; Heims, *Cybernetics Group*; Kay, "Cybernetics, Information, Life"; Pickering, *Cybernetic Brain*; Turner, *From Counterculture to Cyberculture.*

57. Amadae, *Rationalizing Capitalist Democracy*, 37–57; Fortun and Schweber, "Scientists and World War II"; Hounshell, "Cold War, RAND"; Stockfish, *Intellectual Foundations of Systems Analysis*; Thomas, *Rational Action.*

58. Amadae, *Rationalizing Capitalist Democracy*, 54–83; Anser, "Linear Model"; Johnson, "From Concurrency to Phased Planning," 98–102; David Novick, ed., *Program Budgeting; Program Analysis and Federal Budget* (Cambridge, MA: Harvard University Press, 1965).

59. Beniger, *Control Revolution*, 390–425; Edwards, *Closed World*, 125–33; Palmer, *McNamara Strategy and Vietnam War*, 39–77.

60. Jardini, "Out of the Blue Yonder," 334–43; Light, *From Warfare to Welfare.*

61. Lambright, *Powering Apollo*, 93–116.

62. Parker, "Space Age Management"; James E. Webb, *Space Age Management: The Large-Scale Approach* (New York: McGraw-Hill, 1969).

63. "Program Planning for Viruses and Cancer Research Expansion," memo by Carl Baker (?), July 10, 1961, in Baker papers, Virus and Cancer Programs, Backup Memoranda 1950–1962, 1961.

64. Baker draft memo, "Outline of NCI Requirements," October 23, 1963, in Baker papers, Virus and Cancer Programs, Backup Memoranda 1963–1964, 1963.

65. Carl Baker to Members of Human Cancer Virus Task Force, October 16, 1962, in Baker papers, Virus and Cancer Programs, Backup Memoranda 1950–1962, 1962.

66. Briefing Memo, "Comparison of the Polio Story with Developing Cancer Virus Research," March 21, 1963, p. 8, in Baker papers, Virus and Cancer Programs Backup Memoranda 1963–1964, 1963.

67. "Special Report: Virus Cancer Program" (for director, NCI), January 1962, in Baker papers, Virus and Cancer Programs, Backup Memoranda 1950–1962, 1962.

68. Draft of remarks on program planning, October 11, 1963 (Carrese or Baker), in Baker papers, Virus and Cancer Programs Backup Memoranda 1963–1964, 1963. Baker also drew attention to the importance of this new language as he wrote "Guidelines for Presentation of Institute Program Plans" in the fall of 1963. "For the purposes of this document," Baker began, "definitions given herein will have the meanings ascribed, irrespective of definitions given to these words in other sources." Coordinating the use of this new language was important for the discussion of planning goals, especially as the institute attempted to undertake a "full spectrum of research" ranging from "the exploratory" to the "large-scale." Carl Baker, draft memo, "Guidelines for Presentation of Institute Program Plans," September 30, 1963, in Baker papers, Virus and Cancer Programs Backup Memoranda 1963–1964, 1963.

69. Macfarlane Burnet, "Leukemia as a Problem in Preventive Medicine," *New England Journal of Medicine* 259 (August 28, 1958): 423–31, quote on 428.

70. P. R. J. Burch, "Human Cancer: Mendelian Inheritance or Vertical Transmission?," *Nature* 197, no. 4872 (March 16, 1963): 1042–45; Jerome T. Syverton and John D. Ross, "The Virus Theory of Leukemia," *American Journal of Medicine* 28, no. 5 (May 1960): 683–98.

71. In particular, Baker described program planning as encompassing "(a) the consideration of the total spectrum of available knowledge . . . (b) the interrelationship of this knowledge with . . . major problem areas . . . (c) the use of the objectives as the basis for action." Carl Baker draft memo, "Basic and Programmatic Research—Two Systems," November 29, 1963, pp. 2–3, in Baker papers, Virus and Cancer Programs Backup Memoranda 1963–1964, 1963.

72. Louis Mario Carrese, "Application for Federal Employment," in Baker papers, Virus and Cancer Programs Backup Memoranda 1950–1962, 1962.

73. Memo to members of Committee for Developing Guidelines for Presentation of Institute Program Plans, October 7, 1963, pp. 3–4, in Baker papers, Virus and Cancer Programs Backup Memoranda 1963–1964, 1963.

74. Suspending the question of *whether* PERT had made a decisive contribution to the Polaris program, it drew widespread attention. More than one thousand books and articles appeared on PERT by 1964. Sapolsky, *Polaris System Development*, 110–11. The air force adopted a similar set of techniques for developing its own intercontinental ballistic missiles. There was good reason to suspect that the link between the technique itself and the success of the development program may have been as much a product of public relations by the navy's Special Projects Office as of planning. Johnson, "From Concurrency to Phased Planning," 96–98.

75. However, it did not address "quantity, quality, or cost" aspects—as one introductory text cautioned. Robert W. Miller, *Schedule, Cost, and Profit Control with PERT: A Comprehensive Guide for Program Management* (New York: McGraw-Hill, 1963), 1–3.

76. Harry F. Evarts, *Introduction to PERT* (Boston: Allyn and Bacon, 1964), 1–6.

77. "Graphs and Scheduling," translated from *Revue Française de Recherche Opérationnelle* 6, no. 25 (1962), in Baker papers, Virus and Cancer Programs Backup Memoranda 1963–1964, 1963.

78. Chandler, *Visible Hand*; Yates, *Control through Communication*.

79. Bud, "Strategy in American Cancer Research."

80. Evarts, *Introduction to PERT*, 14–20.

81. Chart A, attached to memo entitled "Program Planning for Viruses and Cancer Research Expansion," July 10, 1961, in Baker papers, Virus and Cancer Programs, Backup Memoranda 1950–1962, 1961.

82. *Special Virus Leukemia Program Progress Report* 4 (1967), 7.

83. Gilman, *Mandarins of the Future*; Latham, *Modernization as Ideology*; Ekbladh, *Great American Mission*.

84. Louis Carrese and Carl G. Baker, "The Convergence Technique: A Method for the Planning and Programming of Research Efforts," *Management Science* 13, no. 8 (April 1, 1967): B424–28.

85. "Research Report #13: Research and Development," produced by Research Information Branch of the NCI for the National Association of Educational Broadcasters, 1966, in LION, AR003888.

86. Carrese and Baker, "Convergence Technique," B429–30.

87. Ibid., B435.

88. Ibid., B436–37.

89. Minutes, NACC subcommittee on carcinogenesis, July 23, 1964, p. 2, in R. Lee Clark papers, MS 070, series III, carton 8, folder 6, John P. McGovern Historical Collections and Research Center, Texas Medical Center Library (hereafter "Clark papers," with series, carton, and folder information).

90. *Departments of Labor and Health, Education, and Welfare Appropriations for 1964: Hearings on Part 3: National Institutes of Health, Day 3, Before Subcomm. on Departments of Labor and Health, Education, and Welfare and Related Agencies Appropriations*, 88th Cong. 352–54 (1963) (statement of Kenneth Endicott, director of the National Cancer Institute).

91. Director, NCI, to National Advisory Cancer Council, September 10, 1964, p. 1, in Clark papers, series III, carton 8, folder 6.

92. *Departments of Labor and Health, Education, and Welfare Appropriations for 1965: Hearings on Part 3: National Institutes of Health, Day 1, Before Subcomm. on Departments of Labor and Health, Education, and Welfare and Related Agencies Appropriations*, 88th Cong. 192–93 (1964) (statement of Kenneth Endicott, director of the National Cancer Institute).

93. Rosenfeld, "Superplan," 111.

94. Robert C. Toth, "Hope Raised for Cure of Leukemia: Recent Discoveries Indicate Viruses as Cause of Trouble," *Los Angeles Times*, November 29, 1964; Robert C. Toth, "Virus Link Found: Huge Leukemia Project Pushed," *Los Angeles Times*, November 30, 1964; "Leukemia Cure Near?," *Science News-Letter* 86, no. 14 (October 3, 1964): 215.

95. Thomas T. Fenton, "Leukemia's Cure Seen: Scientist Predicts Vaccine within 5 to 7 Years," *Baltimore Sun*, January 27, 1965.

96. Frank Rauscher, interviews for "Research Report #13: Research and Development," produced by Research Information Branch of the NCI for the National Association of Educational Broadcasters, 1966, in LION, AR003888.

97. A similar point about flowcharts for programming has been made in Ensmenger, "Multiple Meanings of a Flowchart." Historians of science have drawn attention to the importance of visual representation languages for "reifying" immaterial or inaccessible ideas. I thank David Kaiser for bringing this connection to my attention. For example, see Rudwick, "Emergence of a Visual Language"; Kaiser, *Drawing Theories Apart*, 318–55.

98. Stevenson, "Interview with Dr. Carl Baker," 8.

99. NIH Study Committee, *Biomedical Science and Its Administration: A Study of the National Institutes of Health* (Washington, DC: US Government Printing Office, 1965), 208. For more on the impact that the Wooldridge Report had on administration and organization within the National Institutes of Health, see Patel, "Methods and Management."

100. NIH Study Committee, *Biomedical Science and Its Administration*, 40.

101. Ibid., 43.

102. James Shannon memo, "Strengthening the Program Planning and Analysis Function within NIH Institutes and Program Divisions," June 25, 1965, in Baker papers, Virus and Cancer Programs, Backup Memoranda 1965–1966, 1965.

103. Baker memo, "Program Planning and Analysis Functions within the NCI," July 9, 1965, in Baker papers, Virus and Cancer Programs, Backup Memoranda 1965–1966, 1965.

104. "Minutes on Planning Conference," July 8, 1965, p. 2, in Baker papers, Virus and Cancer Programs Backup Memoranda 1965–1966, 1965.

105. Kenneth M. Endicott, "The Human Cancer Virus Task Force: A New Approach to Virus-Cancer Research," *CA: A Cancer Journal for Clinicians* 13, no. 4 (July 1, 1963): 156.

106. "Collaborative Programs, NCI," Committees—Wooldridge, filed March 31, 1964, pp. 70–72, in LION, 6400[illegible]04.

107. Erik Rau has made a similar point about operations research during the Second World War. Rau, "Combat Science."

108. Zubrod et al., "Chemotherapy Program."

109. Carl G. Baker, "Cancer Research Program Strategy and Planning—the Use of Contracts for Program Implementation," *Journal of the National Cancer Institute* 59, no. 2 (August 1977): 656, 668.

110. Mary Lasker to James Shannon, July 17, 1964, in Lasker papers, box 104, folder "Statement to Runia Committee (1966)."

111. Kenneth Endicott to Howard Skipper, May 4, 1965, in Baker papers, Virus and Cancer Programs Backup Memoranda 1965–1966, 1965.

112. Endicott also had assistance from the leadership of the NIH, which was also resistant to the idea of having its discretion over contracts curtailed by peer review. Shannon to Secretary of DHEW, October 25, 1965, in RG 443, UDoD6-1, box 93, RES 9-7 National Cancer Institute (NCI) 1937–1968, NARA II.

113. "Report of the Secretary's Advisory Committee on the Management of National Institutes of Health Research Contracts and Grants," March 1966, US Department of Health, Education, and Welfare, Washington, DC, 16.

114. Ibid., 2, 30.

115. Carl Baker, responses to "Survey of Scientist-Administrators at the National Institutes of Health, January 20, 1969, question 56, in Baker papers, unsorted.

116. This recalls another point by Derrida, that institutions, in their "founding acts" as well as in "the act as performance," must maintain themselves within the linguistic structures that they lay out. Derrida, "Declarations of Independence," 8. See also Giraudeau, "Performing Physiocracy."

117. Creager, *Life Atomic*; Kay, *Who Wrote Book of Life?*; Rader, "Hollaender's Postwar Vision for Biology"; Kevles, "Big Science and Big Politics."

118. Edwards, *Closed World*, 7–15.

119. Timothy Mitchell has raised a similar point as to how we should think about the governance of "the economy." Mitchell, *Rule of Experts*, 82.

Chapter Six

1. Special Virus Cancer Program, *SVCP Progress Report* 6 (1969), 24.

2. Scientists are quoted anonymously in Nicolas Wade's 1971 article criticizing the SVCP. Most of the critics, according to Wade, were unwilling to speak

on record for fear of having their funding cut by the NCI. Wade, "Special Virus Cancer Program," 1306.

3. ACS Historical Data on Cancer Virology Grants, 1946–1980, courtesy of ACS National Cancer Information Center, See p. 279 note 3.

4. This was all the more impressive considering that most cancer virus studies did not pay for clinical work, whereas the budget for chemotherapy (the leading field of expenditure) was high because it had to pay for the treatment of patients as well as for experimental investigations. Figures adjusted to 2017 dollars by the biomedical price research index. NCI Memo for the Record: Support for the Virus Cancer Program at the NCI, August 10, 1983, in LION, DC8308; *National Cancer Institute 1972 Fact Book* (Washington, DC: Department of Health, Education, and Welfare, 1971), 17.

5. Chubin and Studer, "Politics of Cancer"; Rettig, *Cancer Crusade*, 63–70; Wade, "Special Virus Cancer Program"; John Walsh, "The Nixon Administration: End of a Long Campaign," *Science*, 185, no. 4152 (1974): 675–78.

6. This resonates with later work on the construction of model or experimental systems in biology. For example, see Creager, Lunbeck, and Wise, *Science without Laws*; Daston, "The Coming into Being of Scientific Objects"; Kohler, *Lords of the Fly*, 1–17; Rheinberger, "Experimental Systems," 67–72.

7. Carl Baker, "Summary Report: Office of the Scientific Director for Etiology, NCI" (August 24, 1968?), pp. 4–6, in Baker papers, Virus and Cancer Programs, Backup Memoranda 1967–1969, 1968.

8. *Departments of Labor and Health, Education, and Welfare Appropriations for 1963: Hearings on Part 3: National Institutes of Health, Day 2, Before Subcomm. on Departments of Labor and Health, Education, and Welfare and Related Agencies Appropriations*, 87th Cong. 1062 (1962) (statement of Kenneth Endicott, director of National Cancer Institute).

9. Kenneth Endicott, interview transcript, October 30, 1964, p. 50, in Center for Oral History Archives, Columbia University Rare Books and Manuscript Library.

10. Summary Information on Research Contracts—Special Virus Leukemia Program, attachment 1 to memo entitled "Program Materials," from Frank Rauscher and Louis Carrese, August 3, 1965, in Baker papers, Virus and Cancer Programs Backup Memoranda 1965–1966, 1965. These 1965 dollar amounts, if adjusted by the 2017 Biomedical Research and Development Price Index, would be about eleven times greater—e.g., $800,000 would be $8.8 million in 2017 dollars.

11. Ibid.

12. "Ground Broken for New NCI Facility," *NIH Record* 19, no. 16 (August 9, 1966): 1, 5.

13. "Fact Sheet: Fort Detrick," by the NCI (January 1972), NIH Office of the Director Files, INTRA 10-2 #335872. For press coverage, see Peter Osnos, "Mathias Urges Ft. Detrick Be Given 'National Mission,'" *Washington Post*, January 10, 1970, sec. general; John H. Douglas, "Ft. Detrick: From 'Doomsday Bug' to

Cancer," *Science News* 104, no. 4 (July 28, 1973): 56–57; Lawrence Meyer, "Cancer Research Lab Planned at Ft. Detrick," *Washington Post*, October 19, 1971.

14. Harold Schmeck, "Litton to Run Cancer Research Lab," *New York Times*, June 25, 1972. See also Litton's response to the call for management teams, which focused on its capacity to work with PERT, in Baker papers, Viruses and Cancer Backup Memoranda 1970–1971, 1971.

15. Report on Litton Bionetics Inc. operation of Frederick Cancer Research Facility, compiled by Surveys and Investigations Staff of the House Appropriations Subcommittee on Labor, Health, Education, and Welfare, 1978, 8–10, contained within *Department of Labor and Health, Education, and Welfare Appropriations Act for 1980: Hearings, Day 1, Before H.R. Sub. Comm. on Appropriations*, 96th Cong. 360–424 (1979) (statement of Arthur Upton, director of NCI).

16. *Special Leukemia Virus Program Progress Report 4* (FY 1967), 17, 22, 24, 39.

17. This did not mean that there were not such attempts, only that they became more controversial, as demonstrated by a 1964 suit by patients at New York's Jewish Chronic Disease Hospital against Chester Southam, an immunologist who had attempted to transplant cancer cells into terminally ill patients. Langer, "Human Experimentation." On the use of cancer patients as experimental subjects, and the unexpectedly long tenure of these practices, see Kutcher, *Contested Medicine*.

18. The SVLP also continued to engage in widespread tissue collection from "inbred" human groups and isolated human populations. Account of contract NCI-VCL-68-42 for CDC as well as accounts of contract PH43-64-9247 to Tulane University, contract PH43-65-97 to the Hospital for Sick Children, and contract PH43-68-1389 to the University of Padua. *Special Virus Cancer Program Progress Report 6* (1969), 39, 135, 142, 145, 160–61.

19. Denis Burkitt, "A Sarcoma Involving the Jaws in African Children," *British Journal of Surgery* 46, no. 197 (1958): 218–23; Denis Burkitt, "A 'Tumour Safari' in East and Central Africa," *British Journal of Cancer* 16, no. 3 (September 1962): 379–86; Denis Burkitt, "A Children's Cancer Dependent on Climatic Factors," *Nature* 194, no. 4825 (April 1962): 234. The Rockefeller Foundation had sponsored extensive studies of yellow fever epidemiology in East Africa during the 1930s. *Rockefeller Foundation Annual Report* (1949), 95–98, in Rockefeller Foundation Governance Reports, accessed August 1, 2018, https://assets.rockefellerfoundation.org/app/uploads/20150530122119/Annual-Report-1933.pdf.

20. M. A. Epstein, B. G. Achong, and Y. M. Barr, "Virus Particles in Cultured Lymphoblasts from Burkitt's Lymphoma," *Lancet* 283, no. 7335 (March 28, 1964): 702–3.

21. Gilbert Dalldorf, "Lymphomas of African Children: With Different Forms or Environmental Influences," *Journal of the American Medical Association* 181, no. 12 (September 22, 1962): 1026–28; G. Dalldorf, C. A. Linsell, F. E. Barnhart, and R. Martyn, "An Epidemiologic Approach to the Lymphomas of African

Children and Burkitt's Sarcoma of the Jaws," *Perspectives in Biology and Medicine* 7 (1964): 435–49. These findings were disputed later in the 1960s.

22. Peyton Rous to Gilbert Dalldorf, December 17, 1962, in Rous papers, series I, Sloan Kettering Institute for Cancer Research 2. This criticism reflects long-standing debates in American biology over the uses of "the lab" and "the field" as sites for knowledge production. Kuklick and Kohler, "Introduction"; Kohler, *Landscapes and Labscapes.*

23. Masnyk, "Activities of National Cancer Institute."

24. At this time there was still a widespread attitude that cancer in general was not a problem in "under-developed" continents such as Africa; it was a product of theories of cancer as a "disease of civilization," biological theories of racial "primitive immunity," and the varied reach of cancer registries. Livingston, *Improvising Medicine*, 38–40; Wailoo, *How Cancer Crossed Color Line*, 40–46.

25. Denis Burkitt, "Chemotherapy of African (Burkitt) Lymphoma—Clinical Evidence Suggesting an Immunological Response," *British Journal of Surgery* 54, no. 10 (1967): 817–19; R. H. Morrow, M. C. Pike, and A. Kisuule, "Survival of Burkitt's Lymphoma Patients in Mulago Hospital, Uganda," *British Medical Journal* 4, no. 5575 (November 11, 1967): 323–27; Bernard Glemser, *Mr. Burkitt and Africa* (New York: World, 1970), 132–40.

26. I will explore the development of this program in a future article. Julia Cummiskey and Marissa Mika have already examined how Uganda, in particular Makerere Hospital (and later the Uganda Cancer Institute) and the East African Virus Research Institute (now the Uganda Virus Research Institute), operated as sites of biomedical research. See Mika, "Fifty Years of Creativity"; Mika, "Research Is Our Resource"; Stultiens and Mika, *Staying Alive*; Cummiskey, "Placing Global Science in Africa."

27. "Summary Report: Associate Scientific Director for Clinical Trials—Chemotherapy," in NCI Annual Report, 1967–68, 1:154.

28. The annual value of the contract was roughly $250,000. Olweny, "Uganda Cancer Institute."

29. John Ziegler, interview by author, April 17, 2013 (Global Health Program, University of California, San Francisco).

30. Peter Gunvén, George Klein, Gertrude Henle, Werner Henle, and Peter Clifford, "Epstein-Barr Virus in Burkitt's Lymphoma and Nasopharyngeal Carcinoma: EBV DNA in Biopsies of Burkitt Tumours and Anaplastic Carcinomas of the Nasopharynx," *Nature* 228 (December 12, 1970): 1054.

31. Gertrude Henle, Werner Henle, Peter Clifford, Volker Diehl, George W. Kafuko, Barnabas G. Kirya, George Klein, et al., "Antibodies to Epstein-Barr Virus in Burkitt's Lymphoma and Control Groups," *Journal of the National Cancer Institute* 43, no. 5 (November 1, 1969): 1148.

32. "Report of the Associate Scientific Director for Viral Oncology," in NCI Annual Report, 1967–68, 1:912–13.

33. Frank Rauscher, "Memo: Meeting to Discuss the Possible Role of a Herpes-Type Virus (EBV) as an Inducer of Burkitt Lymphoma and Other Neoplasms of Man," May 24, 1968, in Albert Sabin papers, Cancer series, box 2, NCI Documents 1967–68, Harry R. Winkler Center for the History of the Health Professions, University of Cincinnati.

34. John Higginson, "International Research: Its Role in Environmental Biology," *Science* 170, no. 3961 (November 27, 1970): 935, 937, 939.

35. *Special Cancer Virus Program Progress Report* 7 (1970), 33.

36. *Proceedings of a Planning Conference for Epidemiological Studies of Burkitt Lymphoma and Infectious Mononucleosis* (1968), 12, 16, 17–18, 20, in Werner and Gertrude Henle papers, box 18, Burkitt Lymphoma Prospective Studies 1968–69, National Library of Medicine—History of Medicine Division (hereafter "Henle papers," with box and folder information).

37. Summary Sheet for IARC Contract, p. 7, in Joseph Melnick papers (#15), NACC box 3, NCI Mtg. April 21, 1969, John P. McGovern Historical Collections and Research Center, Texas Medical Center Library (hereafter "Melnick papers," with box and folder information).

38. Minutes of 46th Regularly Scheduled Meeting of the Program Segment Chairmen, January 9, 1969, pp. 9–10, in RG 443, UDUP-7, container 1 (FRC), Segment Chair Meeting Minutes, NARA II.

39. *Special Virus Cancer Program Progress Report* 9 (1971), 131–32. Several years later, the contract amounted to $3.6 million (in 2017 USD) annually. Report for Contract No1-CP4-3273, in NCI Annual Report, 1976, 3:1658.

40. Gertrude Henle and Werner Henle, "Immunofluorescence in Cells Derived from Burkitt's Lymphoma," *Journal of Bacteriology* 91, no. 3 (March 1966): 1249.

41. Alfred S. Evans, James C. Niederman, and Robert W. McCollum, "Sero-epidemiologic Studies of Infectious Mononucleosis with EB Virus," *New England Journal of Medicine* 279, no. 21 (1968): 1121–27; Gertrude Henle, Werner Henle, and Volker Diehl, "Relation of Burkitt's Tumor-Associated Herpes-Type Virus to Infectious Mononucleosis," *Proceedings of the National Academy of Sciences of the United States of America* 59, no. 1 (January 15, 1968): 94–101; "Does EB Virus Cause Infectious Mononucleosis?," *New England Journal of Medicine* 279, no. 21 (1968): 1168–69.

42. The SVLP was renamed the Special Cancer Virus Program in 1968, to reflect its interest in tumors responsible for solid tumors as well as blood cancers such as leukemia. *Special Virus Cancer Program Progress Report* 6 (1969), 127, in RG 443, UD0D6-1, box 96, folder RES 9-7-D-1, NARA II.

43. Guy Newell, Memorandum for the Record, February 20, 1969, in RG 443, UDUP-7, container 1 (FRC), Segment Chair Meeting Minutes, NARA II; A. Geser, N. E. Day, G. B. de-Thé, B. K. Chew, R. J. Freund, H. C. Kwan, M. F. Lavoue, D. Simkovic, and R. Sohier, "The Variability in Immunofluorescent Viral Capsid

Antigen Antibody Tests in Population Surveys of Epstein-Barr Virus Infections," *Bulletin of the World Health Organization* 50, no. 5 (1974): 389–90.

44. See records held in the Henles' correspondence. Henle papers, box 18, Burkitt lymphoma Patient Cases, 1968–1974.

45. Harald zur Hausen, Heinrich Schulte-Holthausen, Werner Henle, Gertrude Henle, Peter Clifford, and Lars Santeson, "Epstein-Barr Virus in Burkitt's Lymphoma and Nasopharyngeal Carcinoma: EBV DNA in Biopsies of Burkitt Tumours and Anaplastic Carcinomas of the Nasopharynx," *Nature* 228 (December 12, 1970): 1056–58.

46. Werner Henle to Frank Rauscher, July 15, 1970, in Henle papers, box 14, Rauscher, Frank J.

47. Harald zur Hausen, "Biochemical Approaches to Detection of Epstein-Barr Virus in Human Tumors," *Cancer Research* 36, no. 2, part 2 (February 1, 1976): 678–80; Matthias Durst, Lutz Gissmann, Hans Ikenberg, and Harald zur Hausen, "A Papillomavirus DNA from a Cervical Carcinoma and Its Prevalence in Cancer Biopsy Samples from Different Geographic Regions," *Proceedings of the National Academy of Sciences of the United States of America* 80, no. 12 (June 15, 1983): 3812–15; Harald zur Hausen, "Papillomaviruses in Anogenital Cancer as a Model to Understand the Role of Viruses in Human Cancers," *Cancer Research* 49, no. 17 (September 1, 1989): 4677–81; zur Hausen, "Viruses in Human Tumors."

48. Some recent scholars to consider, in addition to those working on the study of cancer and Burkitt lymphoma, are Radin, "Latent Life"; Crane, *Scrambling for Africa*; Kowal, Radin, and Reardon, "Indigenous Body Parts"; Tilley, *Africa as a Living Laboratory*.

49. One active researcher termed this a "phenomenological" view of viruses, in contrast to a mechanistic or chemical approach. Ray Gilden, interview by author, January 3, 2014 (Skype).

50. Maurice Hilleman, "Problems and Potentials for Human Viral Cancer Vaccines," *Preventive Medicine* 1, no. 3 (August 1972): 352–70.

51. *Special Virus Cancer Program Progress Report* 7 (1970), 28–29.

52. Robert J. Huebner, "70 Newly Recognized Viruses in Man," *Public Health Reports* 74, no. 1 (January 1, 1959): 6–12; W. P. Rowe, R. J. Huebner, L. K. Gilmore, R. H. Parrott, and T. G. Ward, "Isolation of a Cytopathogenic Agent from Human Adenoids Undergoing Spontaneous Degeneration in Tissue Culture," *Proceedings of the Society for Experimental Biology and Medicine* 84, no. 3 (December 1953): 570–73.

53. Wallace P. Rowe, Janet W. Hartley, John D. Estes, and Robert J. Huebner, "Studies of Mouse Polyoma Virus Infection. I. Procedures for Quantitation and Detection of Virus," *Journal of Experimental Medicine* 109, no. 4 (April 1, 1959): 379–91; Wallace P. Rowe, Janet W. Hartley, Isadore Brodsky, and Robert J. Huebner, "Complement Fixation with a Mouse Tumor Virus (S.E. Polyoma)," *Science*, 128, no. 3335 (November 28, 1958): 1339–40.

54. Andrew Lewis, interview transcript, July 30, 1975, p. 7, in Recombinant DNA Collection (MC 100), Massachusetts Institute of Technology Archives and Special Collections, Cambridge, MA.

55. Robert J. Huebner, Wallace P. Rowe, and William T. Lane, "Oncogenic Effects in Hamsters of Human Adenovirus Types 12 and 18," *Proceedings of the National Academy of Sciences of the United States of America* 48, no. 12 (December 15, 1962): 2051–58.

56. Robert J. Huebner, Gary J. Kelloff, Padman S. Sarma, William T. Lane, Horace C. Turner, Raymond V. Gilden, Stephen Oroszlan, Hans Meyier, David D. Myers, and Robert L. Peters, "Group-Specific Antigen Expression during Embryogenesis of the Genome of the C-Type RNA Tumor Virus: Implications for Ontogenesis and Oncogenesis," *Proceedings of the National Academy of Sciences* 67, no. 1 (September 1, 1970): 366–76; Raymond V. Gilden and Stephen Oroszlan, "Group-Specific Antigens of RNA Tumor Viruses as Markers for Subinfectious Expression of the RNA Virus Genome," *Proceedings of the National Academy of Sciences* 69, no. 4 (April 1, 1972): 1021–25.

57. Ludwik Gross, "The 'Vertical' Transmission of Mouse Mammary Carcinoma and Chicken Leukemia: Its Possible Implications for Human Pathology," *Cancer* 4, no. 3 (May 1, 1951): 626–33.

58. Robert Huebner to Ray Gilden, memo, March 29, 1971, in Beeman/Huebner papers (Acc. 2006-001, unprocessed), box 1, folder 11, National Library of Medicine—History of Medicine Division.

59. The mechanism of oncogene "repression" drew upon the "operon" model of genetic regulation proposed by Jacques Monod and François Jacob. Robert J. Huebner, and George J. Todaro, in "Oncogenes of RNA Tumor Viruses as Determinants of Cancer," *Proceedings of the National Academy of Sciences of the United States of America* 64, no. 3 (November 15, 1969): 1087–90.

60. Boveri, *Origin of Malignant Tumors*.

61. Kevles, *In the Name of Eugenics*, 241–45. For a recent account of the rise of medical genetics, see Hogan, *Life Histories of Genetic Disease*.

62. Wapner, *Philadelphia Chromosome*, 22.

63. Cantor, "Frustrations of Families," 295–301.

64. Keyes, "Prion Challenge to 'Central Dogma.'"

65. Robert Huebner, "Memo: A Return to Galenic Thinking?," March 11, 1971, in Baker papers, Virus and Cancer Programs Backup Memoranda 1970–1971, 1971.

66. "A Cure for Cancer May Be Found in Gene," *New Castle News*, February 2, 1970. This was from a wire report, indicating that the pronouncement appeared in other papers as well.

67. Kevles, "Howard Temin," 250–58.

68. Howard M. Temin, "Nature of the Provirus of Rous Sarcoma," *National Cancer Institute Monograph* 17 (1964): 557, 561–66; Kay, *Who Wrote Book of Life?*, 128–92.

69. Howard M. Temin, "Genetic and Possible Biochemical Mechanisms in Viral Carcinogenesis," *Cancer Research* 26, no. 2, part 1 (February 1, 1966): 215.

70. Howard M. Temin, "Homology between RNA from Rous Sarcoma Virus and DNA from Rous Sarcoma Virus-Infected Cells," *Proceedings of the National Academy of Sciences of the United States of America* 52, no. 2 (August 15, 1964): 323–29.

71. Galperin, "Virus, provirus et cancer," 36–46; Kevles, "Howard Temin." Other authors, however, have disputed the claim that the heretical nature of Temin's claim led to its rejection, asserting, respectively, that the central dogma was not firmly established in virology and that Temin's experiments demonstrating the existence of reverse transcriptase were in fact inconclusive. These authors suggest that the story of Temin's "heresy" owes more to the disciplinary aspirations of retrovirus research in the 1970s. Fisher, "Not beyond Reasonable Doubt"; Marcum, "From Heresy to Dogma," 173–74. For more on the technique of RNA-DNA hybridization, see Giacomoni, "Origin of DNA:RNA Hybridization"; S. Spiegelman and I. Haruna, "Problems of an RNA Genome Operating in a DNA-Dominated Biological Universe," *Journal of General Physiology* 49, no. 6 (July 1, 1966): 263–304.

72. Howard Temin's remarks upon receiving the US Steel Award for Molecular Biology at the National Academy of Sciences, April 24, 1972, in Howard Temin papers, Acc. 2013-234, box "Seminars, 1971–1972," folder "National Academy of Sciences April 24, 1972," University of Wisconsin–Madison University Archives.

73. In a 1978 interview, David Baltimore claimed that Temin's announcement that he had identified the provirus in 1964 had done his work an "enormous disservice" by prematurely claiming evidence that did not support the theory. Chubin and Studer, *Cancer Mission*, 139.

74. James D. Watson, *Molecular Biology of the Gene* (New York: W. A. Benjamin, 1965), 298.

75. David Baltimore, interview by author, January 6, 2014 (Skype); Chubin and Studer, *Cancer Mission*, 136–37.

76. David Baltimore, "Viral RNA-Dependent DNA Polymerase: RNA-Dependent DNA Polymerase in Virions of RNA Tumour Viruses," *Nature* 226 (June 27, 1970): 1209–11; Howard M. Temin and Satoshi Mizutani, "Viral RNA-Dependent DNA Polymerase: RNA-Dependent DNA Polymerase in Virions of Rous Sarcoma Virus," *Nature* 226 (June 27, 1970): 1211–13.

77. "Summary Report, Office of Viral Oncology," in NCI Annual Report, 1972, 2:617.

78. S. Spiegelman, A. Burny, M. R. Das, J. Keydar, J. Scholm, M. Travnicek, and K. Watson, "Characterization of the Products of RNA-Directed DNA Polymerases in Oncogenic RNA Viruses," *Nature* 227, no. 5258 (August 8, 1970): 567.

79. "Cancer Viruses: More of the Same," *Nature* 227 (August 29, 1970): 887.

80. NCI Annual Report, 1971, 2:1200–1201.

81. Gilden interview; Murray Gardner, interview by author, January 14, 2014 (Skype); Robert Gallo, interview by author, November 13, 2013 (telephone).

82. James Watson, "Getting Realistic about Cancer," *Cold Spring Harbor Laboratory Director's Report* (1974), 1, in James Watson papers, section 2, series 4, subseries 1, Cold Spring Harbor Laboratory Library and Archives. Online Repository, accessed April 9, 2013, http://libgallery.cshl.edu/items/show/52313 (hereafter, "Watson papers Online," with section, series, subseries, and URL).

83. NCI Annual Report, 1972, 2:1133–79.

84. "Cancer Viruses," 887.

85. "Happy Birthday, Reverse Transcriptase?," *Nature New Biology* 231 (June 9, 1971): 161.

86. Kaiser, "Booms, Busts"; Rasmussen, "Mid-Century Biophysics Bubble."

87. Peck, *Reinventing Free Labor*, 8, 33, 47.

Chapter Seven

1. See correspondence in Richard Nixon Presidential Papers, White House Central Files—Subject Files, Federal Aid 6, box 29; and White House Central Files—Subject Files, HE 1–3, box 9. Richard Nixon Presidential Library and Museum (hereafter "Nixon papers," with collection, box, and folder information).

2. "National Cancer Program: The Strategic Plan," January 1973, Department of Health, Education, and Welfare, Public Health Service, Bethesda, MD, III-6.

3. See the discussion in chapters 4 and 5 as well as Hollinger, "Free Enterprise and Free Inquiry."

4. I borrow this idea of molecular biologists as a nascent political community from Marks, *Progress of Experiment*, 3–4.

5. Cairns, Stent, and Watson, *Phage and Molecular Biology*. For a broader discussion of how biochemists and molecular biologists viewed themselves, see a discussion of many of the autobiographical materials they produced in the 1980s: Abir-Am, "Review." For secondary sources on how molecular biology became a discipline in the 1960s, see Abir-Am, "Legitimation of New Scientific Disciplines"; de Chadarevian, *Designs for Life*, 199–259; Strasser, "Institutionalizing Molecular Biology."

6. Gunther S. Stent, "That Was the Molecular Biology That Was," *Science*, 160, no. 3826 (April 26, 1968): 390–95.

7. Rheinberger, "Recent Science and Its Exploration"; Yi, "Cancer, Viruses, and Mass Migration."

8. Vettel, *Biotech*, 134–36.

9. President Lyndon B. Johnson, remarks following a meeting with directors of the National Institutes of Health, June 27, 1966, in Subcomm. on Government

Research, Research in the Service of Man: Biomedical Knowledge, Development, and Use, S. Doc. No. 90-55, at 4–5 (1966) (remarks entitled "Application of Biomedical Research: The White House View").

10. See, for example, Stokes, *Pasteur's Quadrant*. The meanings of terms such as "basic," "pure," and "fundamental" change depending on the context. Clarke, "Pure Science with a Practical Aim"; Gooday, "'Vague and Artificial'"; Bud, "'Applied Science.'"

11. Salvador Luria to Senator Edward M. Kennedy, September 24, 1969, in Salvador Luria papers, series I, Kennedy, Edward M. (1965–1977), American Philosophical Library. For the "study life" movement, see Vettel, *Biotech*, 99–129.

12. D. S. Greenberg, "NIH: Fountain Committee Issues Bitter Attack on Programs," *Science*, 158, no. 3801 (November 3, 1967): 611–14.

13. Dr. W. C. Davidson, as quoted in H.R. Rep. No. 90-800, at 45 (1967) (The Administration of Research Grants in the Public Health Service).

14. Chalmers W. Sherwin and Raymond S. Isenson, "Project Hindsight," *Science* 156, no. 3782 (June 23, 1967): 1577.

15. The results of this study were not completed and released until nearly a decade later, too late to influence these debates. Julius H. Comroe and Robert D. Dripps, "Scientific Basis for the Support of Biomedical Science," *Science* 192, no. 4235 (April 9, 1976): 105–11.

16. In these comparisons, public relations triumphed over substance. The management of the "moonshot" was also beset by controversies and setbacks, as was the Program Evaluation Review Technique used in the Polaris missile program that inspired Carl Baker when he designed the convergence technique. However, these flaws were not apparent to many observers at the time. Lambright, *Powering Apollo*; Levine, *Managing NASA in Apollo Era*; Parker, "Space Age Management"; Slotkin, *Doing the Impossible*; James E. Webb, *Space Age Management: The Large-Scale Approach* (New York: McGraw-Hill, 1969).

17. James A. Shannon, "Science and Social Purpose," *Science*, 163, no. 3869 (February 21, 1969): 769, 771, 773.

18. As quoted and related in Vettel, *Biotech*, 138–44.

19. Drew, "Health Syndicate"; Rettig, *Cancer Crusade*, 29–30.

20. Quinn was also arguing against the official ACS stance that regulations issued by the IRS threatened the tax-exempt status of organizations that engaged in direct legislative lobbying. Another factor was that the ACS, a doctor's organization, had always been warier of government intervention than Lasker and her allies were. Luke Quinn to Mary Lasker, January 2, 1969, in Lasker papers, carton 494, unlabeled folder; briefing memo from Quinn to ACS Legislative Committee, February 24, 1969, in Lasker papers, carton 493, folder Col Luke Quinn (i).

21. Garb, *Cure for Cancer*, 13, 303.

22. Ibid., 296–97.

23. See correspondence in Lasker papers, carton 185, folder Dr. Garb.

24. More recent scholarship has revealed the presence of earlier grassroots activism around federal cancer policy, especially that of the Food and Drug Administration and NCI's policies on the testing and regulation of "alternative" cancer cures such as the compound Krebizoen in the 1960s. This activism started to articulate the idea of a citizen's right to expect federal action on the problem of cancer, but unlike the effort that Lasker and Garb launched in the late 1960s, those involved in this earlier wave of protest focused on how the federal government was blocking promising cures from wider distribution. Cantor, "Cancer, Quackery and Hope"; Wailoo, *How Cancer Crossed Color Line*, 122–29.

25. Quinn to Lasker, October 7, 1970, in Lasker papers, carton 494, folder unlabeled.

26. R. Shay to [??] Hopkins, May 5, 1971, in Nixon papers, series Federal Aid 6, box 30, folder Public Health 5-1-71–8-31-71.

27. Rettig, *Cancer Crusade*, 77–83.

28. Oppel, "Schmidt, Financier, Dead at 86."

29. Rettig, *Cancer Crusade*, 90–97.

30. Robert J. Bazell, "Cancer Research: Senate Consultants Likely to Push for Planned Assault," *Science*, 170, no. 3955 (October 16, 1970): 304–5; Rettig, *Cancer Crusade*, 88–97.

31. Bazell, "Cancer Research," 304.

32. Cairns, Stent, and Watson, *Phage and Molecular Biology*. Also see Helvoort, "Construction of Bacteriophage"; Summers, "How Bacteriophage Came to Be Used."

33. Renato Dulbecco, "The Plaque Technique and the Development of Quantitative Animal Virology," in Cairns, Stent, and Watson, *Phage and Molecular Biology*, 287–91; Renato Dulbecco and Marguerite Vogt, "Plaque Formation and Isolation of Pure Lines with Poliomyelitis Viruses," *Journal of Experimental Medicine* 99, no. 2 (February 1, 1954): 167–82; Renato Dulbecco, "Production of Plaques in Monolayer Tissue Cultures by Single Particles of an Animal Virus," *Proceedings of the National Academy of Sciences of the United States of America* 38, no. 8 (August 15, 1952): 747–52; Kevles, "Dulbecco and New Animal Virology," 434–39; Kevles and Geison, "Experimental Life Sciences," 118–19.

34. Harry Rubin, "Quantitative Tumor Virology," in Cairns, Stent, and Watson, *Phage and Molecular Biology*, 290–300; Howard M. Temin and Harry Rubin, "Characteristics of an Assay for Rous Sarcoma Virus and Rous Sarcoma Cells in Tissue Culture," *Virology* 6, no. 3 (December 1958): 669–88; Creager and Gaudillière, "Experimental Technologies and Techniques," 226–29; Kevles, "Howard Temin," 250–53.

35. Galperin, "Virus, provirus et cancer," 36–46; Peyrieras and Morange, "Study of Lysogeny."

36. S. E. Luria, "Viruses, Cancer Cells, and the Genetic Concept of Virus Infection," *Cancer Research* 20 (1960): 686.

37. James D. Watson, *Molecular Biology of the Gene* (New York: W. A. Benjamin, 1965), 468.

38. F. H. Crick, "Molecular Biology and Medical Research," *Journal of the Mount Sinai Hospital* 36, no. 3 (1969): 178–88.

39. Paul Berg to Sol Spiegelman, May 13, 1969, in Paul Berg papers, box 2, Correspondence Folder 1969 P–Z, Stanford University Special Collections; Spiegelman to Senator Charles Mathias, February 22, 1971, in *Conquest of Cancer Act: Hearings on S.34 and S.1828, Day 2, Before Subcomm. on Health of Committee on Labor and Public Welfare*, 92nd Cong. 180–81 (1971) (letter attached to statement by Senator Charles Mathias).

40. Appel, *Shaping Biology*, 252–55.

41. J. D. Watson to Mathilde Krim, July 9, 1970, in Clark papers, series III, box 70, folder 3.

42. Dr. Maurice Black to Senator Jacob K. Javits (R-NY), May 10, 1971, in Office of the President: R. Lee Clark papers [Microfilm], "Panel of Consultants May 1971," Historical Resources Center, Research Medical Library, MD Anderson Cancer Center (hereafter "Clark papers, microfilm").

43. Harry Rubin to Krim, July 8, 1970; Frank L. Horsfall to Krim, July 22, 1970; Albert Sabin to Krim, July 10, 1970; all three in Clark papers, series III, box 70, folder 3. Sabin had made similar points earlier in the 1960s: Albert B. Sabin, "Collaboration for Accelerating Progress in Medical Research," *Science*, 156, no. 3782 (June 23, 1967): 1568–71.

44. S. Rep. No. 91-1402, at 1–4, 31 (1970) (Report of the National Panel of Consultants on the Conquest of Cancer).

45. Yarborough, a liberal Democrat, had been defeated in a primary challenge by Lloyd Bentsen, a member of the Texas Democratic Party's conservative wing. As a result, his term of office ended in January 1971, when his seat was filled by Bentsen, who had prevailed over George H. W. Bush in the general election. Rettig, *Cancer Crusade*, 87, 102.

46. James Cavanaugh (White House) to Jonathan Moore (DHEW), April 23, 1971, in Nixon papers, White House Central Files—SMOF, William Timmons, box 56, Cancer-Cure Program, 1 of 2.

47. Richard Nixon, "Cancer-Cure Research," *Weekly Compilation of Presidential Documents* 7, no. 22 (May 31, 1971): 822. For political background, see "Nixon and Kennedy Peddle Rival Cancer Cures," *Nature* 229 (February 26, 1971): 590–92; Perlstein, *Nixonland*, 543–44; Rettig, *Cancer Crusade*, 117–21; MacLean, *Freedom Is Not Enough*, 95–97.

48. Quoted in *Conquest of Cancer Act: Hearings, Day 3*, 92nd Cong. 307 (1971) (Nixon Budget Signing Remarks, May 11, 1971).

49. "Conquest of Cancer" speech by Benno C. Schmidt at Krebs Auditorium, Memorial Sloan Kettering Cancer Center, April 21, 1971, in Clark papers, microfilm, "Panel of Consultants May 1971," emphasis in original.

50. "Senate Resolution 376—An Editorial," *Cancer Research* 31, no. 2 (February 1, 1971): i.

51. *Conquest of Cancer Act: Hearings on S.34 and S.1828, Day 1, Before Sub-comm. on Health of Committee on Labor and Public Welfare*, 92nd Cong. 30–31 (1971) (statement of Jesse Leonard Steinfeld, surgeon general of the US Public Health Service).

52. *National Cancer Attack Act of 1971: Hearings on H.R. 8343, H.R. 10681, Day 10, Before Subcomm. on Public Health and the Environment of Committee on Interstate and Foreign Commerce*, 92nd Cong. 678 (1971) (statement of Robert Hill, American Society of Biological Chemists).

53. Rettig, *Cancer Crusade*, 194–201; Lasker Oral History, part 2, sec. 10, p. 325.

54. Robert Huebner to Michael Gorman, December 4, 1969, in Lasker papers, carton 111, folder Cancer Commission 1969.

55. Huebner to Lloyd Shearer, November 18, 1969, in Lasker papers, carton 111, folder Cancer Commission 1969.

56. George J. Todaro and Robert J. Huebner, "The Viral Oncogene Hypothesis: New Evidence," *Proceedings of the National Academy of Sciences* 69, no. 4 (April 1, 1972): 1009–15.

57. Elizabeth S. Priori, L. Dmochowski, Brooks Myers, and J. R. Wilbur, "Constant Production of Type C Virus Particles in a Continuous Tissue Culture Derived from Pleural Effusion Cells of a Lymphoma Patient," *Nature New Biology* 232, no. 28 (July 14, 1971): 61–62. For media coverage sampling, see the press clippings books maintained by the Memorial Sloan Kettering Cancer Center, in Memorial Sloan Kettering Cancer Center papers, 375.1.3, microfilm reel 12, vol. 1971, Rockefeller Archive Center.

58. "First Virus Seen Likely to Cause Cancer in Humans Is Being Studied by Scientists," *Wall Street Journal*, December 6, 1971.

59. "Next Step Stalking Cancer Virus Will Take Months and $1 Million," *Houston Chronicle*, December 6, 1971, in Clark papers, microfilm, "Type-C Virus (ESP-1) 6.30.1(P)"; Spiegelman to Mathias, February 22, 1971, in *Conquest of Cancer Act: Hearings, Day* 2, 92nd Cong. 180–181 (1971) (letter attached to statement by Senator Charles Mathias); Berg to Spiegelman, May 13, 1969, in Paul Berg papers, box 2, Correspondence Folder 1969 P–Z, Stanford University Special Collections.

60. "Means Sought to Stop Cancer Virus Scramble," *Eugene Register-Guard* (OR), December 16, 1971; "More Candidates for a Human Cancer Virus," *Science News* 100, no. 24 (December 11, 1971): 388–89.

61. "Of Mice and Men," *Nature* 233 (September 10, 1971): 88; Raymond V. Gilden, Wade P. Parks, Robert J. Huebner, and George J. Todaro, "Murine Leukaemia Virus Group-Specific Antigen in the C-Type Virus–Containing Human Cell Line, ESP-1," *Nature* 233 (September 10, 1971): 102–3; Olson, *Making Cancer History*, 137–38.

62. J. Schlom, S. Spiegelman, and D. Moore, "RNA-Dependent DNA Polymerase Activity in Virus-Like Particles Isolated from Human Milk," *Nature* 231, no. 5298 (May 14, 1971): 97–100.

63. Ludwik Gross, "Is Cancer a Communicable Disease?," *Cancer Research* 4 (May 1944): 293–303.

64. Nicholas Wade, "Scientists and the Press: Cancer Scare Story That Wasn't," *Science* 174, no. 4010 (November 12, 1971): 679–80.

65. Nicholas Wade, "The Great Race: Virus Find Awakens Hope for Sufferers," *Science* 174, no. 4014 (December 10, 1971): 1112.

66. Nicholas Wade, "Cancer Politics: NIH Backers Mount Late Defense in House," *Science* 174, no. 4005 (October 8, 1971): 127–31; Rettig, *Cancer Crusade*, xi–xii.

67. Rettig, *Cancer Crusade*, 98.

Chapter Eight

1. For background on the Center for Cancer Research, see Herbert Black, "Biologist Luria to Direct Efforts: MIT to Get $6m Cancer Research Unit," *Boston Globe*, December 5, 1972.

2. Manuscript copy of speech, "The Academic Community and Cancer Research," given at dedication of Seeley G. Mudd Building, March 6, 1975, pp. 6–8, 13. Indeed, it appears that Watson spoke too honestly to what he assumed was a small gathering of like-minded molecular biologists. On March 13, following headlines that he had declared the War on Cancer a "sham," he wrote to Frank Rauscher, director of the National Cancer Institute, apologizing for not realizing that the dedication ceremony would have national media coverage. Of course, this professed surprise may have also been part of Watson's pattern of confrontation and accommodation with the NCI. James Watson to Frank Rauscher, March 13, 1975. See speech and associated correspondence in Watson papers online, sec. 2, series 8, subseries 1, accessed April 9, 2013, http://libgallery.cshl.edu/items/show /53179.

3. Amenta, *Bold Relief*; Quadagno, "Social Movements and State Transformation"; Self, *American Babylon*, 177–213.

4. See my discussion of molecular biology as a distinct community within biomedicine in chapter 7.

5. For this idea of expert political culture, see Balogh, *Chain Reaction*; Zelizer, *Taxing America*, 10–12. For a broader discussion of how culture and expertise interact in the creation of spheres for state action, see Mitchell, "Society, Economy, and State Effect"; O'Connor, *Poverty Knowledge*.

6. McElheny, *Watson and DNA*, 184–85. For more on the Eugenics Record Office, see Allen. "Eugenics Record Office."

7. James Watson, "Molecular Biological Approach to the Cancer Problem," December 28, 1970 (presented at Social Impact of Modern Biology Conference sponsored by British Society for Social Responsibility in Science), in MIT

Office of the President and Chancellor, Records of the President, 1960–1984 (Jerome B. Weisner) (AC 8), box 194, folder 1, Massachusetts Institute of Technology Archives.

8. Baltimore's defense of scientific inquiry invoked the libertarian idea of a marketplace in ideas, finding comfort in the hope that "if you don't get support one place, you can go another place." *National Cancer Attack Act of 1971: Hearings on H.R. 8343, H.R. 10681, Day 3, Before Subcomm. on Public Health and the Environment of Committee on Interstate and Foreign Commerce*, 92nd Cong. 339–40 (1971) (statement of David Baltimore, Massachusetts Institute of Technology).

9. David Baltimore interview, in Chubin and Studer, *Cancer Mission*, 136–37. Baltimore did acknowledge the SVCP in his later Nobel address. David Baltimore, "Viruses, Polymerases, and Cancer," *Science*, 192, no. 4240 (May 14, 1976): 632–36.

10. Wade, "Special Virus Cancer Program," 1309–11. Wade's article is the only one on the history of cancer virus research at the NCI during the War on Cancer that most other historical sources cite. It drew criticism from many other molecular biologists at the time, and there are reasons to suspect that the sources quoted are not representative. The molecular biologist Renato Dulbecco wrote to *Science* in defense of the program, stating, "The lack of objectivity is evident in Wade's failure to report more favorable opinions or to compare the scientific output of the SVCP to other programs." Similarly, a doctor from the University of Southern California wrote, "Grant programs have contributed greatly to cancer research but have not yet found the solutions. As urgent as the cancer problem is, other approaches are necessary to obtain rapid answers; the contract system is one such means." Fred Rapp, Renato Dulbecco, and J. Earle Officer, "Special Virus Cancer Program," *Science*, 175, no. 4026 (March 10, 1972): 1061–63. In a far longer letter, which *Science* did not publish, Maurice Green, a molecular biologist at Washington University and a SVCP contractor, wrote, "Wade interviewed me for about thirty minutes by telephone, but none of my remarks were quoted, perhaps because they were laudatory, and he has selected those which were derogatory." Maurice Green to Philip H. Abelson (editor, *Science*), February 4, 1972, in Sol Spiegelman papers (MSC 561), box 4, folder 11 (Green), National Library of Medicine—History of Medicine Division.

11. Carl Baker, "An Expanded Research and Development Program on Cancer [draft]," March 1971, pp. 34–35, 43, in LEGAL 1-10-A-1 NIH 32042, Central Files, Office of the Director, National Institutes of Health (hereafter "NIH Office of the director central files").

12. Louis M. Carrese, "National Cancer Institute Program Planning Approach," in *Planning Biomedical Research Programs*, ed. Louis M. Carrese (Bethesda, MD: National Cancer Institute, 1973), 3-6–3-7.

13. "National Cancer Program: The Strategic Plan," January 1973, Department of Health, Education, and Welfare, Public Health Service, Bethesda, MD, III-v.

14. "Follow Up on Review of NCI Resident Project Officers," August 11, 1976, in NIH Office of the director central files, EXT RES 1-2-E NIH 33515.

15. "Top 10 NCI Research Contracts," October 3, 1972, in NIH Office of the director central files, EXT RES 1-2-E NIH 33511.

16. Chubin and Studer, *Cancer Mission*, 143–95.

17. Chief, Research Contracts Branch (NCI), to associate director for administration (NIH), December 15, 1972, in RG 443, UDoD6-1, carton 94, RES-9-7-A National Cancer Program, 1973–1974, NARA II.

18. Chubin and Studer, *Cancer Mission*, 143–95.

19. As discussed in chapter 6.

20. Lee, *Nixon's Super-Secretaries*; McKenna, *World's Newest Profession*, 80–85.

21. Nathan, *Plot That Failed*, 40–41.

22. Barbara J. Culliton, "Office of Management and Budget: Skeptical View of Scientific Advice," *Science*, 183, no. 4123 (February 1, 1974): 392–96; Mandel, "Division of Research Grants," 145.

23. L. D. Longo, "Some Problems Facing Biomedical Research," *Federation Proceedings* 32, no. 11 (November 1973): 2078–85.

24. David Goldthwait to Mary Lasker, February 11, 1974, in Lasker papers, carton 118, folder NCAB, 2 of 3.

25. Ward Pigman, "Government Support of Biomedical Research," *Federation Proceedings* 32, no. 7 (July 1973): 1733.

26. James Cavanaugh (White House) to Edmund L. Keeney (Scripps Clinic), March 5, 1973, in Nixon papers, White House Central Files—SMOF, James Cavanaugh, Correspondence File 1972–1974, box 2, March Chron. 1973, folder 1 of 3. This sense of precariousness resonates with the experience of other scientific communities on "soft" money. See Mukerji, *Fragile Power*, 6–14.

27. Geiger, *Research and Relevant Knowledge*, 270–96.

28. For the widespread nature of the American Medical Association's campaign against national health insurance, see Starr, *Social Transformation of American Medicine*, 284–85.

29. A similar dynamic arose around community participation and the technocratic vision of the War on Poverty. McKee, " 'Government Is with Us.' "

30. James Watson, "Getting Realistic about Cancer," in *Cold Spring Harbor Laboratory Director's Report* (1974), pp. 1–2, 6–7, in Watson papers online, sec. 2, series 4, subseries 1, accessed April 9, 2013, http://libgallery.cshl.edu/items/show/52313.

31. James Watson, "Tumor Viruses: A Route to the Mt. Everest of Cancer," in *Cold Spring Harbor Director's Report* (1973), pp. 6–7, in Watson papers online, sec. 2, series 4, subseries 1, accessed April 9, 2013, http://libgallery.cshl.edu/items/show/52310.

32. Watson, "Getting Realistic about Cancer," pp. 1–2, 6–7.

33. James D. Watson, "When Worlds Collide: Research and Know-Nothingism," *New York Times*, March 22, 1973.

34. For comments on Baker's personality see Rettig, *Cancer Crusade*, 72–73.

35. Frank Rauscher, memo, September 8, 1972, in John Moloney papers (Acc. 2011-020), box 1, folder General Information SVCP, National Library of Medicine—History of Medicine Division.

36. Hixson, *Patchwork Mouse*, 127–28.

37. The committee members included Robert Good, a leading researcher in immunology, Wallace Rowe, a researcher at the National Institute of Allergy and Infectious Diseases as well as an adviser to the SVCP; Aaron Shatkin, a virologist at the Roche Institute of Molecular Biology (sponsored by the pharmaceutical firm Hoffman-LaRoche); James Darnell, a cell biologist at Albert Einstein Medical College and soon at Rockefeller University; Keith Porter, a Nobel Prize–winning electron microscopist; Vittorio Defendi, a leukemia specialist and tumor virus researcher from the Wistar Institute; and Chandler Stetson, chair of the New York University Department of Pathology. Zinder's leadership of this ad hoc committee is one of the major professional events mentioned in his obituary, alongside his pioneering work in bacterial genetics and later involvement in AIDS research and the Human Genome Project. Lodish and Fedoroff, "Norton Zinder"; Nicholas Wade, "Norton D. Zinder, Researcher in Molecular Biology, Dies at 83," *New York Times*, February 7, 2012; Schwartz, "Fine Playground." On Joshua Lederberg as a public intellectual for biology, see Crowe, "Joshua Lederberg's 'Euphenics.'"

38. Norton Zinder to Norton Zinder, September 12, 1973, in Norton Zinder papers, series Government Activities, box 135, folder (2? Labeled "Cancer"), Cold Spring Harbor Laboratory Library and Archives.

39. Norton Zinder, Minutes of Meeting of *ad hoc* Committee for Special Virus Cancer Program, June 8, 1973 (NYC), p. 1, in LION, 73606-011080 (hereafter "Zinder Committee Transcript," with page).

40. Comments of Sol Spiegelman and Robert Good, in Zinder Committee Transcript, p. 45.

41. Zinder Committee Transcript, p. 50.

42. Comments of Keith Porter, in Zinder Committee Transcript, p. 43.

43. Comments of Jerald Hurwitz, in Zinder Committee Transcript, p. 10.

44. Comments of Robert Gallo, p. 37; comments of Wallace Rowe, p. 14, both in Zinder Committee Transcript.

45. Report of Ad Hoc Committee to Review the Special Virus Cancer Program, p. 7, in LION, AR001108 (hereafter "Zinder report").

46. Zinder report, p. 18.

47. Zinder report, pp. 30–31.

48. Report on Virus Cancer Program, November 11, 1974, in Clark papers, series III, box 2, folder 3.

49. Howard Skipper to Harold Amos, April 15, 1975, in Clark papers, series III, box 21, folder 3, emphasis in original.

50. Stephen Hazen, memo for the record, August 10, 1983, in NCI LION, DC8308.

51. Frank J. Rauscher, "Budget and the National Cancer Program," *Science*, 184, no. 4139 (May 24, 1974): 872, 875.

52. Barbara J. Culliton, "Virus Cancer Program: Review Panel Stands By Criticism," *Science*, 184, no. 4133 (April 12, 1974): 144.

53. Howard Temin, "Decision Making in Science," address to University of Wisconsin Honors Convocation, June 1972, in Howard Temin papers, Acc. 2013-23, box General Correspondence, 1972, folder July–September 1972, University of Wisconsin–Madison University Archives, (hereafter "Temin papers," with box and folder information); Arthur Kornberg to Benno Schmidt, June 24, 1974, in Sol Spiegelman papers (MSC 561), box 11, folder 22, National Library of Medicine— History of Medicine Division.

54. Barbara J. Culliton, "Cancer: Select Committee Calls Virus Program a Closed Shop," *Science*, 182, no. 4117 (December 14, 1973): 1110–12.

55. *National Cancer Attack Act of 1974: Hearings on S. 2893, Day 1, Before Subcomm. Health of Committee on Labor and Public Welfare*, 93rd Cong. 170–171 (1974) (comments by Norton Zinder in National Cancer Advisory Board meeting minutes of March 19, 1974).

56. "A Rap at Cancer Research Spending," *Medical World News*, April 12, 1974, 21.

57. Carl Baker, "An Expanded Research and Development Program on Cancer" (draft), March 1971, pp. 34–35, 43, in NIH Office of the director central files, LEGAL 1-10-A-1 NIH 32042.

58. J. B. Moloney (associate director, Viral Oncology), memo to chief of NCI Research Contracts Branch re: "Viral Oncology Resident Program Officers," November 20, 1973, in NIH Office of the director central files, EXT RES 1-2-E, NIH 33517.

59. "Report of the Secretary's Advisory Committee on the Management of National Institutes of Health Research Contracts and Grants," March 1966, US Department of Health, Education, and Welfare, Washington, DC, 16. For a discussion of efforts to shape scientific identity, see Daston and Sibum, "Introduction"; Shapin, *Scientific Life*, 165–208; Traweek, *Beamtimes and Lifetimes*, 74–105.

60. *National Cancer Attack Act of 1974: Hearings, Day 1*, 93rd Cong. 170–171 (1974) (comments by Norton Zinder in National Cancer Advisory Board meeting minutes of March 19, 1974).

61. Paul Nieman (University of Washington) to J. Thomas August (Albert Einstein College of Medicine), February 22, 1974, in Temin papers, box Inactive Correspondence L–N, folder N 1966–1983.

62. Mandel, Hirshberg, and Burchenal, "Funding," 1113–14.

63. Rauscher to Schmidt, July 17, 1974, in RG 443, UDoD6-1, carton 94, RES-9-7-A, National Cancer Program, 1973–1974, NARA II.

64. Sweek, "Management Transfer," 21, 122, 159. For biographical information on Robert Sweek, see Rettig, *Cancer Crusade*, 89–90.

65. Sweek, "Management Transfer," 159.

66. Burawoy, *Manufacturing Consent*, 77–94, 161–77. In many ways, this resembles the function of what Stephen Hilgartner has called a "vanguard" group of scientists in managing the sociotechnical imaginaries of large biomedical projects—these small groups, because of their specialized knowledge of a scientific process, have immense leverage over the trajectory of projects. Hilgartner, *Reordering Life*, 6.

67. "A Report to the Director of the Office of Contracts and Grants, Associate Director for Administration on Researching Contracting Operations of the National Cancer Institute" (January 1974), Tab L, in Alex Smallberg, memo, February 27, 1974, in NIH Office of the director central files, EXT RES 1-2-E, NIH 33514.

68. George Klein to Schmidt, July 22, 1977, in LION, AR7707.

69. Arthur Pardee to Schmidt, July 20, 1977, in Watson papers online, sec. 2, series 17, subseries 44, accessed April 9, 2013, http://libgallery.cshl.edu/items/show/54189.

70. L. Thomas, "On the Planning of Cancer Science," *Yale Journal of Biology and Medicine* 50, no. 3 (1977): 257.

71. *Basic Issues in Biomedical and Behavioral Research, 1976: Hearings, Day 1, Before Subcomm. on Health of the Committee on Labor and Public Welfare*, 94th Cong. 344–45 (1976) (Report of the President's Biomedical Research Panel, Appendix A: The Place of Biomedical Science in Medicine and the State of the Science).

72. James D. Watson, *Molecular Biology of the Gene*, 3rd ed. (Menlo Park, CA: W. A. Benjamin, 1976), 685.

73. Beeman, "Robert J. Huebner," 483; Rettig, *Cancer Crusade*, 300.

74. For discussion of the regulatory debates about recombinant DNA, see Wright, *Molecular Politics*; Emrich, "Dr. Genelove"; Krimsky, *Genetic Alchemy*.

75. Angier, *Natural Obsessions*, 12–13. This sense of "grantsmanship" has also been discussed in relation to oceanography by Mukerji, *Fragile Power*, 85–110.

76. Baruch S. Blumberg and W. Thomas London, "Hepatitis B Virus and the Prevention of Primary Cancer of the Liver," *Journal of the National Cancer Institute* 74, no. 2 (February 1, 1985): 267–73.

77. Harold Varmus to John Cole (program director, RNA Virus Studies, NCI), May 11, 1982, in Harold Varmus papers, National Library of Medicine Profiles in Science, accessed January 6, 2014, http://profiles.nlm.nih.gov/ps/retrieve/Resource Metadata/MVBBFS.

Chapter Nine

1. Howard Temin, as quoted by Sugden in "Beyond the Provirus," 161.

2. "Symposium Participants," *Cold Spring Harbor Symposia on Quantitative Biology* 39 (January 1, 1974): v–xi.

3. "Cancer Viruses: More of the Same," *Nature* 227 (August 29, 1970): 887.

4. Technically, the genes that Bishop and Varmus discovered were called "proto-oncogenes" because the presence of the gene in a cell did not directly cause cancer, but I will use "oncogene" as a term of convenience to denote the study of viral, and later cellular, genes that if activated were thought to result in cancer. J. Michael Bishop, "The Molecular Genetics of Cancer," *Science*, 235, no. 4786 (January 16, 1987): 305–11.

5. "Press Release: The Nobel Prize in Physiology or Medicine 1989," Nobel Media AB, October 9, 1989.

6. Harold Varmus, "Retro Viruses and Oncogenes I," Nobel Prize Address, Stockholm, 1989, 517.

7. Rheinberger, "From Microsomes to Ribosomes," 50–52; Rheinberger, *Toward a History of Epistemic Things*, 2, 27–29; Rheinberger, "Consistency of Experimental Systems Approach."

8. Khot, Park, and Longstreth, "Vietnam War and Medical Research," 503.

9. Bishop, *How to Win Nobel Prize*, 52.

10. Varmus, *Art and Politics of Science*, 44.

11. Bishop and Varmus completed their research at the same time UCSF faculty members Herbert Boyer (recombinant DNA) and Stanley Prusiner (prions) made their major discoveries. All were heavily dependent on NIH support. Bourne, *Paths to Innovation*, 44.

12. Bishop, *How to Win Nobel Prize*, 52.

13. Report on contract NIH-NCI-E-71-2147, NCI Annual Report, 1972, 2: 1136–37.

14. J. M. Bishop, interview by author, November 25, 2013, at UCSF Hooper Foundation, quoted by permission of J. M. Bishop.

15. Of the $87,500 in the laboratory's budget for RSV studies, $75,000 was from the SVCP. It's unclear why this amount doesn't match the $77,000 amount listed in the contract, but overhead fees or other accounting features are likely explanations. See Harold Varmus, grant application for Special Grant from the American Cancer Society, California Division, "Detection of Virus-Specific Nucleic Acids in Normal and Transformed Avian and Mammalian Cells," March 1, 1972, p. 3a, in Harold Varmus papers (MSS 84-25), box 3, "NCI-NIH Grants JM Bishop: Viron-associated DNA polymerase," University of California, San Francisco Special Collections (hereafter "Varmus papers UCSF," with box and folder information).

16. J. M. Bishop and W. E. Levinson Cancer Research Coordination Committee (University of California), Research Grant Application, February 1970, pp. 5, 7, in Varmus papers UCSF, box 3, "NCI-NIH Grants JM Bishop: Viron-associated DNA polymerase."

17. Varmus, *Art and Politics of Science*, 31.

18. Harold Varmus, interview, NCI, July 17, 2013.

19. Varmus, *Art and Politics of Science*, 48–49.

20. See listing of contracts in *Special Virus Cancer Program Progress Report* 8 (1971), 362–63, 379–82.

21. Varmus, *Art and Politics of Science*, 79. Varmus appears to be the only person who uses the term "West Coast Tumor Virus Cooperative" in print, raising the question of to what degree it represents a rechristening of the Pacific Tumor Virus Group. The "Cooperative," however, had a much smaller membership.

22. Robin Weiss to Harold Varmus, J. Michael Bishop, and Warren Levinson, April 9, 1971, in Varmus papers UCSF, box 1, correspondence 1971. I thank Angela Creager for noting that "Watson and Crick" are also occasionally used as shorthand for the individual strands of the DNA helix—making Weiss's comment all the more playful.

23. Rheinberger and Gaudillière, introduction, 1–3.

24. Report on Contract NIH-NCI-E-72-2032, in NCI Annual Report, 1972, 2:1169–72.

25. G. S. Martin, "Rous Sarcoma Virus: A Function Required for the Maintenance of the Transformed State," *Nature* 227 (September 5, 1970): 1021–23.

26. Howard Temin to Howard K. Schachman (tenure promotion letter to chair of UCB Molecular Biology Department), October 22, 1970, in Temin papers, Inactive Correspondence S–T, S 197-1972.

27. Report on Contract NIH-NCI-E-71-2173, in NCI Annual Report, 1972, 2:1139–41.

28. P. H. Duesberg and Peter K. Vogt, "Differences between the Ribonucleic Acids of Transforming and Nontransforming Avian Tumor Viruses," *Proceedings of the National Academy of Sciences of the United States of America* 67, no. 4 (December 15, 1970): 1673–80. For more on the gel electrophoresis machine, see Chiang, "Laboratory Technology of Molecular Separation."

29. G. S. Martin and P. H. Duesberg, "The a Subunit in the RNA of Transforming Avian Tumor Viruses: I. Occurrence in Different Virus Strains. II. Spontaneous Loss Resulting in Nontransforming Variants," *Virology* 47, no. 2 (February 1972): 494–97.

30. "Tumour Viruses: Transforming Subunit," *Nature* 236, no. 5344 (March 31, 1972): 203.

31. Varmus, *Art and Politics of Science*, 39–41.

32. Bishop interview.

33. DNA is made of two strands of nucleotides, each containing a series of four different nitrogen-containing "bases" (A, G, T, C) attached to a spine of carbon molecules, and the insight of molecular biologists in the 1950s was that when two strands joined together to form a DNA double helix, the bases bonded only in certain pairs—A with T and C with G. As Watson and Crick suggested in 1953, this pattern of bonding explained the stability of the DNA molecule as a double-stranded helix formed by the attraction between complementary base pairs on two strands of DNA and the chemical process by which DNA could be copied. What other physical chemists soon determined, however, was that under

the right chemical or thermal conditions, this double helix could be unwound into its two single strands of DNA. Given that the single strands of DNA behaved differently, the winding and unwinding of DNA hybrids provided a very accurate way of comparing the sequence of base pairs in individual strands of DNA. The fuller a double-helix that formed between single strands of DNA, the higher the number of complementary bases. RNA, which only came in single strands and with a slightly different series of bases (A, G, U, C) but otherwise behaved in a similar manner to DNA during hybridization reactions, could also be compared to DNA strands. If the strand of DNA or RNA that a researcher wanted to study could be marked in advance—for example, through incorporating radioactive isotopes into the base pairs of a DNA or RNA sequence of interest—then the marked segment could act as a probe for similar sequences of base pairs among fragments of DNA. B. J. McCarthy and R. B. Church, "The Specificity of Molecular Hybridization Reactions," *Annual Review of Biochemistry* 39, no. 1 (1970): 131–50; J. G. Wetmur, "Hybridization and Renaturation Kinetics of Nucleic Acids," *Annual Review of Biophysics and Bioengineering* 5, no. 1 (1976): 337–61.

34. J. O. Bishop, "Molecular Hybridization of Ribonucleic Acid with a Large Excess of Deoxyribonucleic Acid," *Biochemical Journal* 126, no. 1 (January 1972): 171. For a discussion of these reactions more generally, see Giacomoni, "Origin of DNA:RNA Hybridization."

35. This procedure used the association kinetics of a mixture of single-stranded cellular DNA and radioactively labeled viral DNA to assess the presence of the viral genome in the original cellular DNA. Harold E. Varmus, Warren E. Levinson, and J. Michael Bishop, "Extent of Transcription by the RNA-Dependent DNA Polymerase of Rous Sarcoma Virus," *Nature* 233, no. 35 (September 1, 1971): 19–21; Harold E. Varmus, Robin A. Weiss, Robert R. Friss, Warren Levinson, and J. Michael Bishop, "Detection of Avian Tumor Virus-Specific Nucleotide Sequences in Avian Cell DNAs," *Proceedings of the National Academy of Sciences of the United States of America* 69, no. 1 (January 1, 1972): 20; Lawrence D. Gelb, David E. Kohne, and Malcolm A. Martin, "Quantitation of Simian Virus 40 Sequences in African Green Monkey, Mouse and Virus-Transformed Cell Genomes," *Journal of Molecular Biology* 57, no. 1 (April 14, 1971): 129–45.

36. Sol Spiegelman, A. Burny, M. R. Das, J. Keydar, J. Scholm, M. Travnicek, and K. Watson, "Characterization of the Products of RNA-Directed DNA Polymerases in Oncogenic RNA Viruses," *Nature* 227, no. 5258 (August 8, 1970): 567.

37. "Summary Report: Viral Carcinogenesis Branch," in NCI Annual Report, 1972, 2:920, 932–33.

38. Tri-Annual Report for Contract NIH 71-2147 (October 1, 1971), in Varmus papers UCSF, box 3, "NCI-NIH Grants JM Bishop: Viron Associated DNA Polymerase."

39. Ibid., October 1972, p. 4.

40. Ramareddy Guntaka, interview by author, June 17, 2015 (telephone).

41. Application for Renewal of NIH Contract 71-2147: Study on the Role of Viron Associated DNA Polymerase, p. 7, in Varmus papers UCSF, "NCI-NIH Grants JM Bishop: Viron Associated DNA Polymerase."

42. The imprecision arose from the way the laboratory created their (vDNA) probes. The laboratory started with RSV RNA as the template for the synthesis of vDNA by means of reverse transcriptase and radioactively labeled DNA fragments. Questions about the purity of the reverse transcriptase enzyme and the extent of the transcription of RNA into vDNA left considerable uncertainty about the nature of the cDNA that the probe revealed. Varmus et al., "Detection of Avian Tumor," 25–26.

43. Harold E. Varmus, Suzanne Heasley, and J. Michael Bishop, "Use of DNA-DNA Annealing to Detect New Virus-Specific DNA Sequences in Chicken Embryo Fibroblasts after Infection by Avian Sarcoma Virus," *Journal of Virology* 14, no. 4 (October 1974): 899.

44. Varmus to Paul Nieman (oncologist, University of Washington), February 11, 1972, in Harold Varmus papers, National Library of Medicine Profiles in Science, accessed January 6, 2014, http://profiles.nlm.nih.gov/ps/retrieve/ResourceMetadata/MVBBBJ.

45. Varmus to Norman Davison (Caltech), June 25, 1974, in Varmus papers UCSF, carton 1, Correspondence 1973.

46. Bishop and Varmus to Robert Huebner, February 22, 1972, in Varmus papers UCSF, carton 1, Correspondence 72.

47. Minutes of Virus Cancer Program Chairs Meeting, April 4, 1974, p. 2, in RG 443, UDUP-7, container 1 (FRC), Segment Chair Meeting Minutes, NARA II.

48. NCI Annual Report, 1974, p. 980.

49. J. M. Bishop, Semi-Annual Progress Report for NIH Contract No1 CP 33293 (from January 2 to July 2, 1974), p. 1, in Varmus papers UCSF, box 3, "NCI-NIH Grants JM Bishop: Viron Associated DNA Polymerase."

50. Application for Renewal of NIH Contract 71-2167, January 21, 1972, pp. 4–7, quote on p. 6, in Varmus papers UCSF, box 3, "NCI-NIH Grants JM Bishop: Viron Associated DNA Polymerase." Bishop envisioned the further purification of reverse transcriptase at UCSF using gel electrophoresis and the use of this highly purified reverse transcriptase to construct single-stranded vDNA hybridization probes for the presence of vDNA in chicken, mouse, and human cells along with viral RNA in human tumor cells. Tri-Annual Progress Report for NIH Contract 71-2147 (1972), pp. 3–5, in Varmus papers UCSF, carton 3, NCI-NIH Grants: JM Bishop.

51. J. M. Bishop, NIH Grant Application, "Rous Sarcoma Virus Replication and Cellular Transformation, September 1975, pp. 5, 7, 9–10, 13, in Varmus papers UCSF, box 3, "October 1975—Rous Sarcoma Virus and Cellular Transformation."

52. H. E. Varmus, ACS Grant Application (1973–1974?), p. 2, in Varmus papers UCSF, box 3, "ACS CA Division Special Grants Comm. 72–73 (#2/3)."

53. Guntaka interview; R. V. Guntaka, "Antecedents of a Nobel Prize," *Nature* 343, no. 6256 (January 25, 1990): 302.

54. Report on Contract NIH-NCI-E-72-2032, in NCI Annual Report, 1973, 3B:1400–1401; Peter K. Vogt, Robin A. Weiss, and Hidesaburo Hanafusa, "Proposal for Numbering Mutants of Avian Leukosis and Sarcoma Viruses," *Journal of Virology* 13, no. 2 (February 1, 1974): 551–54.

55. Dominique Stehelin, Ramareddy V. Guntaka, Harold E. Varmus, and J. Michael Bishop, "Purification of DNA Complementary to Nucleotide Sequences Required for Neoplastic Transformation of Fibroblasts by Avian Sarcoma Viruses," *Journal of Molecular Biology* 101, no. 3 (March 5, 1976): 349–65.

56. P. E. Neiman, S. E. Wright, C. McMillin, and D. MacDonnell, "Nucleotide Sequence Relationships of Avian RNA Tumor Viruses: Measurement of the Deletion in a Transformation-Defective Mutant of Rous Sarcoma Virus," *Journal of Virology* 13, no. 4 (April 1, 1974): 838.

57. Stehelin recalled in an open letter to the Nobel Committee in 1989 "the intensity of the emotion I experienced and the intellectual clarity induced by the situation. The fantastic results came out in the night of Saturday, October 26th, 1974: Normal DNA contained sequences related to the *src* gene of the transforming virus." Quoted in Bishop, *How to Win Nobel Prize*, 162–63.

58. For an influential appellation of the discovery of cellular *src* as a "revolutionary" event, see Weinberg, *Racing to Beginning of Road*, 96. For a more measured assessment of what the discovery meant, see Morange, *History of Molecular Biology*, 222.

59. D. Stehelin, H. E. Varmus, J. M. Bishop, and P. K. Vogt, "DNA Related to Transforming Gene(s) of Avian Sarcoma Viruses Is Present in Normal Avian DNA," *Nature* 260, no. 5547 (March 11, 1976): 170–73; Bishop, *How to Win Nobel Prize*, 163; Varmus, *Art and Politics of Science*, 79, 116.

60. Robert J. Huebner and George J. Todaro, "Oncogenes of RNA Tumor Viruses as Determinants of Cancer," *Proceedings of the National Academy of Sciences of the United States of America* 64, no. 3 (November 15, 1969): 1090.

61. If this had been the case, of course, the hopes of developing a vaccine would have been slim, but neither Todaro nor Huebner dwelled on this point. "Summary Report: Viral Carcinogenesis Branch," NCI Annual Report, 1972, 2:920, 932–33.

62. E. Zuckerklandl and L. Pauling, "Molecular Disease, Evolution and Genetic Heterogeneity," in *Horizons in Biochemistry*, ed. Michael Kasha and Bernard Pullman (New York: Academic Press, 1962), 189–225.

63. A. C. Wilson, L. R. Maxson, and V. M. Sarich, "Two Types of Molecular Evolution: Evidence from Studies of Interspecific Hybridization," *Proceedings of the National Academy of Sciences of the United States of America* 71, no. 7 (July 1, 1974): 2843–47.

64. Stehelin et al., "DNA Related to Transforming Gene(s)," 173. Others have noted the importance of this evolutionary interpretation for Bishop and Varmus's

paper. Fujimura, *Crafting Science*, 120–21; Kevles, "Pursuing the Unpopular," 103–5.

65. In particular, Spector was able to alter the concentration and composition of the salt solutions used for hybridization reactions. Changing these conditions allowed the creation of v-*src* hybrids with cellular DNA of lower and lower percentages of overlap. Deborah Spector, interview by author, June 9, 2015 (telephone); Deborah H. Spector, Harold E. Varmus, and J. Michael Bishop, "Nucleotide Sequences Related to the Transforming Gene of Avian Sarcoma Virus Are Present in DNA of Uninfected Vertebrates," *Proceedings of the National Academy of Sciences of the United States of America* 75, no. 9 (September 1, 1978): 4102–6.

66. Bishop interview; Spector interview.

67. Robin Weiss, "Molecular Analysis of Oncogene," *Nature* 260, no. 5547 (March 11, 1976): 93.

68. David Baltimore, "Viruses, Polymerases, and Cancer," *Science*, 192, no. 4240 (May 14, 1976): 635.

69. J. Michael Bishop and Harold Varmus, "Functions and Origins of Retroviral Transforming Genes," in *RNA Tumor Viruses: Supplements and Appendixes*, ed. Robin Weiss, Natalie M. Teich, Harold Varmus, and John M. Coffin, 2nd ed. (Cold Spring Harbor, NY: Cold Spring Harbor Laboratory, 1982), 308.

70. Bishop interview; Thomas H. Maugh, "RNA Viruses: The Age of Innocence Ends," *Science* 183, no. 4130 (March 22, 1974): 1185; Jean L. Marx, "RNA Tumor Viruses: Getting a Handle on Transformation," *Science* 199, no. 4325 (January 13, 1978): 161–64.

71. Jean L. Marx, "Cancer Gene Research Wins Medicine Nobel," *Science*, 246, no. 4928 (October 20, 1989): 326.

72. Gina Kolata, "Frenchman Says Nobel Panel Overlooked His Contribution," *New York Times*, October 11, 1989, sec. U.S.; "French Researcher Asks for Share," *Nature* 341, no. 6243 (October 19, 1989): 556; Jean L. Marx, "Stehelin Persists in Nobel Protest," *Science*, 246, no. 4934 (December 1, 1989): 1121.

73. Shapin, "Invisible Technician." Of course, there are many other examples of how laboratory organization and the circulation of credit interact in biology and other sciences. Kohler, *Lords of the Fly*, 91–132; Todes, "Pavlov's Physiology Factory."

74. "Conduct Unbecoming," *Nature* 342, no. 6248 (November 23, 1989): 328.

75. Weinberg, *Racing to Beginning of Road*, 83–84; Klein, "Tale of Great Cuckoo Egg."

76. Rheinberger writes, "The experimental ensemble of technical objects transcends the identity condition of its parts. . . . Established tools can acquire new functions in the process of their reproduction. Their insertion into a productive or consumptive process beyond their intended use may reveal characters other than the original functions they were designed to perform." Rheinberger, *Toward History of Epistemic Things*, 32.

77. Rheinberger, "Experimental Systems" (1999), 421.

Chapter Ten

1. Quoted in Arthur M. Crocker, "Elusive Quest: Cancer Research Drive, Begun with Fanfare, Hits Disillusionment," *Wall Street Journal*, October 10, 1978.

2. Kushner, *Breast Cancer*, 114.

3. Howard Temin, "Banquet Speech," December 10, 1975, accessed February 8, 2014, www.nobelprize.org/nobel_prizes/medicine/laureates/1975/temin-speech.html. For a sense of the reception of this speech, see Kevles, "Howard Temin," 248.

4. Belich, *Replenishing the Earth*, 548–60; Kaiser, "Booms, Busts."

5. Barbara J. Culliton, "Science's Restive Public," *Daedalus* 107, no. 2 (April 1, 1978): 147–56.

6. George A. Keyworth, "The Role of Science in a New Era of Competition," *Science* 217, no. 4560 (1982): 607.

7. Cantor, "Between Prevention and Therapy"; Proctor, *Cancer Wars*, 48–64; Patterson, *Dread Disease*, 280–300.

8. Harold Schmeck, "On Trail of Cancer: Discoveries Reach a Crescendo," *New York Times*, September 13, 1983.

9. For a review of these events, see Robert A. Weinberg, "The Genetic Origins of Human Cancer," *Cancer* 61, no. 10 (May 15, 1988): 1963–68.

10. On the extension of molecular biology into higher organisms, see Rheinberger, "Recent Science and Its Exploration," 7–8. On the political and therapeutic stakes of oncogenes, see Gaudillière, "Oncogenes as Metaphors for Cancer." The historian Michel Morange and the sociologist Joan Fujimura have engaged in a spirited debate regarding the origins of the "oncogene paradigm." Morange places the roots of this transformation in the adaptation of new technologies, such as recombinant DNA sequencing, to solve the outstanding intellectual challenge that previous experimental methods had been unable to solve: how the techniques of molecular biology could explain cellular regulation and development. Oncogenes provided a vital window into this process. Fujimura emphasizes that the infrastructure of the VCP, coupled with the development of standard tools, created a "package" of theory and method around oncogenes that appealed to both the professional interests of scientists and the needs of government administrators. Morange, "Transformation of Molecular Biology"; Fujimura, "Molecular Biological Bandwagon"; Fujimura, *Crafting Science*, 141–54. Morange, in turn, argues that Fujimura's emphasis on the sociological dimensions of oncogene research is an insufficient explanation for the convergence on this paradigm, especially given the persistent issues posed by proteins and cell regulation; he stresses the intellectual tempo of research as an important dimension as well. Morange, "Regulatory Vision to Oncogene Paradigm," 11–13.

11. Lewis Thomas, foreword to *NIH: An Account of Research in Its Laboratories and Clinics*, ed. DeWitt Stetten and W. T. Carrigan (Orlando, FL: Academic Press, 1984), xvii.

12. Löwy, *Between Bench and Bedside*, 125–27; Patterson, *Dread Disease*, 286–98.

13. Bix, "Diseases Chasing Money and Power"; Lerner, *Breast Cancer Wars*, 170–95, 209–22; Wailoo, *How Cancer Crossed Color Line*, 142–47.

14. On changes in the committee structure, see Deering and Smith, *Committees in Congress*, 33–42.

15. Proctor, *Cancer Wars*, 48–53.

16. "Special Report: Environmental Cancers: Humans as the Experimental Model?," *Environmental Science & Technology* 10, no. 13 (December 1, 1976): 1190.

17. Jasanoff, "Science, Politics, and EPA"; Lazarus, *Making of Environmental Law*, 70–89.

18. Proctor, *Cancer Wars*, 54–62.

19. Although the use of screening tests had inspired enthusiasm among environmentalists, the regulation of carcinogens ignited new controversies between public health authorities and theories of the biological mechanisms of cancer, especially the question of DNA repair. One notable scientist, Bruce Ames, switched from being an advocate of carcinogenesis screening to opposing regulation. Creager, " 'EAT: DIE,' " 29–32.

20. *National Cancer Program, 1979: Hearings, Day 2, Before Subcomm. on Health and Scientific Research of the Committee on Labor and Human Resources*, 96th Cong. 459 (1979) (statement of Stephen Epstein).

21. Allen Kneese and William D. Schulze, "Environment, Health, and Economics—The Case of Cancer," *American Economic Review* 67, no.1 (February 1977): 328. I thank Angela Creager for drawing my attention to this comparison.

22. Umberto Stafotti, April 23, 1976, memo to NCI director, DCCP director, staff, and advisers, in Clark papers, microfilm, Division of Cancer Cause and Prevention (4.5) 1976.

23. Comments by George Miller and David Baltimore recorded in the transcript of NCI director's Group Consultation on the Prevention of Cancer, February 4, 1978, pp. 26–27, in Baruch Blumberg papers (Ms. Coll. 144), series III, box 205, folder 6, NCI directors' Group Consultation on the Prevention of Cancer (1978), American Philosophical Society Library (hereafter "NCI director consultation").

24. *National Cancer Program, 1979: Hearings*, 96th Cong. 459, 467 (1979) (statement of Stephen Epstein).

25. Through Oak Ridge's mouse colonies, Upton had in fact worked on the conformation of some of Ludwik Gross's observations of murine leukemia in the 1950s, but his overall focus remained on radiation and cancer. Michael R. Greenberg and Karen W. Lowrie, "Arthur C. Upton: Let the Evidence Speak," *Risk Analysis* 28, no. 6 (December 2008): 1493–96; A. C. Upton, "The Dose-Response Relation in Radiation-Induced Cancer," *Cancer Research* 21, no. 6 (July 1, 1961): 717–29.

26. Marvin A. Schneiderman (NCI DCCP Field Studies and Statistics) to Watson, October 17, 1978, in Watson papers online, sec. 2, series 2, subseries 17, accessed April 9, 2013, http://libgallery.cshl.edu/items/show/54198.

27. "Moloney Out as Viral Oncology Chief," *Cancer Letter* 4, no. 1 (January 6, 1978): 1, 2–4. See also Arthur Upton, "Improvements to NCI Organization," January 27, 1978, in Watson papers online, Personal Papers, Government Relations, Upton File (51), accessed April 9, 2013, http://libgallery.cshl.edu/items/show/54196.

28. "Annual Report: Division of Cancer Cause and Prevention," October 1979–September 1980, in NCI Annual Report, 1980, 3A:1–2.

29. On the important role of the NCI relative to other branches of the NIH during the 1970s, see Kalberer and Newell, "National Cancer Act and Beyond," 4280. The Cold Spring Harbor Laboratory 1977 budget worksheet highlights the important role of NCI funding. Of the $1,083,000 in direct costs on the 1976 NCI Cancer Research Center Grant to CSHL, $423,000 was from tumor virus studies, versus $105,000 for protein synthesis, the next-largest segment of the grant. CSHL anticipated that support for tumor virus studies would rise to $541,260 by 1981 as the grant overall rose to $1,886,309. Meanwhile, in 1977 other Public Health Services, charitable, and National Science Foundation grant support to individual investigators working at the laboratory amounted to $547,130—including $140,365 from other programs of the NCI. Lewis Thomas papers, series II, carton 34, folder 7, Princeton University Library Special Collections and Archives.

30. Daniel S. Greenberg, "Whatever Happened to the War on Cancer?," *Discover*, March 1986, 60.

31. Mahlon B. Hoagland (president and scientific director, Worcester Institute for Experimental Biology) to Arthur C. Upton, February 15, 1978, in Watson papers online, Personal Papers, Government Relations—NCI, Arthur C. Upton, accessed April 9, 2013, http://libgallery.cshl.edu/items/show/54196.

32. This point has been raised by Robert Proctor as a species of "agnotology"—it is possible to produce ignorance of one set of causes of cancer (e.g., tobacco or environmental exposures) by sponsoring a surfeit of knowledge regarding other possible causes (e.g., molecular genetics). Proctor, *Cancer Wars*, 8, 102. See also Proctor and Schiebinger, *Agnotology*.

33. Quote from Harry D. Williams (American Business Cancer Foundation) to Michael H. Wigler (head, Mammalian Genetics Section, Cold Spring Harbor Laboratory), February 9, 1982, in Watson papers online, JDW/2/2/47/5, accessed January 16, 2018, http://libgallery.cshl.edu/items/show/29993; "Group Taps Business for Cancer Research Funds," *Chemical & Engineering News Archive* 58, no. 9 (March 3, 1980): 24–26.

34. Proctor, *Cancer Wars*, 75–80; Kamlet, Mowery, and Su, "Upsetting National Priorities?"

35. Janet D. Rowley, Denman Hammond, Maureen M. Henderson, Joseph Gale Katterhagen, Rose Kushner, Henry C. Pitot, Sheldon W. Samuels, and Irving J. Selikoff, "Cancer Advisory Board," *Science*, 217, no. 4560 (1982): 585.

36. James Watson, "Cancer Is a Solvable Problem," *Cold Spring Harbor Laboratory Director's Report* (1982), p. 8, in Watson papers online, JDW/2/4/1/45, accessed June 27, 2017, http://libgallery.cshl.edu/items/show/52332.

37. President of NAS, as quoted in essay, "On Adequately Funding Medical Research in 1982 and Beyond," sent to Howard Temin by Arthur Pardee (1981? 1982?), in Temin papers, box "Seminars, 1981–1982," folder "National Academy of Sciences April 24, 1982."

38. Howard M. Temin, "On the Origin of the Genes for Neoplasia: G. H. A. Clowes Memorial Lecture," *Cancer Research* 34, no. 11 (November 1, 1974): 2836.

39. John Tooze, ed., *The Molecular Biology of Tumor Viruses* (Cold Spring Harbor, NY: Cold Spring Harbor Laboratory, 1973), 648.

40. My focus on these two discoveries follows from the major developments classified in Michel Morange's account of the intellectual transformation of the oncogene paradigm along the lines of genes as "privileged targets of transformation" and establishing their "functional" relevance. He acknowledges the importance of virus research for the first development while calling the latter "fortuitous," but I will address how oncoprotein research also reflected the infrastructure and ambitions of the VCP. The third major development—experimental work indicating that more than one oncogene is required for cancer—came somewhat later in the 1980s. Morange, "Discovery of Cellular Oncogenes," 48–50.

41. Scolnick worked in Todaro's section, which was not on the NCI campus but in a facility in Virginia operated by contractor Meloy Laboratories. Edward M. Scolnick, Stuart A. Aaronson, and George J. Todaro, "DNA Synthesis by RNA-Containing Tumor Viruses," *Proceedings of the National Academy of Sciences of the United States of America* 67, no. 2 (October 15, 1970): 1034–41; Scolnick interview, NCI Oral History Project, 2–6.

42. Edward M. Scolnick, Richard S. Howk, Anthony Anisowicz, Paul T. Peebles, Charles D. Scher, and Wade P. Parks, "Separation of Sarcoma Virus-Specific and Leukemia Virus-Specific Genetic Sequences of Moloney Sarcoma Virus," *Proceedings of the National Academy of Sciences of the United States of America* 72, no. 11 (November 1, 1975): 4650–54; Edward M. Scolnick, R. J. Goldberg, and W. P. Parks, "A Biochemical and Genetic Analysis of Mammalian RNA-Containing Sarcoma Viruses," *Cold Spring Harbor Symposia on Quantitative Biology* 39 (January 1, 1974): 885–95; Edward M. Scolnick, Elaine Rands, David Williams, and Wade P. Parks, "Studies on the Nucleic Acid Sequences of Kirsten Sarcoma Virus: A Model for Formation of a Mammalian RNA-Containing Sarcoma Virus," *Journal of Virology* 12, no. 3 (September 1973): 458–63.

43. E. M. Scolnick, "Transformation by Rat-Derived Oncogenic Retroviruses," *Microbiological Reviews* 45, no. 1 (March 1981): 2.

44. Scolnick interview, NCI Oral History Project, 17–21, 28–29; L. Evans, M. Nunn, P. H. Duesberg, D. Troxler, and E. Scolnick, "RNAs of Defective and Non-defective Components of Friend Anemia and Polycythemia Virus Strains Identified and Compared," *Cold Spring Harbor Symposia on Quantitative Biology* 44, (January 1, 1980): 823–35; Cha-Mer Wei, Douglas R. Lowy, and Edward M. Scolnick, "Mapping of Transforming Region of the Harvey Murine Sarcoma Virus Genome by Using Insertion-Deletion Mutants Constructed in Vitro," *Proceedings of the National Academy of Sciences of the United States of America* 77, no. 8 (August 1, 1980): 4674–78.

45. Varmus interview. For the recombinant DNA restrictions controversy, see Paul Berg, David Baltimore, Herbert W. Boyer, Stanley N. Cohen, Ronald W. Davis, David S. Hogness, Daniel Nathans, et al., "Potential Biohazards of Recombinant DNA Molecules," *Science*, 185, no. 4148 (July 26, 1974): 303; Krimsky, *Genetic Alchemy*; Wright, *Molecular Politics*.

46. Incidentally, these restrictions did not slow work in commercial laboratories, which were not bound by the NIH guidelines. Wright, *Molecular Politics*, 281–311; Durant, "'Refrain from Using the Alphabet.'"

47. The discovery of this chemical treatment, a calcium salt solution, was originally developed by cancer virus researchers seeking to enhance the rate at which their viruses infected cultured cells. F. L. Graham and A. J. van der Eb, "A New Technique for the Assay of Infectivity of Human Adenovirus 5 DNA," *Virology* 52, no. 2 (April 1, 1973): 456–67; Graham, "This Week's Citation Classic."

48. G. M. Cooper and S. Okenquist, "Mechanism of Transfection of Chicken Embryo Fibroblasts by Rous Sarcoma Virus DNA," *Journal of Virology* 28, no. 1 (October 1978): 45–52; David Smotkin, Alessandro M. Gianni, Shmuel Rozenblatt, and Robert A. Weinberg, "Infectious Viral DNA of Murine Leukemia Virus," *Proceedings of the National Academy of Sciences of the United States of America* 72, no. 12 (1975): 4910–13.

49. NCI Annual Report, 1972, 3:69–71; Massachusetts Institute of Technology Bulletin 110, no. 4 (November, 1974): 78–79, 273–74; Herbert Black, "Biologist Luria to Direct Efforts: MIT to Get $6m Cancer Research Unit," *Boston Globe*, December 5, 1972; Selya, "Salvador Luria's Unfinished Experiment, 254–60. On Watson's speech, see chapter 8.

50. Smotkin et al., "DNA of Murine Leukemia Virus."

51. Kathleen Danna and Daniel Nathans, "Specific Cleavage of Simian Virus 40 DNA by Restriction Endonuclease of Hemophilus Influenzae," *Proceedings of the National Academy of Sciences of the United States of America* 68, no. 12 (December 1, 1971): 2913–17.

52. Weinberg, *Racing to Beginning of Road*, 131–33. In fact, this mouse line had been developed by George Todaro before he left to work at the NCI. George J. Todaro and Howard Green, "Quantitative Studies of the Growth of Mouse Embryo Cells in Culture and Their Development into Established Lines," *Journal of*

Cell Biology 17, no. 2 (May 1, 1963): 299–313; Tadao Aoki and George J. Todaro, "Antigenic Properties of Endogenous Type-C Viruses from Spontaneously Transformed Clones of BALB/3T3," *Proceedings of the National Academy of Sciences of the United States of America* 70, no. 5 (May 1, 1973): 1598–1602; George J. Todaro and Howard Green, "High Frequency of SV40 Transformation of Mouse Cell Line 3T3," *Virology* 28, no. 4 (April 1966): 756–59; Littlefield, "NIH 3T3 Cell Line."

53. Chiaho Shih, Ben-Zion Shilo, Mitchell P. Goldfarb, Ann Dannenberg, and Robert A. Weinberg, "Passage of Phenotypes of Chemically Transformed Cells via Transfection of DNA and Chromatin," *Proceedings of the National Academy of Sciences of the United States of America* 76, no. 11 (November 1, 1979): 5714–18. In seeking to identify a source of possible transforming genes, Weinberg was inspired by work on chemical carcinogenesis, in particular that of UC–Berkeley biochemist Bruce Ames's assay for chemical carcinogens. The Ames assay equated the ability of chemicals to cause mutations in *Salmonella typthimirium* bacteria with their ability to cause cancer. The so-called Ames Test for carcinogens had offered one of the first instances in which it seemed possible to bridge laboratory and environmental perspectives on cancer causation. Environmental organizations embraced the test in the 1970s as a means of rapidly screening for industrial carcinogens. Creager, "Political Life of Mutagens."

54. G. M. Cooper, Sharon Okenquist, and Lauren Silverman, "Transforming Activity of DNA of Chemically Transformed and Normal Cells," *Nature* 284, no. 5755 (April 3, 1980): 418–21; G. M. Cooper, N. G. Copeland, A. D. Zelenetz, and T. Krontiris, "Transformation of NIH-3T3 Mouse Cells by Avian Retroviral DNAs," *Cold Spring Harbor Symposia on Quantitative Biology* 44 (1980): 1169–76.

55. Morange, "Regulatory Vision to Oncogene Paradigm," 4.

56. John Wyke, "Fishing for Oncogenes," *Nature* 284, no. 5755 (1980): 398.

57. Clifford J. Tabin, Scott M. Bradley, Cornelia I. Bargmann, Robert A. Weinberg, Alex G. Papageorge, Edward M. Scolnick, Ravi Dhar, Douglas R. Lowy, and Esther H. Chang, "Mechanism of Activation of a Human Oncogene," *Nature* 300, no. 5888 (November 11, 1982): 143–49; G. M. Cooper, "Cellular Transforming Genes," *Science* 217, no. 4562 (August 27, 1982): 801–6; Cooper, Okenquist, and Silverman, "Transforming Activity of DNA."

58. Committee members included Harold Varmus, J. Michael Bishop, G. Steven Martin, Edward M. Scolnick, Robert A. Weinberg, and Peter Vogt. John M. Coffin, Harold E. Varmus, J. Michael Bishop, Myron Essex, William D. Hardy Jr., G. Steven Martin, Naomi E. Rosenberg, Edward M. Scolnick, Robert A. Weinberg, and Peter K. Vogt, "Proposal for Naming Host Cell-Derived Inserts in Retrovirus Genomes," *Journal of Virology* 40, no. 3 (December 1981): 953–54.

59. Scolnick interview, NCI Oral History Project, 29.

60. Luis F. Parada, Clifford J. Tabin, Chiaho Shih, and Robert A. Weinberg, "Human EJ Bladder Carcinoma Oncogene Is Homologue of Harvey Sarcoma

Virus Ras Gene," *Nature* 297, no. 5866 (June 10, 1982): 474–78; Esther H. Chang, Mark E. Furth, Edward M. Scolnick, and Douglas R. Lowy, "Tumorigenic Transformation of Mammalian Cells Induced by a Normal Human Gene Homologous to the Oncogene of Harvey Murine Sarcoma Virus," *Nature* 297, no. 5866 (June 10, 1982): 479–83.

61. George Vande Woude, interview by author, January 15, 2014 (telephone).

62. Roger Watson, Marianne Oskarsson, and George F. Vande Woude, "Human DNA Sequence Homologous to the Transforming Gene (Mos) of Moloney Murine Sarcoma Virus," *Proceedings of the National Academy of Sciences of the United States of America* 79, no. 13 (July 1, 1982): 4078–82; Ravi Dhar, William L. McClements, Lynn W. Enquist and George F. Vande Woude, "Nucleotide Sequences of Integrated Moloney Sarcoma Provirus Long Terminal Repeats and Their Host and Viral Junctions," *Proceedings of the National Academy of Sciences of the United States of America* 77, no. 7 (July 1, 1980): 3937–41; M. Oskarsson, William L. McClements, Donald G. Blair, and George F. Vande Woude, "Properties of a Normal Mouse Cell DNA Sequence (Sarc) Homologous to the Src Sequence of Moloney Sarcoma Virus," *Science* 207, no. 4436 (March 14, 1980): 1222–24.

63. Simonetta Pulciani, Eugenio Santos, Anne V. Lauver, Linda K. Long, Stuart A. Aaronson, and Mariano Barbacid, "Oncogenes in Solid Human Tumours," *Nature* 300, no. 5892 (December 9, 1982): 539–42.

64. Jean Yin Jen Wang, "Evolution of Oncogenes: From c-Abl to v-Abl," *Nature* 304, no. 5925 (August 4, 1983): 400; John Wyke, "Evolution of Oncogenes: From c-Src to v-Src," *Nature* 304, no. 5926 (August 11, 1983): 491–92; Inder M. Verma, "Oncogenic Intelligence: From c-Fos to v-Fos," *Nature* 308, no. 5957 (March 22, 1984): 317.

65. Weinberg, *Racing to Beginning of Road*, 194–95.

66. "Oncogenes in Solid Human Tumours," *Nature* 300, no. 5892 (December 9, 1982): 477.

67. Takis Papas and George Vande Woude, preface to *Oncogenes*, ed. Takis S. Papas and George F. Vande Woude (New York: Elsevier, 1986), 4:i. This provides an example in support of Joanna Radin's observation that the technology of freezing has changed the temporality of biomedical research—in this instance the existence of human tissue samples from many years earlier was an important element of the progression of oncogene research. Radin, *Life on Ice*, 10–12.

68. Robert A. Weinberg, "A Molecular Basis of Cancer," *Scientific American* 249, no. 5 (November 1983): 140.

69. M. Strand and J. T. August, "Structural Proteins of RNA Tumor Viruses as Probes for Viral Gene Expression," *Cold Spring Harbor Symposia on Quantitative Biology* 39 (January 1, 1974): 1109–16.

70. NCI Annual Report, 1968, 1:1087–88; Stuart Aaronson, interview by author, April 26, 2017 (telephone).

71. *Special Virus Cancer Program Progress Report* 8 (1971), 6, 54, 199, 208–9, 255–56.

72. Aaronson oversaw contractors working for the VCP, such as Hazelton Laboratories, as they designed standardized immunological assays using radioactively labeled antibodies to detect proteins or protein fragments associated with type-C retroviruses. *Virus Cancer Program Progress Report* 14 (1977), 73, quote on 78; Report on Contract NO1-CP-6-1024, in *Virus Cancer Program Progress Report* 15 (1978), 290.

73. See, for example, Bowker and Star, *Sorting Things Out.*

74. Gathering at the Sloan-Kettering Institute, a group of retrovirologists proposed a common naming convention, commenting that "uniform nomenclature for oncornavirus proteins is essential in order to avoid confusion and to permit rapid communication of data." J. T. August, Dani P. Bolognesi, Erwin Fleissner, Raymond V. Gilden, and Robert C. Nowinski, "A Proposed Nomenclature for the Virion Proteins of Oncogenic RNA Viruses," *Virology* 60, no. 2 (August 1, 1974): 595–600. In this regard, the project to systematize the infrastructure for viral proteins actually moved faster than the machinery to classify the cascade of retroviral oncogenes revealed during the 1970s, which started in 1980. Peter Duesberg to Harold Varmus and John Coffin, July 21, 1980, in Harold Varmus papers, National Library of Medicine Profiles in Science, accessed April 17, 2016, https://profiles .nlm.nih.gov/ps/access/MVBBGV.pdf; Coffin et al., "Naming Host Cell-Derived Inserts."

75. *Virus Cancer Program Progress Report* 14 (1977), 43–44.

76. Report on Project Z01-CP-04940-13-LCMB, NCI Annual Report, 1980, 3A:221–28.

77. While a new host could produce the sequence of amino acids—the building blocks of proteins—which corresponded to the sequence in an oncogene, the biochemical function of the oncoproteins itself was governed by a set of three-dimensional folds, which occurred after the sequence of amino acids was in place. The synthesis of proteins by different organisms, involving elements of messenger RNA chemistry that were not fully understood, could produce incomplete or different patterns of folding than the proteins produced in their original context. For discussion of the challenges of expressing eukaryotic genes for proteins in the early 1980s, see Leonard Guarente, Thomas M. Roberts, and Mark Ptashne, "A Technique for Expressing Eukaryotic Genes in Bacteria," *Science* 209, no. 4463 (1980): 1428–30; Tom Maniatis, E. F. Fritsch, and Joseph Sambrook, *Molecular Cloning: A Laboratory Manual* (Cold Spring Harbor, NY: Cold Spring Harbor Laboratory, 1982), 422–33. This was a challenge common to efforts to use recombinant DNA techniques for the production of other molecules such as insulin and interferon, as Nicolas Rasmussen has discussed; the problem of calibrating activity and amino acid sequence, though solvable, could be a laborious process in the early 1980s and was certainly not a routine technique. Moreover, some of these results were not published, and thus they would have been unavailable to researchers in retrovirology working on the periphery of recombinant DNA. Rasmussen, *Gene Jockeys*, 113–15. For a contemporary discussion of recombinant DNA and protein

synthesis, see Germán L. Rosano and Eduardo A. Ceccarelli, "Recombinant Protein Expression in Escherichia Coli: Advances and Challenges," *Frontiers in Microbiology* 5 (April 2014): 1–17.

78. Keith C. Robbins, Sushilkumar G. Devare, E. Premkumar Reddy, and Stuart A. Aaronson, "In Vivo Identification of the Transforming Gene Product of Simian Sarcoma Virus," *Science* 218, no. 4577 (1982): 1131–33.

79. R. F. Doolittle, Michael W. Hunkapiller, Leory E. Hood, Sushilkumar G. Devare, Keith C. Robbins, Stuart A. Aaronson, and Harry N. Antoniades, "Simian Sarcoma Virus Onc Gene, v-Sis, Is Derived from the Gene (or Genes) Encoding a Platelet-Derived Growth Factor," *Science* 221, no. 4607 (July 15, 1983): 275–77; Sushilkumar G. Devare, E. P. Reddy, K. C. Robbins, P. R. Andersen, S. R. Tronick, and S. A. Aaronson, "Nucleotide Sequence of the Transforming Gene of Simian Sarcoma Virus," *Proceedings of the National Academy of Sciences of the United States of America* 79, no. 10 (May 15, 1982): 3179–82; Sushilkumar G. Devare, E. Premkumar Reddy, J. Doria Law, Keith C. Robbins, and Stuart A. Aaronson, "Nucleotide Sequence of the Simian Sarcoma Virus Genome: Demonstration That Its Acquired Cellular Sequences Encode the Transforming Gene Product P28sis," *Proceedings of the National Academy of Sciences of the United States of America* 80, no. 3 (1983): 731–35.

80. Michael D. Waterfield, Geoffery T. Scrace, Nigel Whittle, Paul Stroobant, Ann Johnson, Åke Wasteson, Bengt Westermark, Carl-Henrik Heldin, Jung San Huang, and Thomas F. Deuel, "Platelet-Derived Growth Factor Is Structurally Related to the Putative Transforming Protein P28sis of Simian Sarcoma Virus," *Nature* 304 (July 1, 1983): 35–39.

81. Beverly Merz, "Publishing Flurry Attends Oncogene-Growth Factor Link," *Journal of the American Medical Association* 250, no. 9 (September 2, 1983): 1127–31; Newmark, "Oncogenic Intelligence."

82. Robin Weiss, "Cancer: Oncogenes and Growth Factors," *Nature* 304, no. 5921 (July 7, 1983): 12.

83. Both quoted in Newmark, "Oncogene Discovery."

84. Matt Clark and Deborah Witherspoon, "Cancer: The Enemy Within," *Newsweek*, March 5, 1984.

85. Watson, "Cancer Is Solvable Problem, p. 8, in Watson papers online.

86. *The Status and Implications of Oncogene Research: Hearings before Subcomm. on Investigations and Oversight of the Comm. on Science and Technology, Day 1*, 98th Cong. 4 (1984) (testimony of Peter Fischinger, deputy director, NCI).

87. Manaker, Sibal, and Moloney, "Activities at National Cancer Institute," 631.

88. Bishop interview; Bishop, "Oncogenes," 92.

89. Also honored were Hidesaburo Hanafusa (for work on the genetics of RSV) and Robert Gallo (for his isolation of HTLV-I, the first human retrovirus). Harold Schmeck, "Seven Medical Experts Win Lasker Prizes," *New York Times*,

November 18, 1982; Beverly Merz, "Seven Honored for Tracking Disorders to Genetic Source," *Journal of the American Medical Association* 248, no. 21 (December 3, 1982): 2799–2801.

90. Howard M. Temin, "Oncogenes: We Still Don't Understand Cancer," *Nature* 302, no. 5910 (April 21, 1983): 656.

91. "Retroviral *onc* genes [oncogenes] are the only proof that altered proto-*onc* genes [proto-oncogenes] can be cancer genes. There is as yet no conclusive evidence that an unaltered proto-*onc* gene can function as a cancer gene simply through enhanced transcription or gene amplification." Peter H. Duesberg, "Retroviral Transforming Genes in Normal Cells?," *Nature* 304, no. 5923 (July 21, 1983): 223. Two years later, Duesberg cautioned that the effort to identify cellular cancer genes had been frustrated, "despite fierce efforts." Retroviral v-*onc* genes remained the only known transforming genes, and the function of their associated cellular proto-oncogenes remained unclear. Proto-oncogenes were not consistently associated with tumors. Duesberg also questioned the suitability of the 3T3 mouse cell line used by Weinberg and others to assess the carcinogenicity of various proto-oncogenes, on the grounds that these cells were prone to a high rate of spontaneous transformation. Peter H. Duesberg, "Activated Proto-Onc Genes: Sufficient or Necessary for Cancer?," *Science* 228, no. 4700 (May 10, 1985): 669, 676. While later commentators have recalled Duesberg's caution through the prism of the eventual acceptance of the oncogene theory and his denial of the link between another retrovirus, human immunodeficiency virus, and AIDS, in the mid-1980s Duesberg was one of the preeminent authorities on the biochemistry of retroviral genomes. The points that he raised indicated ongoing uncertainties for oncogene research. Weinberg, *Racing to Beginning of Road*, 113. See Duesberg's later criticism of the oncogene theory in P. H. Duesberg, "Cancer Genes: Rare Recombinants Instead of Activated Oncogenes (a Review)," *Proceedings of the National Academy of Sciences of the United States of America* 84, no. 8 (April 1987): 2117–24.

92. Harry Rubin, "Understanding Cancer," *Science*, 219, no. 4589 (March 11, 1983): 1170.

93. The procedural critiques of contract research by Zinder and others had also served to strengthen the grant-making mechanism, which the director of the NIH controlled rather than the director of the NCI. Vincent DeVita, interview by author, April 17, 2017 (telephone).

94. Vincent DeVita, interview by Gretchen A. Chase, June 5, 1997, 12–18, NIH Office of History.

95. Meeting Summary, National Cancer Advisory Board Minutes for October 4–6, 1982, accessed June 29, 2017, https://deainfo.nci.nih.gov/advisory/ncab/archive/43_1082/04oct82mins.pdf.

96. DeVita and DeVita-Raeburn, *Death of Cancer*, 174–75.

97. DeVita, "Governance at National Cancer Institute," 3972.

98. Peter J. Fischinger and Vincent T. DeVita, "Governance of Science at the National Cancer Institute: Perceptions and Opportunities in Oncogene Research," *Cancer Research* 44, no. 10 (October 1, 1984): 4696.

99. Ibid., 4693.

100. *Status and Implications of Oncogene Research: Hearings*, 98th Cong. 33–34 (1984) (prepared statement of William F. Feller, president-elect, American Cancer Society, Washington, DC).

101. D. Shapley, "Oncogenes Cause Cancer Institute to Change Tack," *Nature* 301, no. 5895 (January 6, 1983): 5.

102. NCI associate director Peter Fischinger to Charlotte Friend, September 9, 1982 (form letter apparently sent to all FCRF Advisory Committee Members), in Friend papers, carton 16, NCI Frederick Research Facility Reports 1982.

103. Fischinger to James Liverman (acting PI, Basic Research Program, FCRF), December 23, 1982, in Friend papers, carton 16, NCI Frederick Research Facility Reports.

104. Fischinger and DeVita, "Governance at National Cancer Institute," 4693.

105. DeVita, "Governance at National Cancer Institute," 3969.

106. Watson, "Cancer Is Solvable Problem," p. 6, in Watson papers online.

107. Watson to Vincent DeVita, March 17, June 3, 1982 (quote), in Watson papers online, JDW/2/2/2098/114, JDW2/2/2099/55, accessed March 17, 2017, http://libgallery.cshl.edu/items/show/78420, http://libgallery.cshl.edu/items/show/78477.

108. Hunter and Simon, "History of the Oncogene Meeting"; Frederick Cancer Research Facility and National Cancer Institute (U.S.), *Annual Report: Frederick Cancer Research Facility* (Frederick, MD: FCRF, 1993).

109. *Departments of Labor, Health and Human Services, Education, and Related Agencies Appropriations for 1990 Part 4A: Hearings before Subcomm. of the Committee on Appropriations, Day 1*, 101st Cong. 304 (1989) (testimony of Dr. Samuel Broder, director, NCI).

110. J. W. Berg, "Implications for the Control of Human Cancer," in *RNA Tumor Viruses, Oncogenes, Human Cancer and AIDS: On the Frontiers of Understanding*, ed. Philip Furmanski, Jean Carol Hager, and Marvin A. Rich (New York: Springer US, 1985), 392.

111. Robert Weinberg, preface to *Oncogenes and the Molecular Origins of Cancer*, ed. Robert A. Weinberg (Cold Spring Harbor, NY: Cold Spring Harbor Laboratory, 1989), vii–viii; Renato Dulbecco, "A Turning Point in Cancer Research: Sequencing the Human Genome," *Science*, 231, no. 4742 (March 7, 1986): 1055–56.

112. Michel Morange has surveyed the relative rates of articles published in *Nature* from 1975 to 1985, and the most conspicuous decline in publications is for those on the "biological and medical aspects of cancer." The rising profile of the molecular genetics of cancer came with its success in superseding other therapeutic or biological studies of cancer, while publications on cancer viruses remained strong. Morange, "Discovery of Cellular Oncogenes," 53.

113. Joan Fujimura discusses this as part of her analysis of how different groups came to align their different interests behind the "oncogene bandwagon," but she tends to treat the resources available to these different groups as a matter of fact. My analysis contributes to further historicizing these starting conditions, and it suggests the important role that the sense of acceleration they produced played in getting the "bandwagon" rolling. Fujimura, "Ecologies of Action."

Conclusion

1. Greer Williams, *Virus Hunters* (New York: Knopf, 1959), 497.

2. Daniel S. Greenberg, "Whatever Happened to the War on Cancer?," *Discover*, March 1986, 47.

3. Proctor, *Cancer Wars*, 5–6.

4. Nass and Stillman, *Large-Scale Biomedical Science*, 44–45.

5. Mukherjee, *Emperor of All Maladies*, 175. This assessment is in fact based on an article published midway into the search for a cancer virus. See Wade, "Special Virus Cancer Program."

6. This point has been raised in relationship to sociology by Spector and Kitsuse, *Constructing Social Problems*.

7. Centers for Disease Control, "Kaposi's Sarcoma and Pneumocystis Pneumonia among Homosexual Men—New York City and California," *Morbidity and Mortality Weekly Report* 30, no. 25 (July 3, 1981): 305–8; "Epidemiologic Aspects of the Current Outbreak of Kaposi's Sarcoma and Opportunistic Infections," *New England Journal of Medicine* 306, no. 4 (January 28, 1982): 248–52.

8. Epstein, *Impure Science*.

9. Harden, "NIH and Research on AIDS," 31–33.

10. AIDS was at first colloquially known as a "gay cancer." Shilts, *And the Band Played On*, 96.

11. Beeman, "Robert J. Huebner," 480; Gallo, *Virus Hunting*, 35–37.

12. Gallo interview.

13. Robert C. Gallo, "RNA Tumor Viruses and Leukemia: Evaluation of Present Results Supporting Their Presence in Human Leukemias," in *Modern Trends in Human Leukemia II. Biological, Immunological, Therapeutical and Virological Aspects*, ed. Rolf Neth, Robert C. Gallo, Klaus Mannweiler, and William C. Moloney (Munich: J. F. Lehmanns, 1976), 431–50.

14. Bernard J. Poiesz, Francis W. Ruscetti, Adi F. Gazdar, Paul A. Bunn, John D. Minna, and Robert C. Gallo, "Detection and Isolation of Type C Retrovirus Particles from Fresh and Cultured Lymphocytes of a Patient with Cutaneous T-Cell Lymphoma," *Proceedings of the National Academy of Sciences of the United States of America* 77, no. 12 (December 1, 1980): 7415–19; Gallo, "Discoveries of First Human Retroviruses."

15. Robert C. Gallo, Prem S. Sarin, E. P. Gelmann, Marjorie Robert-Guroff, Ersell Richardson, V. S. Kalyanaraman, Dean Mann, et al., "Isolation of Human T-Cell Leukemia Virus in Acquired Immune Deficiency Syndrome (AIDS)," *Science* 220, no. 4599 (1983): 865–67.

16. M. G. Sarngadharan, Mikulas Popovic, Lilian Bruch, Jörg Schüpbach, and Robert C. Gallo, "Antibodies Reactive with Human T-Lymphotropic Retroviruses (HTLV-III) in the Serum of Patients with AIDS," *Science* 224, no. 4648 (1984): 506–8.

17. Montagnier, *Virus*, 21–48.

18. This dispute was eventually resolved in Montagnier's favor. Gallo and Montagnier, "HIV as Cause of AIDS"; Grmek, *History of AIDS*, 70–77.

19. DeVita interview; Barbara J. Culliton, "Crash Development of AIDS Test Nears Goal," *Science* 225, no. 4667 (1984): 1128–31.

20. P. J. Fischinger, W. G. Robey, H. Koprowski, R. C. Gallo, and D. P. Bolognesi, "Current Status and Strategies for Vaccines against Diseases Induced by Human T-Cell Lymphotropic Retroviruses (HTLV-I, -II, -III)," *Cancer Research* 45, no. 9, suppl. (September 1985): 4694s–99s.

21. Epstein, *Impure Science*, 80.

22. NIH historian Victoria Harden notes that most of the retrovirologists active in the United States on AIDS research had spent time working with the Virus Cancer Program. Harden, *AIDS at 30*, 44–45.

23. P. H. Duesberg, "Human Immunodeficiency Virus and Acquired Immunodeficiency Syndrome: Correlation but Not Causation," *Proceedings of the National Academy of Sciences of the United States of America* 86, no. 3 (February 1, 1989): 755–64; Peter H. Duesberg, "Retroviruses as Carcinogens and Pathogens: Expectations and Reality," *Cancer Research* 47, no. 5 (March 1, 1987): 1199–1220. For analysis of this discussion, see Fujimura and Chou, "Dissent in Science."

24. Grmek, *History of AIDS*, 47–82; Varmus, *Art and Politics of Science*, 128–30.

25. Berkowitz, *Institute of Medicine and AIDS*, 218–20; Jon Cohen, "A 'Manhattan Project' for AIDS?," *Science* 259, no. 5098 (1993): 1113. Temin quoted in Cohen.

26. Thomas, *Big Shot*; Cohen, *Shots in the Dark*.

27. With the benefit of hindsight, one can see that the promise of a robust social welfare state appears to be less an inevitable development than a possibility rooted in the 1930s and 1940s. Brick, *Transcending Capitalism*; Mitchell, "Society, Economy, and State Effect"; Lichtenstein, ed., *American Capitalism*; Sparrow, *Warfare State*, 10–15. This policy transformation has often been associated with the rise of "neoliberal" approaches to public policy. In this sense, the political consequences of the biomedical settlement represent a moment when the antistatist valorization of the free market in policy emerged from within the operation of state institutions rather than from without. Centeno and Cohen, "Arc of Neoliberalism"; Harvey, *Brief History of Neoliberalism*; Rodgers, *Age of Fracture*, 41–76; Zelizer,

"Reflections." For an example of a framing of the roots of the Reagan-era free-market movement in external terms, see Phillips-Fein, *Invisible Hands*.

28. Berman, *Creating the Market University*; Yi, "Who Owns What?"; Mirowski, *Science-Mart*, 87–138.

29. Joan Hamilton, "The New War on Cancer," *Business Week*, no. 2965 (September 22, 1986): 61.

30. Eric N. Berg, "Small Concerns Battle Cancer," *New York Times*, December 28, 1985, business sec.

31. Shapin, *Scientific Life*, 232–51.

32. Jones, "Entrepreneurial Science."

33. Yi, *Recombinant University*, 177–78. The growth of these new ways of doing research, however, drew upon the legacy of state-centered fights against cancer even as they claimed to make a break with them.

34. For an analysis of the history of science as the politics of memory, see Abir-Am and Elliott, *Commemorative Practices in Science*. On boundary work, see Gieryn, "Boundary-Work and Demarcation of Science." For a particular study of how virologists construct their memories, see Aviles, "Little Death."

35. Klein, *Atheist and the Holy City*; Weinberg, *Racing to Beginning of Road*, 73–84.

36. As quoted in Jerome Groopman, "The Thirty Years' War," *New Yorker*, June 4, 2001, 61–62.

37. Aviles, "Emergence of Ethical Research." On the federal government's role as a major source of innovation in the development of vaccines, see Hoyt, *Long Shot*.

38. Cook-Deegan, *Gene Wars*, 161–85; Hilgartner, "Constituting Large-Scale Biology."

39. C. R. Cantor, "Orchestrating the Human Genome Project," *Science* 248, no. 4951 (April 6, 1990): 49–51; T. D. Yager, D. A. Nickerson, and L. E. Hood, "The Human Genome Project: Creating an Infrastructure for Biology and Medicine," *Trends in Biochemical Sciences* 16 (1991): 454–58; Martin C. Rechsteiner, "The Human Genome Project: Misguided Science Policy," *Trends in Biochemical Sciences* 16, no. 12 (1991): 455–61.

40. Office of Technology Assessment, *Mapping Our Genes: The Genome Projects: How Big, How Fast?* (DIANE, 1988), 117, 127.

41. Loeppky, *Encoding Capital*, 95–134. I am grateful to Robert Kohler for suggesting the difference between the HGP as a "survey" project versus the VCP as a "targeted" project.

42. Leaf, *Truth in Small Doses*. Ironically, one of the voices calling for an industrial-style approach to cancer research was James Watson. "Acclaimed Scientist Watson Discusses Cancer Research," *Yale Daily News*, March 21, 2012.

43. Baruch S. Blumberg, "Australia Antigen and the Biology of Hepatitis B," *Science*, 197, no. 4298 (July 1, 1977): 17–25; Baruch S. Blumberg, "Hepatitis B

Virus, the Vaccine, and the Control of Primary Cancer of the Liver," *Proceedings of the National Academy of Sciences of the United States of America* 94, no. 14 (July 8, 1997): 7121–25; Blumberg, *Hepatitis B*; Beasley, "Control of HBV and HCC"; R. P. Beasley, L. Y. Hwang, C. C. Lin, and C. S. Chien, "Hepatocellular Carcinoma and Hepatitis B Virus. A Prospective Study of 22 707 Men in Taiwan," *Lancet* 2, no. 8256 (November 21, 1981): 1129–33; Cladd E. Stevens, R. Palmer Beasley, Julia Tsui, and Wy-Chan Lee, "Vertical Transmission of Hepatitis B Antigen in Taiwan," *New England Journal of Medicine* 292, no. 15 (April 10, 1975): 771–74.

44. NCI director consultation, Olds, Baltimore, Blumberg, 18–19; Weinstein and Baltimore, 35.

45. Ibid., Blumberg, 20.

46. Ibid., Baltimore, 4.

47. Several anthropologists have recently highlighted that cancer is a disease that takes on different properties in different contexts. In sub-Saharan Africa, for example, the combination of AIDS and malnutrition means that as many as 40 percent of all cancers may have viral origins, versus less than 10 percent in a country like the United States. Livingston, *Improvising Medicine*; Lora-Wainwright, *Fighting for Breath*; Mathews, Burke, and Kampriani, *Anthropologies of Cancer*.

48. Another factor slowing the production of the original vaccine was concern that the supplies of human serum used in its manufacture were vulnerable to contamination with HIV. This provided the basis for American public health authorities to favor an even more expensive hepatitis B vaccine produced using recombinant DNA technology. Conis, "'Hepatitis B on Second Day?'"; Huzair and Sturdy, "Biotechnology and Vaccine Innovation"; Muraskin, "Silent Epidemic"; Steven Schenker, "Alcoholic Liver Disease: Evaluation of Natural History and Prognostic Factors," *Hepatology* 4, no. S1 (January 1, 1984): 36S–43S.

49. Moreover, the promotion of the vaccine attempted to elide the sexual politics associated with HPV's status as a sexually transmitted infection, causing many of the cancers associated with HPV infection to be minimized in favor of the image of the threat of the virus to young women. See Aronowitz, "Gardasil"; Douglas R. Lowy and John T. Schiller, "Reducing HPV-Associated Cancer Globally"; Carpenter and Casper, "Global Intimacies"; Mamo and Epstein, "Pharmaceuticalization of Sexual Risk"; Mamo and Epstein, "New Sexual Politics of Cancer"; Teixeira and Löwy, "Imperfect Tools for Difficult Job."

Bibliography

Manuscript Collections

American Philosophical Society Library, Philadelphia
    Baruch Blumberg papers
    Salvador Luria papers
    James B. Murphy papers
    Peyton Rous papers
Cold Spring Harbor Laboratory Library and Archives, Cold Spring Harbor, NY
    James Watson papers
    Norton Zinder papers
Columbia University Rare Book and Manuscript Library, New York
    Center for Oral History Archives
    Mary Lasker papers
Center for the History of Medicine, Francis A. Countway Library of Medicine, Boston
    Francisco Duran-Reynals papers
Henry R. Winkler Center for the History of the Health Professions, University of Cincinnati, Cincinnati
    Albert Sabin papers
Icahn School of Medicine at Mount Sinai, New York
    Charlotte Friend papers
John P. McGovern Historical Collections and Research Center, Texas Medical Center Library, Houston
    R. Lee Clark papers
    Joseph Melnick papers
March of Dimes Archives, White Plains, NY
    Medical Program Records, Series 14: Poliomyelitis
    Public Relations Records
Massachusetts Institute of Technology Archives, Cambridge
    Recombinant DNA Oral History Collection (MC 100)
    Records of the President, Jerome B. Weisner, 1960–1984 (AC 8)

MD Anderson Cancer Center, Houston, TX
  R. Lee Clark papers (microfilm)
National Archives and Records Administration II, College Park, MD
  Records of the National Institutes of Health, 1912–1990
National Cancer Institute, Bethesda, MD
  LION Database
National Institutes of Health Library, Bethesda, MD
  Annual Report of Program Activities, National Cancer Institute, 1950–1985
National Institutes of Health Office of History, Bethesda, MD
  Carl Baker papers (now transferred to the National Library of Medicine)
  Progress reports and annual reports of the Special Virus Leukemia Program,
    the Special Virus Cancer Program, and the Virus Cancer Program
National Institutes of Health Office of the Director, Bethesda, MD
  Central Files
National Library of Medicine—History of Medicine Division, Bethesda, MD
  Beeman/Huebner papers
  Ludwik Gross papers
  Werner and Gertrude Henle papers
  John Moloney papers
  Profiles in Science (Online)
    John Fogarty papers
    Mary Lasker papers
    Harold Varmus papers
  Sol Spiegelman papers
Princeton University Library Special Collections and Archives, Princeton, NJ
  Lewis Thomas papers
Richard Nixon Presidential Library and Museum, Yorba Linda, CA
  White House Central Files—Subject Files
Rockefeller Archive Center, Sleepy Hollow, NY
  Commonwealth Fund Archive
  Memorial Sloan Kettering Cancer Center papers
  Laurance Rockefeller papers
  Rockefeller University Archives
Stanford University Special Collections, Stanford, CA
  Paul Berg papers
University of California, Berkeley, Bancroft Library, Berkeley
  Wendell Stanley papers
University of California, San Francisco Special Collections, San Francisco
  Harold Varmus papers
University of Wisconsin–Madison University Archives, Madison
  Howard Temin papers
Yale University Archives and Special Collections, New Haven, CT
  John Enders papers

Interviews by Author

Stuart Aaronson, April 26, 2017, telephone

David Baltimore, January 6, 2014, Skype

J. Michael Bishop, November 25, 2013, at Hooper Foundation, University of California, San Francisco

Vincent DeVita, April 19, 2017, telephone

Robert Gallo, November 13, 2013, telephone

Murray Gardner, January 14, 2014, Skype

Raymond Gilden, January 3, 2014, Skype

Ramareddy Guntaka, June 17, 2015, telephone

Paul Levine, July 25, 2013, at National Institutes of Health Office of History

Deborah Spector, June 9, 2015, telephone

George Vande Woude, January 15, 2014, telephone

Harold Varmus, July 17, 2013, at National Cancer Institute Office of the Director

John Ziegler, April 17, 2013, at Global Health Program, University of California, San Francisco

Secondary Sources

Primary Sources, such as scientific publications and newspaper articles, cited fully in the notes, are omitted from the bibliography.

Abir-Am, Pnina G. "The Molecular Transformation of Twentieth Century Biology." In *Science in the Twentieth Century*, edited by John Krige and Dominique Pestre, 495–524. Amsterdam: Harwood Academic, 1997.

———. "Review: Nobelesse Oblige: Lives of Molecular Biologists." *Isis* 82, no. 2 (June 1991): 326–43.

———. "Themes, Genres, and Orders of Legitimation in the Consolidation of New Scientific Disciplines: Deconstructing the Historiography of Molecular Biology." *History of Science* 23 (March 1, 1985): 73–117.

Abir-Am, Pnina G., and Clark A. Elliott, eds. *Commemorative Practices in Science: Historical Perspectives on the Politics of Collective Memory*. Ithaca, NY: Department of Science and Technology Studies, Cornell University, 1999.

Adams, Vincanne, Michelle Murphy, and Adele E. Clarke. "Anticipation: Technoscience, Life, Affect, Temporality." *Subjectivity* 28, no. 1 (September 2009): 246–65. doi:10.1057/sub.2009.18.

Alexander, Jennifer Karns. *The Mantra of Efficiency: From Waterwheel to Social Control*. Baltimore: Johns Hopkins University Press, 2008.

Allen, Garland E. "The Eugenics Record Office at Cold Spring Harbor, 1910–1940: An Essay in Institutional History." *Osiris*, 2 (1986): 225–64.

Amadae, S. M. *Rationalizing Capitalist Democracy: The Cold War Origins of Rational Choice Liberalism*. Chicago: University of Chicago Press, 2003.

Amenta, Edwin. *Bold Relief: Institutional Politics and the Origins of Modern American Social Policy*. Princeton, NJ: Princeton University Press, 1998.

American Capitalism: Social Thought and Political Economy in the Twentieth Century. Lichtenstein Nelson, ed. Philadelphia: University of Pennsylvania Press, 2006.

The American Society for the Control of Cancer: Its Objects and Methods and Some of the Visible Results of Its Work. New York: American Society for the Control of Cancer, 1925.

Angier, Natalie. *Natural Obsessions: The Search for the Oncogene*. Boston: Houghton Mifflin, 1988.

Ankeny, Rachel A., and Sabina Leonelli. "What's So Special about Model Organisms?" *Studies in History and Philosophy of Science Part A* 42, no. 2 (June 2011): 313–23. doi:10.1016/j.shpsa.2010.11.039.

Anser, Glen R. "The Linear Model, the US Department of Defense, and the Golden Age of Industrial Research." In *The Science-Industry Nexus: History, Policy, Implications*, edited by Karl Grandin, Nina Wormbs, and Sven Widmalm, 3–30. Sagamore Beach, MA: Science History Publications/USA, 2004.

Appel, Toby A. *Shaping Biology: The National Science Foundation and American Biological Research, 1945–1975*. Baltimore: Johns Hopkins University Press, 2000.

Aronowitz, Robert A. "Do Not Delay: Breast Cancer and Time, 1900–1970." *Milbank Quarterly* 79, no. 3 (2001): 355–86.

———. "Gardasil: A Vaccine against Cancer and a Drug to Reduce Risk." In Wailoo et al., *Three Shots at Prevention*, 21–38.

———. *Unnatural History: Breast Cancer and American Society*. Cambridge: Cambridge University Press, 2007.

Aviles, Natalie B. "The Little Death: Rigoni-Stern and the Problem of Sex and Cancer in 20th-Century Biomedical Research." *Social Studies of Science* 45, no. 3 (June 1, 2015): 394–415. doi:10.1177/0306312715584402.

———. "Situated Practice and the Emergence of Ethical Research: HPV Vaccine Development and Organizational Cultures of Translation at the National Cancer Institute." *Science, Technology & Human Values* 43, no. 5 (September 2018): 810–33. doi:10.1177/0162243917749728.

Balogh, Brian. *The Associational State: American Governance in the Twentieth Century*. Philadelphia: University of Pennsylvania Press, 2015.

———. *Chain Reaction: Expert Debate and Public Participation in American Commercial Nuclear Power, 1945–1975*. Cambridge: Cambridge University Press, 1991.

Barnes, David S. *The Great Stink of Paris and the Nineteenth-Century Struggle against Filth and Germs*. Baltimore: Johns Hopkins University Press, 2006.

Barnes, Emm. "Cancer Coverage: The Public Face of Childhood Leukaemia in 1960s Britain." *Endeavour* 32, no. 1 (March 2008): 10–15. doi:10.1016/j.endeavour.2008.01.004.

Barrett, Frank A. "Alfred Haviland's Nineteenth-Century Map Analysis of the Geographical Distribution of Diseases in England and Wales." *Social Science & Medicine* 46, no. 6 (March 1, 1998): 767–81. doi:10.1016/S0277-9536(97)00170-6.

Bayne-Jones, Stanhope. *The Advancement of Medical Research and Education through the Department of Health, Education, and Welfare: Final Report of the Secretary's Consultants on Medical Research and Education.* Washington, DC: Office of the Secretary, Department of Health, Education, and Welfare, 1958.

Beasley, R. Palmer. "Rocks along the Road to the Control of HBV and HCC." *Annals of Epidemiology* 19, no. 4 (April 2009): 231–34. doi:10.1016/j.annepidem.2009.01.017.

Beckert, Jens. *Imagined Futures: Fictional Expectations and Capitalist Dynamics.* Cambridge, MA: Harvard University Press, 2016.

Becsei Kilborn, Eva. "Going against the Grain: Francis Peyton Rous (1879–1970) and the Search for the Cancer Virus." PhD diss., University of Illinois–Chicago, 2003.

———. "Scientific Discovery and Scientific Reputation: The Reception of Peyton Rous' Discovery of the Chicken Sarcoma Virus." *Journal of the History of Biology* 43, no. 1 (November 2008): 111–57. doi:10.1007/s10739-008-9171-y.

Beeman, Edward. "Robert J. Huebner, M.D.: A Virologist's Odyssey." NIH Office of History, National Institutes of Health, Bethesda, MD, 2005.

Belich, James. *Replenishing the Earth: The Settler Revolution and the Rise of the Anglo-World, 1783–1939.* Oxford: Oxford University Press, 2009.

Bell, Jonathan. *The Liberal State on Trial: The Cold War and American Politics in the Truman Years.* New York: Columbia University Press, 2004.

Beniger, James R. *The Control Revolution: Technological and Economic Origins of the Information Society.* Cambridge, MA: Harvard University Press, 1986.

Berkowitz, Edward D. *The Institute of Medicine and AIDS.* Washington, DC: National Academies Press, 1998.

Berman, Elizabeth Popp. *Creating the Market University: How Academic Science Became an Economic Engine.* Princeton NJ: Princeton University Press, 2012.

Bessis, M. "How the Mouse Leukemia Virus Was Discovered: A Talk with Ludwik Gross." *Nouvelle Revue Française d'Hématologie* 16, no. 2 (1976): 287–304.

Birch, Kean. "Introduction: Biofutures/Biopresents." *Science as Culture* 15, no. 3 (2006): 173–81. doi:10.1080/09505430600890602.

Bishop, J. Michael. *How to Win the Nobel Prize: An Unexpected Life in Science.* Cambridge, MA: Harvard University Press, 2003.

———. "Oncogenes." *Scientific American* 246 (1982): 81–92.

Bix, Amy Sue. "Diseases Chasing Money and Power: Breast Cancer and Aids Activism Challenging Authority." *Journal of Policy History* 9, no.1 (1997): 5–32. doi:10.1017/S0898030600005807.

Bloor, David. "Toward a Sociology of Epistemic Things." *Perspectives on Science* 13, no. 3 (October 10, 2005): 285–312.

Blumberg, Baruch S. *Hepatitis B: The Hunt for a Killer Virus.* Princeton, NJ: Princeton University Press, 2002.

Bourne, Henry R. *Paths to Innovation: Discovering Recombinant DNA, Oncogenes, and Prions in One Medical School, over One Decade.* San Francisco: University of California Medical Humanities Consortium, 2011.

Boveri, Theodor. *Concerning the Origin of Malignant Tumors.* Translated by Henry Harris. Cambridge, UK: Company of Biologists, 2008.

Bowker, Geoffrey C., and Susan Leigh Star. *Sorting Things Out: Classification and Its Consequences.* Cambridge, MA: MIT Press, 1999.

Brandt, Allan M. *The Cigarette Century: The Rise, Fall, and Deadly Persistence of the Product That Defined America.* New York: Basic Books, 2007.

———. *No Magic Bullet: A Social History of Venereal Disease in the United States since 1880.* Expanded ed. New York: Oxford University Press, 1987.

Brandt, Allan, and Martha Gardner. "The Golden Age of Medicine?" In *Medicine in the Twentieth Century,* edited by Roger Cooter and John V. Pickstone, 21–36. Amsterdam: Harwood Academic, 2000.

Brick, Howard. *Transcending Capitalism: Visions of a New Society in Modern American Thought.* Ithaca, NY: Cornell University Press, 2006.

Brinkley, Alan. *The End of Reform: New Deal Liberalism in Recession and War.* New York: Knopf, 1995.

Broadbent, Alex. "Disease as a Theoretical Concept: The Case of 'HPV-Itis.'" *Studies in History and Philosophy of Science Part C: Studies in History and Philosophy of Biological and Biomedical Sciences* 48, Part B (December 2014): 250–57. doi:10.1016/j.shpsc.2014.07.010.

Brown, E. Richard. *Rockefeller Medicine Men: Medicine and Capitalism in America.* Berkeley: University of California Press, 1979.

Brown, Nik. "Hope against Hype—Accountability in Biopasts, Presents and Futures." *Science Studies* 16, no. 2 (December 2003): 3–21.

Brown, Phil, Sabrina McCormick, Brian Mayer, Stephen Zavestoski, Rachel Morello-Frosch, Rebecca Gasior Altman, and Laura Senier. "'A Lab of Our Own': Environmental Causation of Breast Cancer and Challenges to the Dominant Epidemiological Paradigm." *Science, Technology, & Human Values* 31, no. 5 (2006): 499–536.

Bud, Robert. "'Applied Science': A Phrase in Search of a Meaning." *Isis* 103, no. 3 (September 1, 2012): 537–45. doi:10.1086/667977.

———. *Penicillin: Triumph and Tragedy.* Oxford: Oxford University Press, 2007.

———. "Strategy in American Cancer Research after World War II: A Case Study." *Social Studies of Science* 8, no. 4 (November 1978): 425–59.

Burawoy, Michael. *Manufacturing Consent: Changes in the Labor Process under Monopoly Capitalism.* Chicago: University of Chicago Press, 1979.

Bynum, W. F. *Science and the Practice of Medicine in the Nineteenth Century*. Cambridge: Cambridge University Press, 1994.

Cairns, John, Gunther S. Stent, and James D. Watson, eds. *Phage and the Origins of Molecular Biology*. Cold Spring Harbor, NY: Cold Spring Harbor Laboratory of Quantitative Biology, 1966.

Cambrosio, Alberto, and Peter Keating. "'Going Monoclonal': Art, Science, and Magic in the Day-to-Day Use of Hybridoma Technology." *Social Problems* 35, no. 3 (June 1, 1988): 244–60. doi:10.2307/800621.

Cantor, David. "Between Prevention and Therapy: Gio Batta Gori and the National Cancer Institute's Diet, Nutrition and Cancer Programme, 1974–1978." *Medical History* 56, no. 4 (October 24, 2012): 531–61. doi:10.1017/mdh.2012.73.

———. "Cancer, Quackery and the Vernacular Meanings of Hope in 1950s America." *Journal of the History of Medicine and Allied Sciences* 61, no. 3 (February 2006): 324–68. doi:10.1093/jhmas/jrjo48.

———. "The Frustrations of Families: Henry Lynch, Heredity, and Cancer Control, 1962–1975." *Medical History* 50, no. 3 (July 1, 2006): 279–302.

———. "Introduction: Cancer Control and Prevention in the Twentieth Century." *Bulletin of the History of Medicine* 81, no. 1 (2007): 1–38.

———. "Radium and the Origins of the National Cancer Institute." In *Biomedicine in the Twentieth Century: Practices, Policies, and Politics*, edited by Caroline Hannaway, 95–146. Amsterdam: IOS Press, 2008.

———. "Uncertain Enthusiasm: The American Cancer Society, Public Education, and the Problems of the Movie, 1921–1960." *Bulletin of the History of Medicine* 81, no. 1 (2007): 39–69. doi:10.1353/bhm.2007.0002.

Capshew, James H., and Karen A. Rader. "Big Science: Price to the Present." *Osiris*, 7 (January 1, 1992): 3–25.

Carpenter, Daniel. "Is Health Politics Different?" *Annual Review of Political Science* 15, no. 1 (2012): 287–311. doi:10.1146/annurev-polisci-050409-113009.

———. *Reputation and Power: Organizational Image and Pharmaceutical Regulation at the FDA*. Princeton, NJ: Princeton University Press, 2010.

Carpenter, Laura M., and Monica J. Casper. "Global Intimacies: Innovating the HPV Vaccine for Women's Health." *Women's Studies Quarterly* 37, no. 1 (2009): 80–100. doi:10.1353/wsq.0.0161.

Centeno, Miguel A., and Joseph N. Cohen. "The Arc of Neoliberalism." *Annual Review of Sociology* 38 (January 1, 2012): 317–40. doi:10.2307/23254598.

Chandler, Alfred D. *The Visible Hand: The Managerial Revolution in American Business*. Cambridge, MA: Belknap Press of Harvard University Press, 1977.

Chapin, Christy Ford. *Ensuring America's Health: The Public Creation of the Corporate Health Care System*. New York: Cambridge University Press, 2015.

Chiang, Howard Hsueh-Hao. "The Laboratory Technology of Discrete Molecular Separation: The Historical Development of Gel Electrophoresis and the Material

Epistemology of Biomolecular Science, 1945–1970." *Journal of the History of Biology* 42, no. 3 (November 2008): 495–527. doi:10.1007/s10739-008-9169-5.

Chubin, Daryl E., and Kenneth E. Studer. *The Cancer Mission: Social Contexts of Biomedical Research*. Beverly Hills, CA: Sage, 1980.

———. "The Politics of Cancer." *Theory and Society* 6, no. 1 (1978): 55–74.

Clark, Claudia. *Radium Girls, Women and Industrial Health Reform: 1910–1935*. Chapel Hill, NC: University of North Carolina Press, 1997.

Clarke, Adele E., and Joan H. Fujimura, eds. *The Right Tools for the Job: At Work in Twentieth-Century Life Sciences*. Princeton, NJ: Princeton University Press, 1992.

Clarke, Adele E., Janet K. Shim, Laura Mamo, Jennifer Ruth Fosket, and Jennifer R. Fishman. "Biomedicalization: Technoscientific Transformations of Health, Illness, and U.S. Biomedicine." *American Sociological Review* 68, no. 2 (April 1, 2003): 161–94. doi:10.2307/1519765.

Clarke, Sabine. "Pure Science with a Practical Aim: The Meanings of Fundamental Research in Britain, circa 1916–1950." *Isis* 101, no. 2 (June 1, 2010): 285–311. doi:10.1086/653094.

Cohen, Joe. *Shots in the Dark: The Wayward Search for an AIDS Vaccine*. New York: Norton, 2001.

Cohen, Lizabeth. *A Consumer's Republic: The Politics of Mass Consumption in Postwar America*. New York: Knopf, 2003.

Coleman, William. *Biology in the Nineteenth Century: Problems of Form, Function, and Transformation*. New York: Wiley, 1971.

Collins, Harry. *Gravity's Shadow: The Search for Gravitational Waves*. Chicago: University of Chicago Press, 2004.

———. "The TEA Set: Tacit Knowledge and Scientific Networks." In *The Science Studies Reader*, edited by Mario Biagioli, 95–109. New York: Routledge, 1999.

Comfort, Nathaniel. "Rous's Reception: Tumor Viruses in the Context of the Germ Theory." History of Science Society Annual Meeting presentation, Minneapolis, 1995.

Conis, Elena. " 'Do We Really Need Hepatitis B on the Second Day of Life?': Vaccination Mandates and Shifting Representations of Hepatitis B." *Journal of Medical Humanities* 32, no. 2 (June 1, 2011): 155–66. doi:10.1007/s10912-010-9132-2.

Connolly, Cynthia A. *Saving Sickly Children: The Tuberculosis Preventorium in American Life, 1909–1970*. New Brunswick, NJ: Rutgers University Press, 2014.

Cook-Deegan, Robert M. *The Gene Wars: Science, Politics, and the Human Genome*. New York: W. W. Norton, 1994.

Cowie, Jefferson. *The Great Exception: The New Deal and the Limits of American Politics*. Princeton, NJ: Princeton University Press, 2016.

Crane, Johanna Tayloe. *Scrambling for Africa: AIDS, Expertise, and the Rise of American Global Health Science*. Ithaca, NY: Cornell University Press, 2013.

Creager, Angela N. H. " 'EAT: DIE': The Domestication of Environmental Carcinogens in the 1980s." The Hans Rausing Lecture, Uppsala University, 2015.

———. *Life Atomic: A History of Radioisotopes in Science and Medicine*. Chicago: University of Chicago Press, 2013.

———. *The Life of a Virus: Tobacco Mosaic Virus as an Experimental Model, 1930–1965*. Chicago: University of Chicago Press, 2002.

———. "Mobilizing Biomedicine: Virus Research between Lay Health Organizations and the US Federal Government, 1935–1955." In *Biomedicine in the Twentieth Century: Practices, Policies, and Politics*, edited by Caroline Hannaway, 171–201. Amsterdam: IOS Press, 2008.

———. "Paradigms and Exemplars Meet Biomedicine." In *Kuhn's Structure of Scientific Revolutions—50 Years On*, edited by William J. Devlin and Alisa Bokulich, 151–66. Chicago: University of Chicago Press, 2015.

———. "The Political Life of Mutagens: A History of the Ames Test." In *Making Mutations: Objects, Practices, Contexts*, 285–306. Max Planck Institute for the History of Science Preprints 383. Berlin: Max-Planck-Inst. für Wissenschaftsgeschichte, 2010.

———. "Timescapes of Radioactive Tracers in Biochemistry and Ecology." *History and Philosophy of the Life Sciences* 35, no. 1 (2013): 83–89.

———. "'What Blood Told Dr Cohn': World War II, Plasma Fractionation, and the Growth of Human Blood Research." *Studies in History and Philosophy of Science Part C: Studies in History and Philosophy of Biological and Biomedical Sciences* 30, no. 3 (September 1999): 377–405. doi:10.1016/S1369-8486(99)00017-5.

Creager, Angela N. H., and Jean-Paul Gaudillière. "Experimental Technologies and Techniques of Visualisation: Cancer as a Viral Epidemic, 1930–1960." In Gaudillière and Löwy, *Heredity and Infection*, 203–60.

Creager, Angela N. H., and Hannah Landecker. "Technical Matters: Method, Knowledge and Infrastructure in Twentieth-Century Life Science." *Nature Methods* 6, no. 10 (October 2009): 701–5. doi:10.1038/nmeth1009-701.

Creager, Angela N. H., Elizabeth Lunbeck, and M. Norton Wise, eds. *Science without Laws: Model Systems, Cases, Exemplary Narratives*. Durham, NC: Duke University Press, 2007.

Crowe, Nathan. "Cancer, Conflict, and the Development of Nuclear Transplantation Techniques." *Journal of the History of Biology* 47, no. 1 (2014): 63–105.

———. "Joshua Lederberg's 'Euphenics': The Construction of Human Cloning Narratives in the 1960s." Presented at the History of Science Society Annual Meeting, Chicago, November 7, 2014.

Cummiskey, Julia. "Placing Global Science in Africa: International Networks, Local Places, and Virus Research in Uganda, 1936–2000." PhD diss., Johns Hopkins University, 2017.

Cunningham, Andrew. "Transforming Plague: The Laboratory and the Identity of Infectious Disease." In Cunningham and Williams, *Laboratory Revolution in Medicine*, 209–45.

Cunningham, Andrew, and Perry Williams, eds. *The Laboratory Revolution in Medicine*. Cambridge: Cambridge University Press, 1992.

Daemmrich, Arthur. "BioRisk: Interlukin-2 from Laboratory to Market in the United States and Germany." In *The Risks of Medical Innovation: Risk Perception and Assessment in Historical Context*, edited by Thom Schlich and Ulrich Tröhler, 242–61. London: Routledge, 2006.

Daemmrich, Arthur, and Leah Shaper. "The Gordon Research Conferences as Scientific Infrastructure." *Bulletin of the History of Chemistry* 3, no. 2 (2008): 94–102.

Daston, Lorraine. "The Coming into Being of Scientific Objects." In *Biographies of Scientific Objects*, edited by Lorraine Daston, 1–14. Chicago: University of Chicago Press, 2000.

Daston, Lorraine, and Peter Galison. *Objectivity*. New York: Zone Books, 2007.

Daston, Lorraine, and H. Otto Sibum. "Introduction: Scientific Personae and Their Histories." *Science in Context* 16, nos. 1–2 (2003): 1–8. doi:10.1017/S026988970300067X.

Davies, Gail, Emma Frow, and Sabina Leonelli. "Bigger, Faster, Better? Rhetorics and Practices of Large-Scale Research in Contemporary Bioscience." *BioSocieties* 8, no. 4 (December 1, 2013): 386–96. doi:10.1057/biosoc.2013.26.

De Chadarevian, Soraya. *Designs for Life: Molecular Biology after World War II*. Cambridge: Cambridge University Press, 2002.

———. "Microstudies versus Big Picture Accounts?" *Studies in History and Philosophy of Science Part C: Studies in History and Philosophy of Biological and Biomedical Sciences* 40, no. 1 (March 2009): 13–19. doi:10.1016/j.shpsc .2008.12.003.

———. "Whose Turn? Chromosome Research and the Study of the Human Genome." *Journal of the History of Biology* 51, no. 4 (December 2018): 631–55. doi:10.1007 /s10739-017-9486-7.

De Chadarevian, Soraya, and Harmke Kamminga. Introduction to de Chadarevian and Kamminga, *Molecularizing Biology and Medicine*, 1–16.

———, eds. *Molecularizing Biology and Medicine: New Practices and Alliances, 1910s–1970s*. Amsterdam: Harwood Academic, 1998.

De Chadarevian, Soraya, and Bruno Strasser. "Molecular Biology in Postwar Europe: Towards a 'Glocal' Picture." *Studies in History and Philosophy of Science Part C: Studies in History and Philosophy of Biological and Biomedical Sciences* 33, no. 3 (September 2002): 361–65. doi:10.1016/S1369-8486(02)00009-2.

Deering, Christopher J., and Steven S. Smith. *Committees in Congress*. 3rd ed. Washington, DC: CQ Press, 1997.

DeJong-Lambert, William, and Nikolai Krementsov. "On Labels and Issues: The Lysenko Controversy and the Cold War." *Journal of the History of Biology* 45, no. 3 (2012): 373–88.

Dennis, Michael. "Accounting for Research: New Histories of Corporate Laboratories and the Social History of American Science." *Social Studies of Science* 17, no. 3 (1987): 479–518.

———. "Reconstructing Sociotechnical Order: Vannevar Bush and US Science Policy." In *States of Knowledge: The Co-Production of Science and Social Order*, edited by Sheila Jasanoff, 225–53. London: Routledge, 2004.

Derickson, Alan. *Health Security for All: Dreams of Universal Health Care in America*. Baltimore: Johns Hopkins University Press, 2005.

Derrida, Jacques. "Declarations of Independence." *New Political Science* 7, no. 1 (June 1, 1986): 7–15. doi:10.1080/07393148608429608.

———. "No Apocalypse, Not Now (Full Speed Ahead, Seven Missiles, Seven Missives)." Translated by Catherine Porter and Philip Lewis. *Diacritics* 14, no. 2 (1984): 20–31. doi:10.2307/464756.

DeVita, Vincent. "The Governance of Science at the National Cancer Institute: A Perspective on Misperceptions." *Cancer Research* 43 (1983): 3969–73.

DeVita, Vincent T., and Edward Chu. "A History of Cancer Chemotherapy." *Cancer Research* 68, no. 21 (November 1, 2008): 8643–53. doi:10.1158/0008-5472 .CAN-07-6611.

DeVita, Vincent T., Jr., and Elizabeth DeVita-Raeburn. *The Death of Cancer: After Fifty Years on the Front Lines of Medicine, a Pioneering Oncologist Reveals Why the War on Cancer Is Winnable—and How We Can Get There*. New York: Sarah Crichton Books, 2015.

Drew, Elizabeth. "The Health Syndicate: Washington's Noble Conspirators." *Atlantic Monthly*, December 1967, 75–82.

Dubos, René J., and Jean Dubos. *The White Plague: Tuberculosis, Man and Society*. Boston: Little, Brown, 1952.

Dupree, A. Hunter. "The Great Instauration of 1940: The Organization of Scientific Research for War." In *The Twentieth-Century Sciences: Studies in the Biography of Ideas*, edited by Gerald James Holton, 443–67. New York: Norton, 1972.

Durant, John. "'Refrain from Using the Alphabet': How Community Outreach Catalyzed the Life Sciences at MIT." In *Becoming MIT: Moments of Decision*, edited by David Kaiser, 145–63. Cambridge, MA: MIT Press, 2010.

Edgerton, David. Introduction to *Industrial Research and Innovation in Business*, edited by David Edgerton, x–xvi. Cheltenham, UK: Edward Elgar, 1996.

Edwards, Paul N. *The Closed World: Computers and the Politics of Discourse in Cold War America*. Cambridge, MA: MIT Press, 1996.

———. "Infrastructure and Modernity: Force, Time, and Social Organization in the History of Sociotechnical Systems." In *Modernity and Technology*, edited by Thomas J. Misa, Philip Brey, and Andrew Feenberg, 185–226. Cambridge, MA: MIT Press, 2003.

Edwards, Paul, Geoffrey Bowker, Steven Jackson, and Robin Williams. "Introduction: An Agenda for Infrastructure Studies." *Journal of the Association for Information Systems* 10, no. 5 (May 2009): 364–74. doi:10.17705/1jais .00200.

Ekbladh, David. *The Great American Mission: Modernization and the Construction of an American World Order*. Princeton, NJ: Princeton University Press, 2010.

Emrich, John S. "Dr. Genelove: How Scientists Learned to Stop Worrying and Love Recombinant DNA." PhD diss., George Washington University, 2009.

Endersby, Jim. *A Guinea Pig's History of Biology*. Cambridge, MA: Harvard University Press, 2007.

Engerman, David C. "Introduction: Histories of the Future and the Futures of History." *American Historical Review* 117, no. 5 (December 1, 2012): 1402–10. doi:10.1093/ahr/117.5.1402.

Ensmenger, Nathan. "The Multiple Meanings of a Flowchart." *Information & Culture* 51, no. 3 (2016): 321–51. doi:10.1353/lac.2016.0013.

Epstein, Steven. *Impure Science: AIDS, Activism, and the Politics of Knowledge*. Berkeley: University of California Press, 1996.

Erdey, Nancy Carol. "Armor of Patience: The National Cancer Institute and the Development of Medical Research Policy in the United States, 1937–1971." PhD diss., Case Western Reserve University, 1995.

Ewing, James. *Neoplastic Diseases: A Treatise on Tumors*. Philadelphia: W. B. Saunders, 1919.

———. "Pathological Aspects of Some Problems of Cancer Research." *Journal of Cancer Research* 1, no. 1 (January 1916): 71–86.

Feinstein, Alvan R. "The Intellectual Crisis in Clinical Science: Medaled Models and Muddled Mettle." *Perspectives in Biology and Medicine* 30, no. 2 (1987): 215–30. doi:10.1353/pbm.1987.0047.

Feldberg, Georgina D. *Disease and Class: Tuberculosis and the Shaping of Modern North American Society*. New Brunswick, NJ: Rutgers University Press, 1995.

Field, Robert I. "How the Government Created the Hospital Industry." In *Mother of Invention: How the Government Created Free-Market Health Care*, edited by Robert I. Field, 85–121. New York: Oxford University Press, 2013.

Fischer, Michael M. J. *Anthropological Futures*. Durham, NC: Duke University Press, 2009.

Fisher, Susie. "Not beyond Reasonable Doubt: Howard Temin's Provirus Hypothesis Revisited." *Journal of the History of Biology* 43, no. 4 (September 2009): 661–96. doi:10.1007/s10739-009-9202-3.

Forman, Paul. "Behind Quantum Electronics: National Security as Basis for Physical Research in the United States, 1940–1960." *Historical Studies in the Physical and Biological Sciences* 18, no. 1 (January 1, 1987): 149–229.

Fortun, Michael. "Genomics Scandals and Other Volatiles of Promising." In *Lively Capital: Biotechnologies, Ethics, and Governance in Global Markets*, edited by Kaushik Sunder Rajan, 329–53. Durham, NC: Duke University Press, 2012.

———. "The Human Genome Project and Acceleration of Biotechnology." In *Private Science: Biotechnology and the Rise of the Molecular Sciences*, edited by Arnold Thackray, 182–201. Philadelphia: University of Pennsylvania Press, 1998.

———. *Promising Genomics: Iceland and DeCODE Genetics in a World of Specu-lation*. Berkeley: University of California Press, 2008.

Fortun, Michael, and S. S. Schweber. "Scientists and the Legacy of World War II: The Case of Operations Research (OR)." *Social Studies of Science* 23, no. 4 (November 1, 1993): 595–642.

Fox, Daniel M. "The Politics of the NIH Extramural Program, 1937–1950." *Jour-nal of the History of Medicine and Allied Sciences* 42, no. 4 (October 1, 1987): 447–66. doi:10.1093/jhmas/42.4.447.

Fraser, Steve, and Gary Gerstle. Introduction to *The Rise and Fall of the New Deal Order, 1930–1980*, edited by Steve Fraser and Gary Gerstle, ix–xxv. Princeton, NJ: Princeton University Press, 1989.

Fujimura, Joan H. *Crafting Science: A Sociohistory of the Quest for the Genetics of Cancer*. Cambridge, MA: Harvard University Press, 1996.

———. "Ecologies of Action: Recombining Genes, Molecularizing Cancer, and Transforming Biology." In *Ecologies of Knowledge: Work and Politics in Sci-ence and Technology*, edited by Susan Leigh Star, 302–46. Albany: State Uni-versity of New York Press, 1995.

———. "The Molecular Biological Bandwagon in Cancer Research: Where Social Worlds Meet." *Social Problems* 35, no. 3 (June 1988): 261–83.

Fujimura, Joan H., and Danny Y. Chou. "Dissent in Science: Styles of Scientific Practice and the Controversy over the Cause of AIDS." *Social Science & Medi-cine* 38, no. 8 (April 1994): 1017–36. doi:10.1016/0277-9536(94)90219-4.

Funigiello, Philip J. *Chronic Politics: Health Care Security from FDR to George W. Bush*. Lawrence: University Press of Kansas, 2005.

Galison, Peter. *Image and Logic: A Material Culture of Microphysics*. Chicago: University of Chicago Press, 1997.

———. "The Ontology of the Enemy: Norbert Wiener and the Cybernetic Vi-sion." *Critical Inquiry* 21, no. 1 (Autumn 1994): 228–66.

Galison, Peter, and Bruce William Hevly, eds. *Big Science: The Growth of Large-Scale Research*. Stanford, CA: Stanford University Press, 1992.

Gallo, Robert C. "History of the Discoveries of the First Human Retroviruses: HTLV-1 and HTLV-2." *Oncogene* 24, no. 39 (2005): 5926–30.

———. *Virus Hunting: AIDS, Cancer, and the Human Retrovirus: A Story of Sci-entific Discovery*. New York: Basic Books, 1991.

Gallo, Robert C., and Luc Montagnier. "The Discovery of HIV as the Cause of AIDS." *New England Journal of Medicine* 349, no. 24 (December 11, 2003): 2283–85. doi:10.1056/NEJMp038194.

Galperin, Charles. "Virus, provirus et cancer." *Revue d'Histoire des Sciences* 47, no. 1 (1994): 7–56. doi:10.3406/rhs.1994.1189.

Garb, Solomon. *Cure for Cancer: A National Goal*. New York: Springer, 1968.

Gardner, Kirsten E. *Early Detection: Women, Cancer, and Awareness Campaigns in the Twentieth-Century United States*. Chapel Hill: University of North Carolina Press, 2006.

Gaudillière, Jean-Paul. "Cancer." In *The Modern Biological and Earth Sciences*, edited by Peter J. Bowler and John V. Pickstone, 6:486–503. Cambridge: Cambridge University Press, 2009.

——. "Le cancer entre infection et hérédité: Gènes, virus et souris au National Cancer Institute (1937–1977)." *Revue d'Histoire des Sciences* 47, no. 1 (1994): 57–90. doi:10.3406/rhs.1994.1190.

——. "Circulating Mice and Viruses: The Jackson Memorial Laboratory, the National Cancer Institute, and the Genetics of Breast Cancer, 1930–1965." In *The Practices of Human Genetics*, edited by Michael Fortun and Everett Mendelsohn, 89–124. Dordrecht, Netherlands: Kluwer Academic, 1997.

——. "Genesis and Development of a Biomedical Object: Styles of Thought, Styles of Work and the History of the Sex Steroids." *Studies in History and Philosophy of Science Part C: Studies in History and Philosophy of Biological and Biomedical Sciences* 35, no. 3 (September 2004): 525–43. doi:10.1016/j.shpsc.2004.06.003.

——. *Inventer la biomédecine: La France, l'Amérique et la production des savoirs du vivant, 1945–1965*. Paris: La Découverte, 2002.

——. "The Molecularization of Cancer Etiology in the Postwar United States: Instruments, Politics, and Management." In de Chadarevian and Kamminga, *Molecularizing Biology and Medicine*, 139–70.

——. "Oncogenes as Metaphors for Human Cancer: Articulating Laboratory Practices and Medical Demands." In *Medicine and Change: Historical and Sociological Studies of Medical Innovation*, edited by Ilana Löwy, 213–47. Paris: INSERM, 1993.

——. "Rockefeller Strategies for Scientific Medicine: Molecular Machines, Viruses and Vaccines." *Studies in History and Philosophy of Science Part C: Studies in History and Philosophy of Biological and Biomedical Sciences* 31, no. 3 (September 2000): 491–509. doi:10.1016/S1369-8486(00)00017-0.

Gaudillière, Jean-Paul, and Ilana Löwy. "General Introduction." In *The Invisible Industrialist: Manufactures and the Production of Scientific Knowledge*, edited by Jean-Paul Gaudillière and Ilana Löwy, 3–15. Houndmills, Basingstoke, Hampshire, UK: Macmillan, 1998.

——, eds. *Heredity and Infection: The History of Disease Transmission*. New York: Routledge, 2001.

Geiger, Roger L. *Research and Relevant Knowledge: American Research Universities since World War II*. 2nd ed. New Brunswick, NJ: Transaction, 2004.

Geison, Gerald L. *The Private Science of Louis Pasteur*. Princeton, NJ: Princeton University Press, 1995.

Giacomoni, Dario. "Origin of DNA:RNA Hybridization." *Journal of the History of Biology* 26, no. 1 (1993): 89–107.

Gieryn, Thomas F. "Boundary-Work and the Demarcation of Science from Non-Science: Strains and Interests in the Professional Ideologies of Scientists." *American Sociological Review* 48, no. 6 (1983): 781–95.

Gilman, Nils. *Mandarins of the Future: Modernization Theory in Cold War America*. Baltimore: Johns Hopkins University Press, 2003.

Gilmore, Glenda. *Gender and Jim Crow: Women and the Politics of White Supremacy in North Carolina, 1896–1920*. Chapel Hill: University of North Carolina Press, 1996.

Giraudeau, Martin. "Performing Physiocracy." *Journal of Cultural Economy* 3, no. 2 (July 1, 2010): 225–42. doi:10.1080/17530350.2010.494125.

Gitelman, Lisa. *Paper Knowledge: Toward a Media History of Documents*. Durham, NC: Duke University Press, 2014.

Good, Mary-Jo DelVecchio. "The Practice of Biomedicine and the Discourse on Hope." In *Anthropologies of Medicine*, edited by Beatrix Pfleiderer and Gilles Bibeau, 121–35. Brunswick, Germany: Vieweg, 1991.

Gooday, Graeme. "'Vague and Artificial': The Historically Elusive Distinction between Pure and Applied Science." *Isis* 103, no. 3 (September 1, 2012): 546–54. doi:10.1086/667978.

Goodman, Jordan, and Vivien Walsh. *The Story of Taxol: Nature and Politics in the Pursuit of an Anti-Cancer Drug*. Cambridge: Cambridge University Press, 2001.

Gordon, Colin. *Dead on Arrival: The Politics of Health Care in Twentieth-Century America*. Princeton, NJ: Princeton University Press, 2003.

———. "Why No National Health Insurance in the U.S.? The Limits of Social Provision in War and Peace, 1941–1948." *Journal of Policy History* 9, no. 3 (July 1997): 277–310. doi:10.1017/S0898030600006035.

Gosset, D. P. "L'hôpital des cancérés: Foundation du chanoine Godinot." *Union Medicale du Nord-Est*, no. 2 (February 1926): 17–26.

Gradmann, Christoph. *Laboratory Disease: Robert Koch's Medical Bacteriology*. Translated by Elborg Forster. Baltimore: Johns Hopkins University Press, 2009.

Graham, Frank L. "This Week's Citation Classic." *Current Content*, no. 46 (November 14, 1988): 16.

Grant, Robert P. "National Biomedical Research Agencies: A Comparative Study of Fifteen Countries." *Minerva* 4, no. 4 (September 1, 1966): 466–88. doi:10.1007/BF02207979.

Greenberg, Daniel S. *The Politics of Pure Science*. New York: New American Library, 1968.

Grmek, Mirko Drazen. *History of AIDS: Emergence and Origin of a Modern Pandemic*. Translated by Jacalyn Duffin and Russel Mauitz. Princeton, NJ: Princeton University Press, 1990.

Haber, Samuel. *Efficiency and Uplift: Scientific Management in the Progressive Era, 1890–1920*. Chicago: University of Chicago Press, 1964.

Hacker, Jacob. *The Divided Welfare State: The Battle over Private and Public Social Benefits in the United States*. Cambridge: Cambridge University Press, 2002.

———. "The Historical Logic of National Health Insurance: Structure and Sequence in the Development of British, Canadian, and U.S. Medical Policy." *Studies in American Political Development* 12, no. 1 (April 1998): 57–130.

Hansen, Bert. *Picturing Medical Progress from Pasteur to Polio: A History of Mass Media Images and Popular Attitudes in America*. New Brunswick, NJ: Rutgers University Press, 2009.

Harden, Victoria Angela. *AIDS at 30: A History*. Dulles, VA: Potomac Books, 2012.

———. *Inventing the NIH: Federal Biomedical Research Policy, 1887–1937*. Baltimore: Johns Hopkins University Press, 1986.

———. "The NIH and Biomedical Research on AIDS." In *AIDS and the Public Debate: Historical and Contemporary Perspectives*, edited by Caroline Hannaway, Victoria Angela Harden, and John Parascandola, 30–46. Amsterdam: IOS Press, 1995.

Harris, Bernard. "Public Health, Nutrition, and the Decline of Mortality: The McKeown Thesis Revisited." *Social History of Medicine* 17, no. 3 (December 1, 2004): 379–407. doi:10.1093/shm/17.3.379.

Hart, David M. "Herbert Hoover's Last Laugh: The Enduring Significance of the 'Associative State' in the United States." *Journal of Policy History* 10, no. 4 (October 1998): 419–44. doi:10.1017/S0898030600007156.

Harvey, David. *A Brief History of Neoliberalism*. Oxford: Oxford University Press, 2005.

Hays, Samuel P. *Conservation and the Gospel of Efficiency: The Progressive Conservation Movement, 1890–1920*. College ed. New York: Atheneum, 1972.

Heims, Steve J. *The Cybernetics Group*. Cambridge, MA: MIT Press, 1991.

Heller, John R. "Cornelius Packard Rhoads, Leader in Cancer Research." *Science*, 131, no. 3399 (February 19, 1960): 486–87.

Helvoort, Ton Van. "A Century of Research into the Cause of Cancer: Is the New Oncogene Paradigm Revolutionary?" *History and Philosophy of the Life Sciences* 21 (1999): 293–330.

———. "The Construction of Bacteriophage as Bacterial Virus: Linking Endogenous and Exogenous Thought Styles." *Journal of the History of Biology* 27, no. 1 (April 1, 1994): 91–139.

———. "A Dispute over Scientific Credibility: The Struggle for an Independent Institute for Cancer Research in Pre–World War II Berlin." *Studies in History and Philosophy of Science Part C: Studies in History and Philosophy of Biological and Biomedical Sciences* 31, no. 2 (June 2000): 315–54. doi:10.1016/S1369-8486(99)00030-8.

———. "History of Virus Research in the Twentieth Century: The Problem of Conceptual Continuity." *History of Science* 32, no. 2 (1994): 185–235.

———. "The Start of a Cancer Research Tradition: Peyton Rous, James Ewing, and Viruses as a Cause of Cancer." In *Creating a Tradition of Biomedical Research: Contributions to the History of the Rockefeller University*, edited by Darwin H. Stapleton, 191–210. New York: Rockefeller University Press, 2004.

Hess, David J. "Technology and Alternative Cancer Therapies: An Analysis of Heterodoxy and Constructivism." *Medical Anthropology Quarterly*, n.s., 10, no. 4 (December 1, 1996): 657–74.

Hilgartner, Stephen. "Constituting Large-Scale Biology: Building a Regime of Governance in the Early Years of the Human Genome Project." *BioSocieties* 8, no. 4 (December 1, 2013): 397–416. doi:10.1057/biosoc.2013.31.

———. *Reordering Life: Knowledge and Control in the Genomics Revolution*. Cambridge, MA: MIT Press, 2017.

Hixson, Joseph R. *The Patchwork Mouse*. Garden City, NY: Anchor Press, 1976.

Hoffman, Beatrix Rebecca. *The Wages of Sickness: The Politics of Health Insurance in Progressive America*. Chapel Hill: University of North Carolina Press, 2001.

Hogan, Andrew J. *Life Histories of Genetic Disease: Patterns and Prevention in Postwar Medical Genetics*. Baltimore: Johns Hopkins University Press, 2016.

Hollinger, David A. "The Defense of Democracy and Robert K. Merton's Formulation of the Scientific Ethos." In *Knowledge and Society: Studies in the Sociology of Culture Past and Present*, edited by Robert Alun Jones and Henrika Kuklick, 4:1–15. Greenwich, CT: JAI Press, 1983.

———. "Free Enterprise and Free Inquiry: The Emergence of Laissez-Faire Communitarianism in the Ideology of Science in the United States." *New Literary History* 21, no. 4 (October 1, 1990): 897–919. doi:10.2307/469191.

Hollingsworth, J. Rogers. "Institutionalizing Excellence in Biomedical Research: The Case of the Rockefeller University." In *Creating a Tradition of Biomedical Research: Contributions to the History of the Rockefeller University*, edited by Darwin H. Stapleton, 17–63. New York: Rockefeller University Press, 2004.

Hollingsworth, J. Rogers, Jerald Hage, and Robert Hanneman. *State Intervention in Medical Care: Consequences for Britain, France, Sweden, and the United States, 1890–1970*. Ithaca, NY: Cornell University Press, 1990.

Holloway, Thomas M., and Jane S. Reeb. "A Price Index for Biomedical Research and Development." *Public Health Reports* 104, no. 1 (January 1, 1989): 11–13.

Holmes, Frederic Lawrence. *Reconceiving the Gene: Seymour Benzer's Adventures in Phage Genetics*. Edited by William C. Summers. New Haven, CT: Yale University Press, 2006.

Hounshell, David. "The Cold War, RAND, and the Generation of Knowledge, 1946–1962." *Historical Studies in the Physical and Biological Sciences* 27, no. 2 (January 1, 1997): 237–67.

Hounshell, David A., and John K. Smith. *Science and Corporate Strategy: Du Pont R&D, 1902–1980*. Cambridge: Cambridge University Press, 1988.

Hoyt, Kendall. *Long Shot: Vaccines for National Defense.* Cambridge, MA: Harvard University Press, 2012. doi:10.4159/harvard.9780674063150.

Hughes, Sally Smith. *The Virus: A History of the Concept.* London: Heinemann Educational Books, 1977.

Hughes, Thomas Parke. *American Genesis: A Century of Invention and Technological Enthusiasm, 1870–1970.* New York: Viking, 1989.

Hull, Matthew S. *Government of Paper: The Materiality of Bureaucracy in Urban Pakistan.* Berkeley: University of California Press, 2012.

Hunter, T., and J. Simon. "A Not So Brief History of the Oncogene Meeting and Its Cartoons." *Oncogene* 26, no. 9 (2007): 1260–67. doi:10.1038/sj.onc.1210262.

Hurley, Andrew. *Environmental Inequalities: Class, Race, and Industrial Pollution in Gary, Indiana, 1945–1980.* Chapel Hill: University of North Carolina Press, 1995.

Huzair, Farah, and Steve Sturdy. "Biotechnology and the Transformation of Vaccine Innovation: The Case of the Hepatitis B Vaccines 1968–2000." *Studies in History and Philosophy of Science Part C: Studies in History and Philosophy of Biological and Biomedical Sciences* 64 (August 2017): 11–21. doi:10.1016/j.shpsc.2017.05.004.

Jain, S. Lochlann. *Malignant: How Cancer Becomes Us.* Berkeley: University of California Press, 2013.

Jameson, Fredric. "Utopia as Method, or the Uses of the Future." In *Utopia/Dystopia: Conditions of Historical Possibility*, edited by Michael D. Gordin, Helen Tilley, and Gyan Prakash, 21–44. Princeton, NJ: Princeton University Press, 2010.

Jardini, David Raymond. "Out of the Blue Yonder: The RAND Corporation's Diversification into Social Welfare Research, 1946–1968." PhD diss., Carnegie Mellon University, 1996.

Jasanoff, Sheila. "Future Imperfect." In *Dreamscapes of Modernity: Sociotechnical Imaginaries and the Fabrication of Power*, edited by Sheila Jasanoff and Sang-Hyun Kim, 1–29. Chicago: University of Chicago Press, 2015.

———. "Introduction: Rewriting Life, Reframing Rights." In *Reframing Rights: Bioconstitutionalism in the Genetic Age*, edited by Sheila Jasanoff, 1–22. Cambridge, MA: MIT Press, 2011.

———. "Ordering Knowledge, Ordering Society." In *States of Knowledge: The Co-Production of Science and Social Order*, edited by Sheila Jasanoff, 13–45. London: Routledge, 2004.

———. "Science, Politics, and the Renegotiation of Expertise at EPA." *Osiris* 7 (1992): 194–217.

Jasanoff, Sheila, and Ingrid Metzler. "Borderlands of Life: IVF Embryos and the Law in the United States, United Kingdom, and Germany." *Science, Technology & Human Values*, in press. Accessed April 20, 2018. doi:10.1177/0162243917753990.

Johnson, Stephen B. "From Concurrency to Phased Planning: An Episode in the History of Systems Management." In *Systems, Experts, and Computers: The Systems Approach in Management and Engineering, World War II and After*, edited by Agatha C. Hughes and Thomas Parke Hughes, 93–112. Cambridge, MA: MIT Press, 2000.

Jones, David S., Scott H. Podolsky, and Jeremy A. Greene. "The Burden of Disease and the Changing Task of Medicine." *New England Journal of Medicine* 366, no. 25 (2012): 2333–38. doi:10.1056/NEJMp1113569.

Jones, Mark Peter. "Entrepreneurial Science: The Rules of the Game." *Social Studies of Science* 39, no. 6 (December 1, 2009): 821–51. doi:10.1177/03063127 09104434.

Kaiser, David. "Booms, Busts, and the World of Ideas: Enrollment Pressures and the Challenge of Specialization." *Osiris* 27, no. 1 (January 1, 2012): 276–302. doi:10.1086/667831.

———. *Drawing Theories Apart: The Dispersion of Feynman Diagrams in Postwar Physics*. Chicago: University of Chicago Press, 2005.

Kalberer, John T., and Guy R. Newell. "Funding Impact of the National Cancer Act and Beyond." *Cancer Research* 39, no. 10 (October 1, 1979): 4274–84.

Kamlet, Mark S., David C. Mowery, and Tsai-Tsu Su. "Upsetting National Priorities? The Reagan Administration's Budgetary Strategy." *American Political Science Review* 82, no. 4 (1988): 1293–1307. doi:10.2307/1961761.

Katznelson, Ira. "Was the Great Society a Lost Opportunity?" In *The Rise and Fall of the New Deal Order, 1930–1980*, edited by Steve Fraser and Gary Gerstle, 185–211. Princeton, NJ: Princeton University Press, 1989.

Kay, Lily E. "Conceptual Models and Analytical Tools: The Biology of Physicist Max Delbrück." *Journal of the History of Biology* 18, no. 2 (July 1, 1985): 207–46. doi:10.2307/4330932.

———. "Cybernetics, Information, Life: The Emergence of Scriptural Representations of Heredity." *Configurations* 5, no. 1 (1997): 23–91. doi:10.1353 /con.1997.0004.

———. *The Molecular Vision of Life: Caltech, the Rockefeller Foundation, and the Rise of the New Biology*. New York: Oxford University Press, 1993.

———. *Who Wrote the Book of Life? A History of the Genetic Code*. Stanford, CA: Stanford University Press, 2000.

Keating, Peter, and Alberto Cambrosio. "Cancer Clinical Trials: The Emergence and Development of a New Style of Practice." *Bulletin of the History of Medicine* 81, no. 1 (2007): 197–223.

———. *Cancer on Trial: Oncology as a New Style of Practice*. Chicago: University of Chicago Press, 2012.

———. "Does Biomedicine Entail the Successful Reduction of Pathology to Biology?" *Perspectives in Biology and Medicine* 47, no. 3 (2004): 357–71.

———. "From Screening to Clinical Research: The Cure of Leukemia and the Early Development of the Cooperative Oncology Groups, 1955–1966." *Bulletin of the History of Medicine* 76, no. 2 (2002): 299–334. doi:10.1353/bhm.2002.0074.

Kellehear, Allan. *A Social History of Dying*. Cambridge: Cambridge University Press, 2007.

Kessler-Harris, Alice. *In Pursuit of Equity: Women, Men, and the Quest for Economic Citizenship in 20th-Century America*. Oxford: Oxford University Press, 2001.

Kevles, Daniel J. "Big Science and Big Politics in the United States: Reflections on the Death of the SSC and the Life of the Human Genome Project." *Historical Studies in the Physical and Biological Sciences* 27, no. 2 (January 1, 1997): 269–97.

———. "Genetics in the United States and Great Britain, 1890–1930: A Review with Speculations." *Isis* 71, no. 3 (September 1980): 441–55.

———. "Howard Temin: Rebel of Evidence and Reason." In *Rebels, Mavericks, and Heretics in Biology*, edited by Oren Solomon Harman and Michael Dietrich, 248–64. New Haven, CT: Yale University Press, 2008.

———. *In the Name of Eugenics: Genetics and the Uses of Human Heredity*. New York: Knopf, 1985.

———. "The National Science Foundation and the Debate over Postwar Research Policy, 1942–1945: A Political Interpretation of Science—the Endless Frontier." *Isis* 68, no. 1 (March 1, 1977): 5–26.

———. *The Physicists: The History of a Scientific Community in Modern America*. New York: Knopf, 1977.

———. "Pursuing the Unpopular: A History of Courage, Viruses, and Cancer." In *Hidden Histories of Science*, edited by Robert B. Silvers, 69–114. New York: New York Review of Books, 1995.

———. "Renato Dulbecco and the New Animal Virology: Medicine, Methods, and Molecules." *Journal of the History of Biology* 26, no. 3 (October 1, 1993): 409–42.

Kevles, Daniel J., and Gerald L. Geison. "The Experimental Life Sciences in the Twentieth Century." *Osiris*, 10 (January 1995): 97–121.

Kevles, Daniel J., and Leroy E. Hood, eds. *The Code of Codes: Scientific and Social Issues in the Human Genome Project*. Cambridge, MA: Harvard University Press, 1992.

Keyes, Martha E. "The Prion Challenge to the 'Central Dogma' of Molecular Biology, 1965–1991; Part 1 Prelude to Prions." *Studies in History and Philosophy of Science Part C: Studies in History and Philosophy of Biological and Biomedical Sciences* 30, no. 1 (March 1999): 1–19. doi:10.1016/S1369-8486(98)00028-4.

Khot, Sandeep, Buhm Soon Park, and W. T. Longstreth. "The Vietnam War and Medical Research: Untold Legacy of the U.S. Doctor Draft and the NIH 'Yellow Berets.'" *History of Academic Medicine* 86, no. 4 (April 2011): 502–8.

Kiechle, Melanie A. *Smell Detectives: An Olfactory History of Nineteenth-Century Urban America*. Seattle: University of Washington Press, 2017.

Kimmelman, Barbara, and Diane Paul. "Mendel in America: Theory and Practice, 1900–1919." In *The American Development of Biology*, edited by Ronald Rainger, Keith Rodney Benson, and Jane Maienschein, 281–310. Philadelphia: University of Pennsylvania Press, 1988.

Kirk, Robert G. W. "A Brave New Animal for a Brave New World: The British Laboratory Animals Bureau and the Constitution of International Standards of Laboratory Animal Production and Use, circa 1947–1968." *Isis* 101, no. 1 (March 1, 2010): 62–94.

———. "'Standardization through Mechanization': Germ-Free Life and the Engineering of the Ideal Laboratory Animal." *Technology and Culture* 53, no. 1 (January 2012): 61–93.

———. "'Wanted—Standard Guinea Pigs': Standardization and the Experimental Animal Market in Britain ca. 1919–1947." *Studies in History and Philosophy of Science Part C: Studies in History and Philosophy and Biomedical Sciences* 39, no. 3 (September 2008): 280–91. doi:10.1016/j.shpsc.2008.06.002.

Kisacky, Jeanne Susan. *Rise of the Modern Hospital: An Architectural History of Health and Healing, 1870–1940*. Pittsburgh: University of Pittsburgh Press, 2017.

Klein, George. *The Atheist and the Holy City: Encounters and Reflections*. Translated by Theodore Freidmann and Ingrid Freidmann. Cambridge, MA: MIT Press, 1990.

———. "The Tale of the Great Cuckoo Egg." *Nature* 400, no. 6744 (1999): 515. doi:10.1038/22906.

Klein, Jennifer. *For All These Rights: Business, Labor, and the Shaping of America's Public-Private Welfare State*. Princeton, NJ: Princeton University Press, 2003.

Knorr-Cetina, K. *Epistemic Cultures: How the Sciences Make Knowledge*. Cambridge, MA: Harvard University Press, 1999.

Kohler, Robert E. *From Medical Chemistry to Biochemistry: The Making of a Biomedical Discipline*. Cambridge: Cambridge University Press, 1982.

———. *Landscapes and Labscapes: Exploring the Lab-Field Border in Biology*. Chicago: University of Chicago Press, 2002.

———. *Lords of the Fly: Drosophila Genetics and the Experimental Life*. Chicago: University of Chicago Press, 1994.

———. "The Management of Science: The Experience of Warren Weaver and the Rockefeller Foundation Programme in Molecular Biology." *Minerva* 14, no. 3 (1976): 279–306. doi:10.1007/BF01096274.

———. *Partners in Science: Foundations and Natural Scientists, 1900–1945*. Chicago: University of Chicago Press, 1991.

Korstad, Robert Rodgers. *Civil Rights Unionism: Tobacco Workers and the Struggle for Democracy in the Mid-Twentieth-Century South*. Chapel Hill: University of North Carolina Press, 2003.

Koselleck, Reinhart. *Futures Past: On the Semantics of Historical Time*. Translated by Keith Tribe. Cambridge, MA: MIT Press, 1985.

Kowal, Emma, Joanna Radin, and Jenny Reardon. "Indigenous Body Parts, Mutating Temporalities, and the Half-Lives of Postcolonial Technoscience." *Social Studies of Science* 43, no. 4 (2013): 465–83.

Krige, John. *American Hegemony and the Postwar Reconstruction of Science in Europe*. Cambridge, MA: MIT Press, 2006.

Krimsky, Sheldon. *Genetic Alchemy: The Social History of the Recombinant DNA Controversy*. Cambridge, MA: MIT Press, 1982.

Krueger, Gretchen Marie. "Death Be Not Proud: Children, Families, and Cancer in Postwar America." *Bulletin of the History of Medicine* 78, no. 4 (2004): 836–63. doi:10.1353/bhm.2004.0177.

———. "'For Jimmy and the Boys and Girls of America': Publicizing Childhood Cancers in Twentieth-Century America." *Bulletin of the History of Medicine* 81, no. 1 (2007): 70–93. doi:10.1353/bhm.2007.0004.

———. *Hope and Suffering: Children, Cancer, and the Paradox of Experimental Medicine*. Baltimore: Johns Hopkins University Press, 2008.

Kuklick, Henrika, and Robert E. Kohler. "Introduction." *Osiris*, 11 (1996): 1–14.

Kushner, Rose. *Breast Cancer: A Personal History and an Investigative Report*. New York: Harcourt Brace Jovanovich, 1975.

Kutcher, Gerald. *Contested Medicine: Cancer Research and the Military*. Chicago: University of Chicago Press, 2009.

Lambright, W. Henry. *Powering Apollo: James E. Webb of NASA*. Baltimore: Johns Hopkins University Press, 1995.

Landecker, Hannah. *Culturing Life: How Cells Became Technologies*. Cambridge, MA: Harvard University Press, 2007.

Langer, Elinor. "Human Experimentation: New York Verdict Affirms Patient's Rights." *Science* 151, no. 3711 (February 11, 1966): 663–66.

Langston, Nancy. *Toxic Bodies: Hormone Disruptors and the Legacy of DES*. New Haven, CT: Yale University Press, 2010.

Laszlo, John. *The Cure of Childhood Leukemia: Into the Age of Miracles*. New Brunswick, NJ: Rutgers University Press, 1995.

Latham, Michael E. *Modernization as Ideology: American Social Science and "Nation Building" in the Kennedy Era*. Chapel Hill: University of North Carolina Press, 2000.

Latour, Bruno. "On the Partial Existence of Existing and Nonexisting Objects." In *Biographies of Scientific Objects*, edited by Lorraine Daston, 247–70. Chicago: University of Chicago Press, 2000.

Latour, Bruno, and Steve Woolgar. *Laboratory Life: The Social Construction of Scientific Facts*. Beverly Hills, CA: Sage, 1979.

Law, John, and Marianne Elisabeth Lien. "Slippery: Field Notes in Empirical Ontology." *Social Studies of Science* 43, no. 3 (June 1, 2013): 363–78. doi:10.1177/0306312712456947.

Lazarus, Richard J. *The Making of Environmental Law*. Chicago: University of Chicago Press, 2004.

Leaf, Clifton. *The Truth in Small Doses: Why We're Losing the War on Cancer—and How to Win It*. New York: Simon & Schuster, 2013.

Lean, Christopher, and Anya Plutynski. "The Evolution of Failure: Explaining Cancer as an Evolutionary Process." *Biology & Philosophy* 31, no. 1 (2016): 39–57.

Lee, Mordecai. *Nixon's Super-Secretaries: The Last Grand Presidential Reorganization Effort*. College Station: Texas A&M University Press, 2010.

Lenoir, Timothy, and Marguerite Hays. "The Manhattan Project for Biomedicine." In *Controlling Our Destinies: Historical, Philosophical, Ethical, and Theological Perspectives on the Human Genome Project*, edited by Phillip R. Sloan, 29–61. Notre Dame, IN: University of Notre Dame Press, 2000.

Lerner, Barron H. *The Breast Cancer Wars: Hope, Fear, and the Pursuit of a Cure in Twentieth-Century America*. New York: Oxford University Press, 2001.

Levine, Arnold S. *Managing NASA in the Apollo Era*. Washington, DC: Scientific and Technical Information Branch, National Aeronautics and Space Administration, 1982.

Light, Jennifer S. *From Warfare to Welfare: Defense Intellectuals and Urban Problems in Cold War America*. Baltimore: Johns Hopkins University Press, 2003.

Lindee, M. Susan. *Suffering Made Real: American Science and the Survivors at Hiroshima*. Chicago: University of Chicago Press, 1994.

Lindenmeyer, Kriste. *A Right to Childhood: The U.S. Children's Bureau and Child Welfare, 1912–46*. Urbana: University of Illinois Press, 1997.

Littlefield, John W. "NIH 3T3 Cell Line." *Science* 218, no. 4569 (October 15, 1982): 214–16.

Livingston, Julie. *Improvising Medicine: An African Oncology Ward in an Emerging Cancer Epidemic*. Durham, NC: Duke University Press, 2012.

Lodish, Harvey, and Nina Fedoroff. "Norton Zinder (1928–2012)." *Science* 335, no. 6074 (March 16, 2012): 1316. doi:10.1126/science.1220682.

Loeppky, Rodney. *Encoding Capital: The Political Economy of the Human Genome Project*. New York: Routledge, 2005.

Lora-Wainwright, Anna. *Fighting for Breath: Living Morally and Dying of Cancer in a Chinese Village*. Honolulu: University of Hawai'i Press, 2013.

Lowy, Douglas R., and John T. Schiller. "Reducing HPV-Associated Cancer Globally." *Cancer Prevention Research* 5, no. 1 (January 1, 2012): 18–23. doi:10.1158/1940-6207.CAPR-11-0542.

Löwy, Ilana. *Between Bench and Bedside: Science, Healing, and Interleukin-2 in a Cancer Ward*. Cambridge, MA: Harvard University Press, 1996.

———. "Cancer: The Century of the Transformed Cell." In *Science in the Twentieth Century*, edited by John Krige and Dominique Pestre, 461–77. Amsterdam: Harwood Academic, 1997.

————. "Experimental Systems and Clinical Practices: Tumor Immunology and Cancer Immunotherapy, 1895–1980." *Journal of the History of Biology* 27, no. 3 (October 1, 1994): 403–35.

————. "Historiography of Biomedicine: 'Bio,' 'Medicine,' and In Between." *Isis* 102, no. 1 (March 1, 2011): 116–22.

————. "Immunotherapy of Cancer from Coley's Toxins to Interferons: Molecularization of a Therapeutic Practice." In de Chadarevian and Kamminga, *Molecularizing Biology and Medicine*, 249–71.

————. *Preventive Strikes: Women, Precancer, and Prophylactic Surgery*. Baltimore: Johns Hopkins University Press, 2010.

————. *A Woman's Disease : The History of Cervical Cancer*. Oxford: Oxford University Press, 2011.

Ludmerer, Kenneth M. *Time to Heal: American Medical Education from the Turn of the Century to the Era of Managed Care*. New York: Oxford University Press, 2005.

Lynch, Michael. "Ontography: Investigating the Production of Things, Deflating Ontology." *Social Studies of Science* 43, no. 3 (June 1, 2013): 444–62. doi:10.1177/0306312713475925.

Lynch, Vincent J. "Use with Caution: Developmental Systems Divergence and Potential Pitfalls of Animal Models." *Yale Journal of Biology and Medicine* 82, no. 2 (June 2009): 53–66.

MacLean, Nancy. *Freedom Is Not Enough: The Opening of the American Workplace*. New York: Russell Sage Foundation, 2006.

Mamo, Laura, and Steven Epstein. "The New Sexual Politics of Cancer: Oncoviruses, Disease Prevention, and Sexual Health Promotion." *BioSocieties* 12, no. 3 (September 1, 2017): 367–91. doi:10.1057/biosoc.2016.10.

————. "The Pharmaceuticalization of Sexual Risk: Vaccine Development and the New Politics of Cancer Prevention." *Social Science & Medicine* 101 (January 2014): 155–65. doi:10.1016/j.socscimed.2013.11.028.

Manaker, Robert A., Louis R. Sibal, and John B. Moloney. "Scientific Activities at the National Cancer Institute: Virology." *Journal of the National Cancer Institute* 59, no. 2 (August 1977): 623–31.

Mandel, H. George, Erich Hirshberg, and Joseph H. Burchenal. "Funding: Grants or Contracts? A Survey of Cancer Scientists." *Cancer Research* 35 (May 1975): 1109–15.

Mandel, Richard. "Division of Research Grants: A Half Century of Peer Review, 1946–1996," 2007. Department of Health, Education, and Welfare, Public Health Service, Bethesda, MD.

Marcum, James A. "From Heresy to Dogma in Accounts of Opposition to Howard Temin's DNA Provirus Hypothesis." *History and Philosophy of the Life Sciences* 24 (2002): 165–92.

Marcus, Alan I. *Malignant Growth: Creating the Modern Cancer Research Establishment, 1875–1915*. Tuscaloosa: University of Alabama Press, 2018.

Marks, Harry M. *The Progress of Experiment: Science and Therapeutic Reform in the United States, 1900–1990.* Cambridge: Cambridge University Press, 1997.

Masnyk, Ihor J. "International Activities of the National Cancer Institute." *Journal of the National Cancer Institute* 80, no. 17 (November 2, 1988): 1366–72. doi:10.1093/jnci/80.17.1366.

Mathews, Holly F., Nancy J. Burke, and Eirini Kampriani. *Anthropologies of Cancer in Transnational Worlds.* New York: Routledge, 2015.

Mawdsley, Stephen E. *Selling Science: Polio and the Promise of Gamma Globulin.* New Brunswick, NJ: Rutgers University Press, 2016.

May, Elaine Tyler. *Homeward Bound: American Families in the Cold War Era.* New York: Basic Books, 1988.

McCoy, J. J. *The Cancer Lady: Maud Slye and Her Heredity Studies.* Nashville, TN: T. Nelson, 1977.

McCray, Patrick. *The Visioneers: How a Group of Elite Scientists Pursued Space Colonies, Nanotechnologies, and a Limitless Future.* Princeton, NJ: Princeton University Press, 2013.

McElheny, Victor K. *Watson and DNA: Making a Scientific Revolution.* Cambridge, MA: Perseus, 2003.

McKee, Guian. "'The Government Is with Us': Lyndon Johnson and the Grassroots War on Poverty." In *The War on Poverty: A New Grassroots History, 1964–1980,* edited by Annelise Orleck and Lisa Gayle Hazirjian, 31–62. Athens: University of Georgia Press, 2011.

McKenna, Christopher D. *The World's Newest Profession: Management Consulting in the Twentieth Century.* Cambridge: Cambridge University Press, 2006.

Meckel, Richard A. *"Save the Babies": American Public Health Reform and the Prevention of Infant Mortality, 1850–1929.* Baltimore: Johns Hopkins University Press, 1990.

Mendelsohn, J. Andrew. "'Like All That Lives': Biology, Medicine and Bacteria in the Age of Pasteur and Koch." *History and Philosophy of the Life Sciences* 24, no. 1 (January 1, 2002): 3–36.

———. "Medicine and the Making of Bodily Inequality in Twentieth Century Europe." In Gaudillière and Löwy, *Heredity and Infection,* 21–80.

Mika, Marissa. "Fifty Years of Creativity, Crisis, and Cancer in Uganda." *Canadian Journal of African Studies/Revue Canadienne des Etudes Africaines* 50, no. 3 (September 1, 2016): 395–413. doi:10.1080/00083968.2016.1272061.

———. "Research Is Our Resource: Surviving Experiments and Politics at an African Cancer Institute, 1950 to the Present." PhD diss., University of Pennsylvania, 2015. https://repository.upenn.edu/edissertations/1898.

Milburn, Colin. *Nanovision: Engineering the Future.* Durham, NC: Duke University Press, 2008.

Mintz, Steven. *Huck's Raft: A History of American Childhood.* Cambridge, MA: Belknap Press of Harvard University Press, 2004.

Mirowski, Philip. *Science-Mart: Privatizing American Science*. Cambridge, MA: Harvard University Press, 2011.

Mitchell, Timothy. *Rule of Experts: Egypt, Techno-Politics, Modernity*. Berkeley: University of California Press, 2002.

———. "Society, Economy, and the State Effect." In *State/Culture: State-Formation after the Cultural Turn*, edited by George Steinmetz, 76–97. Ithaca, NY: Cornell University Press, 1999.

Mochales, S. "Forty Years of Screening Programmes for Antibiotics." *Microbiología* 10, no. 4 (1994): 331–42.

Montagnier, Luc. *Virus: The Co-Discoverer of HIV Tracks Its Rampage and Charts the Future*. New York: W. W. Norton, 2000.

Morange, Michel. "The Discovery of Cellular Oncogenes." *History and Philosophy of the Life Sciences* 15, no. 1 (January 1, 1993): 45–58.

———. "From the Regulatory Vision of Cancer to the Oncogene Paradigm, 1975–1985." *Journal of the History of Biology* 30, no. 1 (April 1997): 1–29.

———. *A History of Molecular Biology*. Cambridge, MA: Harvard University Press, 1998.

———. "The Transformation of Molecular Biology on Contact with Higher Organisms, 1960–1980: From a Molecular Description to a Molecular Explanation." *History and Philosophy of the Life Sciences* 19 (1997): 369–93.

Morgan, Gregory J. "Ludwik Gross, Sarah Stewart, and the 1950s Discoveries of Gross Murine Leukemia Virus and Polyoma Virus." *Studies in History and Philosophy of Science Part C: Studies in History and Philosophy of Biological and Biomedical Sciences* 48, Part B (December 2014): 200–209. doi:10.1016/j.shpsc.2014.07.013.

Morris, Andrew J. F. *The Limits of Voluntarism: Charity and Welfare from the New Deal through the Great Society*. Cambridge: Cambridge University Press, 2009.

Mukerji, Chandra. *A Fragile Power: Scientists and the State*. Princeton, NJ: Princeton University Press, 1989.

Mukherjee, Siddhartha. *The Emperor of All Maladies: A Biography of Cancer*. New York: Scribner, 2010.

Müller-Wille, Staffan, and Hans-Jörg Rheinberger. *A Cultural History of Heredity*. Chicago: University of Chicago Press, 2012.

Munns, David P. D. *A Single Sky: How an International Community Forged the Science of Radio Astronomy*. Cambridge, MA: MIT Press, 2013.

Muraskin, William. "The Silent Epidemic: The Social, Ethical, and Medical Problems Surrounding the Fight against Hepatitis B." *Journal of Social History* 22, no. 2 (December 21, 1988): 277–98. doi:10.1353/jsh/22.2.277.

Murphy, Michelle. *Sick Building Syndrome and the Problem of Uncertainty: Environmental Politics, Technoscience, and Women Workers*. Durham, NC: Duke University Press, 2006.

Nass, Sharyl J., and Bruce Stillman, eds. *Large-Scale Biomedical Science: Exploring Strategies for Future Research*. Washington, DC: National Academies Press, 2003.

Nathan, Richard P. *The Plot That Failed: Nixon and the Administrative Presidency.* New York: Wiley, 1975.

Necochea, Raul. "From Cancer Families to HNPCC: Henry Lynch and the Transformations of Hereditary Cancer, 1975–1999." *Bulletin of the History of Medicine* 81, no. 1 (2007): 267–85.

Neushul, Peter. "Science, Government and the Mass Production of Penicillin." *Journal of the History of Medicine and Allied Sciences* 48, no. 4 (October 1, 1993): 371–95. doi:10.1093/jhmas/48.4.371.

Newmark, Peter. "Oncogene Discovery: Priority by Press Release." *Nature* 304, no. 5922 (July 14, 1983): 108. doi:10.1038/304108a0.

———. "Oncogenic Intelligence: The Rasmatazz of Cancer Genes." *Nature* 305, no. 5934 (October 6, 1983): 470–71. doi:10.1038/305470a0.

Noble, David F. *America by Design: Science, Technology, and the Rise of Corporate Capitalism.* New York: Knopf, 1977.

Nolte, Karen. "Carcinoma Uteri and 'Sexual Debauchery'—Morality, Cancer and Gender in the Nineteenth Century." *Social History of Medicine* 21, no. 1 (April 2008): 31–46. doi:10.1093/shm/hkm116.

Oberling, Charles. *The Riddle of Cancer.* Translated by William Henry Woglom. New Haven, CT: Yale University Press, 1944.

O'Connor, Alice. *Poverty Knowledge: Social Science, Social Policy, and the Poor in Twentieth-Century U.S. History.* Princeton, NJ: Princeton University Press, 2001.

Olson, James Stuart. *Making Cancer History: Disease and Discovery at the University of Texas M.D. Anderson Cancer Center.* Baltimore: Johns Hopkins University Press, 2009.

Olweny, Charles L. M. "The Uganda Cancer Institute." *Oncology* 37 (1980): 367–70.

Omran, Abdel R. "The Epidemiologic Transition: A Theory of the Epidemiology of Population Change." *Milbank Memorial Fund Quarterly* 49, no. 4 (October 1, 1971): 509–38. doi:10.2307/3349375.

Oppel, Richard A. "Benno C. Schmidt, Financier, Is Dead at 86." *New York Times,* October 22, 1999, business sec. http://www.nytimes.com/1999/10/22/business/benno-c-schmidt-financier-is-dead-at-86.html.

Oshinsky, David M. *Polio: An American Story.* Oxford: Oxford University Press, 2005.

Palladino, Paolo. "Between Knowledge and Practice: On Medical Professionals, Patients, and the Making of the Genetics of Cancer." *Social Studies of Science* 32, no. 1 (February 2002): 137–65.

Palmer, Gregory. *The McNamara Strategy and the Vietnam War: Program Budgeting in the Pentagon, 1960–1968.* Westport, CT: Greenwood Press, 1978.

Parascandola, Mark. "The Epidemiologic Transition and Changing Concepts of Causation and Causal Inference." *Revue d'Histoire des Sciences* 64, no. 2 (April 15, 2012): 243–62. doi:10.3917/rhs.642.0243.

———. "Uncertain Science and a Failure of Trust: The NIH Radioepidemiologic Tables and Compensation for Radiation-Induced Cancer." *Isis* 93, no. 4 (December 2002): 559–84.

Parker, Martin. "Space Age Management." *Management & Organizational History* 4, no. 3 (August 1, 2009): 317–32. doi:10.1177/1744935909337759.

Parthasarathy, Shobita. "Architectures of Genetic Medicine: Comparing Genetic Testing for Breast Cancer in the USA and the UK." *Social Studies of Science* 35, no. 1 (February 1, 2005): 5–40.

Patel, Sejal S. "Methods and Management: NIH Administrators, Federal Oversight, and the Framingham Heart Study." *Bulletin of the History of Medicine* 86, no. 1 (2012): 94–121.

Patterson, James T. "Cancer, Cancerphobia, and Culture: Reflections on Attitudes in the United States and Great Britain." *Twentieth Century British History* 2, no. 2 (April 1991): 137–49.

———. *The Dread Disease: Cancer and Modern American Culture*. Cambridge, MA: Harvard University Press, 1987.

Paul, John R. *A History of Poliomyelitis*. New Haven, CT: Yale University Press, 1971.

Pauly, Philip J. *Biologists and the Promise of American Life: From Meriwether Lewis to Alfred Kinsey*. Princeton, NJ: Princeton University Press, 2000.

Peck, Gunther. *Reinventing Free Labor: Padrones and Immigrant Workers in the North American West, 1880–1930*. Cambridge: Cambridge University Press, 2000.

Pelling, Margaret. "Contagion/Germ Theory/Specificity," edited by William F. Bynum and Roy Porter, *Companion Encyclopedia of the History of Medicine*, RN 1: 309–34. London: Routledge, 1993.

Perlstein, Rick. *Nixonland: The Rise of a President and the Fracturing of America*. New York: Scribner, 2008.

Peyrieras, Nadine, and Michel Morange. "The Study of Lysogeny at the Pasteur Institute (1950–1960): An Epistemologically Open System." *Studies in History and Philosophy of Science Part C: Studies in History and Philosophy of Biological and Biomedical Sciences* 33, no. 3 (September 2002): 419–30. doi:10.1016/S1369-8486(02)00014-6.

Phillips-Fein, Kim. *Invisible Hands: The Making of the Conservative Movement from the New Deal to Reagan*. New York: W. W. Norton, 2009.

Pickering, Andrew. *The Cybernetic Brain*. Chicago: University of Chicago Press, 2009.

Pickstone, John V. "Contested Cumulations: Configurations of Cancer Treatments through the Twentieth Century." *Bulletin of the History of Medicine* 81, no. 1 (2007): 164–96.

———. *Ways of Knowing: A New History of Science, Technology, and Medicine*. Chicago: University of Chicago Press, 2001.

Pieters, Toine. *Interferon: The Science and Selling of a Miracle Drug*. London: Routledge, 2005.

Pinell, Patrice. *The Fight against Cancer: France, 1890–1940*. Translated by David Mandell. London: Routledge, 2002.

Podolsky, Scott H. *The Antibiotic Era: Reform, Resistance, and the Pursuit of a Rational Therapeutics*. Baltimore: Johns Hopkins University Press, 2015.

Proctor, Robert. *Cancer Wars: How Politics Shapes What We Know and Don't Know about Cancer.* New York: Basic Books, 1995.

———. *Golden Holocaust: Origins of the Cigarette Catastrophe and the Case for Abolition.* Berkeley: University of California Press, 2011.

———. *The Nazi War on Cancer.* Princeton, NJ: Princeton University Press, 1999.

Proctor, Robert, and Londa L. Schiebinger, eds. *Agnotology: The Making and Unmaking of Ignorance.* Stanford, CA: Stanford University Press, 2008.

Quadagno, Jill S. *One Nation, Uninsured: Why the U.S. Has No National Health Insurance.* New York: Oxford University Press, 2005.

———. "Social Movements and State Transformation: Labor Unions and Racial Conflict in the War on Poverty." *American Sociological Review* 57, no. 5 (October 1, 1992): 616–34. doi:10.2307/2095916.

Quinn, Roswell. "Rethinking Antibiotic Research and Development: World War II and the Penicillin Collaborative." *American Journal of Public Health* 103, no. 3 (March 2013): 426–34. doi:10.2105/AJPH.2012.300693.

Rader, Karen A. "Alexander Hollaender's Postwar Vision for Biology: Oak Ridge and Beyond." *Journal of the History of Biology* 39, no. 4 (October 2006): 685–706. doi:10.1007/s10739-006-9109-1.

———. *Making Mice: Standardizing Animals for American Biomedical Research, 1900–1955.* Princeton, NJ: Princeton University Press, 2004.

Radin, Joanna. "Latent Life: Concepts and Practices of Human Tissue Preservation in the International Biological Program." *Social Studies of Science* 43, no. 4 (August 2013): 484–508. doi:10.1177/0306312713476131.

———. *Life on Ice: A History of New Uses for Cold Blood.* Chicago: University of Chicago Press, 2017.

Rajan, Kaushik. *Biocapital: The Constitution of Postgenomic Life.* Durham, NC: Duke University Press, 2006.

Rasmussen, Nicolas. *Gene Jockeys: Life Science and the Rise of Biotech Enterprise.* Baltimore: Johns Hopkins University Press, 2014.

———. "The Mid-Century Biophysics Bubble: Hiroshima and the Biological Revolution in America, Revisited." *History of Science* 35 (1997): 245–93.

———. "Of 'Small Men,' Big Science and Bigger Business: The Second World War and Biomedical Research in the United States." *Minerva* 40 (2002): 115–46.

———. *Picture Control: The Electron Microscope and the Transformation of Biology in America, 1940–1960.* Stanford, CA: Stanford University Press, 1997.

Rather, L. J. *The Genesis of Cancer: A Study in the History of Ideas.* Baltimore: Johns Hopkins University Press, 1978.

Rau, Erik P. "Combat Science: The Emergence of Operational Research in World War II." *Endeavour* 29, no. 4 (December 2005): 156–61. doi:10.1016/j.endeavour.2005.10.002.

Rego, Brianna. "The Polonium Brief: A Hidden History of Cancer, Radiation, and the Tobacco Industry." *Isis* 100, no. 3 (2009): 453–84.

Reich, Leonard S. *The Making of American Industrial Research: Science and Business at GE and Bell, 1876–1926*. Cambridge: Cambridge University Press, 1985.

Reingold, Nathan. "Science and Government in the United States since 1945." *History of Science* 32, no. 4 (1994): 361–85.

———. "Vannevar Bush's New Deal for Research: Or the Triumph of the Old Order." *Historical Studies in the Physical and Biological Sciences* 17, no. 2 (January 1, 1987): 299–344.

Rettig, Richard A. *Cancer Crusade: The Story of the National Cancer Act of 1971*. Princeton, NJ: Princeton University Press, 1977.

Rheinberger, Hans-Jörg. "Consistency from the Perspective of an Experimental Systems Approach to the Sciences and Their Epistemic Objects." *Manuscrito* 34, no. 1 (June 2011): 307–21. doi:10.1590/S0100-60452011000100014.

———. *An Epistemology of the Concrete: Twentieth-Century Histories of Life*. Durham, NC: Duke University Press, 2010.

———. "Experimental Systems: Historiality, Narration, and Deconstruction." *Science in Context* 7 (1994): 65–81.

———. "Experimental Systems: Historiality, Narration, and Deconstruction." In *The Science Studies Reader*, edited by Mario Biagioli, 417–29. New York: Routledge, 1999.

———. "From Microsomes to Ribosomes: 'Strategies' of 'Representation.'" *Journal of the History of Biology* 28, no. 1 (April 1, 1995): 49–89.

———. "Recent Science and Its Exploration: The Case of Molecular Biology." *Studies in History and Philosophy of Science Part C: Studies in History and Philosophy of Biological and Biomedical Sciences* 40, no. 1 (March 2009): 6–12. doi:10.1016/j.shpsc.2008.12.002.

———. *Toward a History of Epistemic Things: Synthesizing Proteins in the Test Tube*. Stanford, CA: Stanford University Press, 1997.

———. "What Happened to Molecular Biology?" *BioSocieties* 3, no. 3 (September 1, 2008): 303–10. doi:10.1017/S1745855208006212.

Rheinberger, Hans-Jörg, and Jean Paul Gaudillière. Introduction to *From Molecular Genetics to Genomics: The Mapping Cultures of Twentieth-Century Genetics*, edited by Jean-Paul Gaudillière and Hans-Jörg Rheinberger, 1–6. London: Routledge, 2004.

Rhodes, Richard. *The Making of the Atomic Bomb*. New York: Simon & Schuster, 1986.

Riles, Annelise, ed. *Documents: Artifacts of Modern Knowledge*. Ann Arbor: University of Michigan Press, 2006.

Rivers, Thomas M. "Some General Aspects of Filterable Viruses." In *Filterable Viruses*, edited by Thomas M. Rivers, 3–54. Baltimore: Williams & Wilkins, 1928.

———. "Viruses and Koch's Postulates." *Journal of Bacteriology* 33, no. 1 (January 1, 1937): 1–12.

Robinson, Judith. *Noble Conspirator: Florence S. Mahoney and the Rise of the National Institutes of Health*. Washington, DC: Francis Press, 2001.

Rodgers, Daniel T. *Age of Fracture*. Cambridge, MA: Belknap Press of Harvard University Press, 2011.

———. *Atlantic Crossings: Social Politics in a Progressive Age*. Cambridge, MA: Belknap Press of Harvard University Press, 1998.

Rogers, Naomi. *Dirt and Disease: Polio before FDR*. New Brunswick, NJ: Rutgers University Press, 1992.

Rose, Nikolas S. *The Politics of Life Itself: Biomedicine, Power, and Subjectivity in the Twenty-First Century*. Princeton, NJ: Princeton University Press, 2007.

Rose, Nikolas, and Carlos Novas. "Biological Citizenship." In *Global Assemblages: Technology, Politics, and Ethics as Anthropological Problems*, edited by Aihwa Ong and Stephen J Collier, 439–63. Malden, MA: Blackwell, 2005.

Rosenberg, Charles E. *Explaining Epidemics and Other Studies in the History of Medicine*. Cambridge: Cambridge University Press, 1992.

———. "Rationalization and Reality in the Shaping of American Agricultural Research, 1875–1914." *Social Studies of Science* 7, no. 4 (November 1977): 401–22.

Rosenkrantz, Barbara Gutmann. *Public Health and the State: Changing Views in Massachusetts, 1842–1936*. Cambridge, MA: Harvard University Press, 1972.

Rothman, Sheila M. *Living in the Shadow of Death: Tuberculosis and the Social Experience of Illness in American History*. Baltimore: Johns Hopkins University Press, 1995.

Rubin, Harry. "Quantitative Tumor Virology." In Cairns, Stent, and Watson, *Phage and Molecular Biology*, 290–300.

Rudwick, Martin J. S. "The Emergence of a Visual Language for Geological Science, 1760–1840." *History of Science* 14, no. 3 (September 1, 1976): 149–95. doi:10.1177/007327537601400301.

Saint-Amour, Paul K. *Tense Future: Modernism, Total War, Encyclopedic Form*. New York: Oxford University Press, 2015.

Sapolsky, Harvey M. *The Polaris System Development: Bureaucratic and Programmatic Success in Government*. Cambridge, MA: Harvard University Press, 1972.

Schaffer, Simon, and Steven Shapin. *Leviathan and the Air-Pump: Hobbes, Boyle, and the Experimental Life*. Princeton, NJ: Princeton University Press, 1985.

Scheffler, Robin W., and Bruno J. Strasser. "Biomedical Sciences, History and Sociology Of." In *International Encyclopedia of the Social & Behavioral Sciences*, edited by James D. Wright, 663–69. 2nd ed. Oxford: Elsevier, 2015. http://www.sciencedirect.com/science/article/pii/B9780080970868850045X.

Schneyer, Solomon, J. Steven Landefeld, and Frank H. Sandifer. "Biomedical Research and Illness: 1900–1979." *Milbank Memorial Fund Quarterly: Health and Society* 59, no. 1 (January 1, 1981): 44–58. doi:10.2307/3349775.

Schwartz, Judith. "A Fine Playground: Dr. Norton Zinder." *Rockefeller University Research Profiles* 27 (Winter 1987): 1–6.

Scolnick, Edward M. NCI Oral History Project: Interview with Edward Scolnick. Transcript, June 28, 1998. NIH Office of History, Bethesda, MD.

Scott, James C. *Seeing like a State: How Certain Schemes to Improve the Human Condition Have Failed.* New Haven, CT: Yale University Press, 1998.

Scotto, Joseph, and John C. Bailar. "Rigoni-Stern and Medical Statistics: A Nineteenth-Century Approach to Cancer Research." *Journal of the History of Medicine and Allied Sciences* 24, no. 1 (January 1, 1969): 65–75. doi:10.1093/jhmas/XXIV.1.65.

Secord, James A. "Knowledge in Transit." *Isis* 95, no. 4 (December 2004): 654–72.

Self, Robert O. *American Babylon: Race and the Struggle for Postwar Oakland.* Princeton, NJ: Princeton University Press, 2003.

Selya, Rena Elisheva. "Salvador Luria's Unfinished Experiment: The Public Life of a Biologist in a Cold War Democracy." PhD diss., Harvard University, 2002.

Shapin, Steven. "Here and Everywhere: Sociology of Scientific Knowledge." *Annual Review of Sociology* 21 (1995): 289–321.

———. "The Invisible Technician." *American Scientist* 77, no. 6 (November 1, 1989): 554–63. doi:10.2307/27856006.

———. *The Scientific Life: A Moral History of a Late Modern Vocation.* Chicago: University of Chicago Press, 2008.

Shaughnessy, Donald F. "The Story of the American Cancer Society." PhD diss., Columbia University, 1957.

Sheingate, Adam. "The Terrain of the Political Entrepreneur." In *Formative Acts: American Politics in the Making,* edited by Stephen Skowronek and Matthew Glassman, 13–31. Philadelphia: University of Pennsylvania Press, 2007.

Shilts, Randy. *And the Band Played On: Politics, People, and the AIDS Epidemic.* 20th anniversary ed., New York: St. Martin's Press, 2007.

Shope, Richard. "Evolutionary Episodes in the Concept of Viral Oncogenesis." *Perspectives in Biology and Medicine* 9, no. 2 (1966): 258–74.

Shryock, Richard Harrison. *National Tuberculosis Association, 1904–1954: A Study of the Voluntary Health Movement in the United States.* New York: National Tuberculosis Association, 1957.

Sklar, Kathryn. *Florence Kelley and the Nation's Work: The Rise of Women's Political Culture.* New Haven, CT: Yale University Press, 1995.

Skocpol, Theda. *Protecting Soldiers and Mothers: The Political Origins of Social Policy in the United States.* Cambridge, MA: Belknap Press of Harvard University Press, 1992.

Skuse, Alanna. *Constructions of Cancer in Early Modern England: Ravenous Natures.* Basingstoke, UK: Palgrave Macmillan, 2015.

Sledge, Daniel. *Health Divided: Public Health and Individual Medicine in the Making of the Modern American State.* Lawrence: University Press of Kansas, 2017.

————. "War, Tropical Disease, and the Emergence of National Public Health Capacity in the United States." *Studies in American Political Development* 26, no. 2 (October 2012): 125–62. doi:10.1017/S0898588X12000107.

Slotkin, Arthur L. *Doing the Impossible: George E. Mueller and the Management of NASA's Human Spaceflight Program.* New York: Springer-Praxis, 2012.

Smith, Jane S. *Patenting the Sun: Polio and the Salk Vaccine.* New York: William Morrow, 1990.

Smith, Susan L. *Toxic Exposures: Mustard Gas and the Health Consequences of World War II in the United States.* New Brunswick, NJ: Rutgers University Press, 2017.

Sontag, Susan. *Illness as Metaphor; and AIDS and Its Metaphors.* New York: Picador, 2001.

Sparrow, James T. *Warfare State: World War II Americans and the Age of Big Government.* New York: Oxford University Press, 2011.

Sparrow, James T., William J. Novak, and Stephen W. Sawyer, eds. *Boundaries of the State in US History.* Chicago: University of Chicago Press, 2015.

Spears, Ellen Griffith. *Baptized in PCBs: Race, Pollution, and Justice in an All-American Town.* Chapel Hill, NC: University of North Carolina Press, 2014.

Spector, Malcolm, and John I. Kitsuse. *Constructing Social Problems.* New Brunswick, NJ: Transaction, 2001.

Star, Susan Leigh, and Geoffrey C. Bowker. "How to Infrastructure." In *Handbook of New Media: Social Shaping and Social Consequences of ICTs.* Updated student ed., 230–45. London: SAGE, 2011.

Starr, Paul. *The Social Transformation of American Medicine.* New York: Basic Books, 1982.

Steere-Williams, Jacob. "Performing State Medicine during Its 'Frustrating' Years: Epidemiology and Bacteriology at the Local Government Board, 1870–1900." *Social History of Medicine* 28, no. 1 (February 1, 2015): 82–107. doi:10.1093/shm/hku064.

Stein, Judith. *Pivotal Decade: How the United States Traded Factories for Finance in the Seventies.* New Haven, CT: Yale University Press, 2010.

Stevenson, Robert. "Interview with Dr. Carl Baker." Oral History Transcript, September 6, 1995. NIH Office of History, Bethesda, MD.

Stockfish, J. A. *The Intellectual Foundations of Systems Analysis.* Santa Monica, CA: RAND, 1987.

Stokes, Donald E. *Pasteur's Quadrant: Basic Science and Technological Innovation.* Washington, DC: Brookings Institution Press, 1997.

Stolberg, Michael. "Metaphors and Images of Cancer in Early Modern Europe." *Bulletin of the History of Medicine* 88, no. 1 (2014): 48–74. doi:10.1353/bhm.2014.0014.

Stolley, Paul D., and Tamar Lasky. "Johannes Fibiger and His Nobel Prize for the Hypothesis That a Worm Causes Stomach Cancer." *Annals of Internal Medicine* 116, no. 9 (May 1, 1992): 765–69. doi:10.7326/0003-4819-116-9-765.

Stolt, Carl-Magnus, George Klein, and Alfred T. R. Jansson. "An Analysis of a Wrong Nobel Prize—Johannes Fibiger, 1926: A Study in the Nobel Archives." *Advances in Cancer Research* 92 (2004): 1–12. doi:10.1016/S0065-230X(04)92001-5.

Storrs, Landon R. Y. "Red Scare Politics and the Suppression of Popular Front Feminism: The Loyalty Investigation of Mary Dublin Keyserling." *Journal of American History* 90, no. 2 (September 2003): 491–524.

Strasser, Bruno J. *La fabrique d'une nouvelle science: La biologie moléculaire a l'âge atomique.* Firenze, Italy: Casa Editrice Leo S. Olschki, 2006. "Institutionalizing Molecular Biology in Post-War Europe: A Comparative Study." *Studies in History and Philosophy of Science Part C: Studies in History and Philosophy of Biological and Biomedical Sciences* 33, no. 3 (September 2002): 515–46. doi:10.1016/S1369-8486(02)00016-X.

———. "Linus Pauling's 'Molecular Diseases': Between History and Memory." *American Journal of Medical Genetics* 115, no. 2 (2002): 83–93. https://doi.org/10.1002/ajmg.10542.

Strickland, Stephen P. *Politics, Science, and Dread Disease: A Short History of United States Medical Research Policy.* Cambridge, MA: Harvard University Press, 1972.

Stultiens, Andrea, and Marissa Mika. *Staying Alive: Documenting the Uganda Cancer Institute.* Edam, NL: Paradox, 2017.

Sturdy, Steve. "Knowing Cases: Biomedicine in Edinburgh, 1887–1920." *Social Studies of Science* 37, no. 5 (October 1, 2007): 659–89.

———. "Looking for Trouble: Medical Science and Clinical Practice in the Historiography of Modern Medicine." *Social History of Medicine* 24, no. 3 (December 1, 2011): 739–57. doi:10.1093/shm/hkq106.

———. "The Political Economy of Scientific Medicine: Science, Education and the Transformation of Medical Practice in Sheffield, 1890–1922." *Medical History* 36, no. 2 (April 1992): 125–59.

Sugden, Bill. "Beyond the Provirus: From Howard Temin's Insights on Rous Sarcoma Virus to the Study of Epstein-Barr Virus, the Prototypic Human Tumor Virus." In *The DNA Provirus: Howard Temin's Scientific Legacy*, edited by Geoffrey M. Cooper, Rayla Greenberg Temin, and Bill Sugden, 161–84. Washington, DC: ASM Press, 1995.

Summers, William C. *Félix d'Herelle and the Origins of Molecular Biology.* New Haven, CT: Yale University Press, 1999.

———. "How Bacteriophage Came to Be Used by the Phage Group." *Journal of the History of Biology* 26, no. 2 (July 1, 1993): 255–67.

Swain, Donald C. "The Rise of a Research Empire: NIH, 1930 to 1950." *Science,* 138, no. 3546 (December 14, 1962): 1233–37.

Swann, John Patrick. *Academic Scientists and the Pharmaceutical Industry: Cooperative Research in Twentieth-Century America.* Baltimore: Johns Hopkins University Press, 1988.

Sweek, Robert Forman. "Management Transfer from the Physical Sciences to Cancer Research." PhD diss., American University, 1975.

Taussig, Karen-Sue, Hoeyer Klaus, and Stefan Helmreich. "The Anthropology of Potentiality in Biomedicine: An Introduction to Supplement 7." *Current Anthropology* 54, no. S7 (October 1, 2013): S3–14. doi:10.1086/671401.

Teixeira, Luiz Antonio, and Ilana Löwy. "Imperfect Tools for a Difficult Job: Colposcopy, 'Colpocytology' and Screening for Cervical Cancer in Brazil." *Social Studies of Science* 41, no. 4 (August 2011): 585–608. doi:10.1177/0306312711408380.

Thomas, Patricia. *Big Shot: Passion, Politics, and the Struggle for an AIDS Vaccine.* New York: Public Affairs, 2001.

Thomas, William. *Rational Action: The Sciences of Policy in Britain and America, 1940–1960.* Cambridge, MA: MIT Press, 2015.

Tilley, Helen. *Africa as a Living Laboratory: Empire, Development, and the Problem of Scientific Knowledge, 1870–1950.* Chicago: University of Chicago Press, 2011.

Timmermann, Carsten. "As Depressing as It Was Predictable? Lung Cancer, Clinical Trials, and the Medical Research Council in Postwar Britain." *Bulletin of the History of Medicine* 81, no. 1 (2007): 312–34. doi:10.1353/bhm.2007.0012.

———. "Chronic Illness and Disease History." In *The Oxford Handbook of the History of Medicine,* edited by Mark Jackson, 393–410. Oxford: Oxford University Press, 2011.

———. *A History of Lung Cancer: The Recalcitrant Disease.* Basingstoke, UK: Palgrave Macmillan, 2013.

Tobbell, Dominique A. *Pills, Power, and Policy: The Struggle for Drug Reform in Cold War America and Its Consequences.* Berkeley: University of California Press; New York: Millbank Memorial Fund, 2012.

Todes, Daniel P. "Pavlov's Physiology Factory." *Isis* 88, no. 2 (June 1997): 205–46.

Tomes, Nancy J. *The Gospel of Germs: Men, Women, and the Microbe in American Life.* Cambridge, MA: Harvard University Press, 1998.

Tomes, Nancy J., and John Harley Warner. "Introduction to Special Issue on Rethinking the Reception of the Germ Theory of Disease: Comparative Perspectives." *Journal of the History of Medicine and Allied Sciences* 52, no. 1 (January 1, 1997): 7–16. doi:10.1093/jhmas/52.1.7.

Toon, Elizabeth. "'Cancer as the General Population Knows It': Knowledge, Fear, and Lay Education in 1950s Britain." *Bulletin of the History of Medicine* 81, no. 1 (2007): 116–38. doi:10.1353/bhm.2007.0013.

Traweek, Sharon. *Beamtimes and Lifetimes: The World of High Energy Physicists.* Cambridge, MA: Harvard University Press, 1988.

Triolo, Victor A. "Nineteenth Century Foundations of Cancer Research: Origins of Experimental Research." *Cancer Research* 24, no. 1, part 1 (January 1, 1964): 4–27.

Turner, Fred. *From Counterculture to Cyberculture: Stewart Brand, the Whole Earth Network, and the Rise of Digital Utopianism*. Chicago: University of Chicago Press, 2006.

Valenčius, Conevery Bolton. *The Health of the Country: How American Settlers Understood Themselves and Their Land*. New York: Basic Books, 2004.

Valier, Helen K. *A History of Prostate Cancer: Cancer, Men, and Medicine*. London: Palgrave Macmillan, 2016.

Varmus, Harold. *The Art and Politics of Science*. New York: W. W. Norton, 2009.

Vermeulen, Niki. *Supersizing Science: On Building Large-Scale Research Projects in Biology*. Maastricht, Netherlands: Maastricht University Press, 2009.

Vettel, Eric James. *Biotech: The Countercultural Origins of an Industry*. Philadelphia: University of Pennsylvania Press, 2006.

Wailoo, Keith. *Dying in the City of the Blues: Sickle Cell Anemia and the Politics of Race and Health*. Chapel Hill: University of North Carolina Press, 2001.

———. *How Cancer Crossed the Color Line*. Oxford: Oxford University Press, 2011.

Wailoo, Keith, Julie Livingston, Steven Epstein, and Robert A. Aronowitz, eds. *Three Shots at Prevention: The HPV Vaccine and the Politics of Medicine's Simple Solutions*. Baltimore: Johns Hopkins University Press, 2010.

Wang, Jessica. "Liberals, the Progressive Left, and the Political Economy of Postwar American Science: The National Science Foundation Debate Revisited." *Historical Studies in the Physical and Biological Sciences* 26, no. 1 (January 1, 1995): 139–66.

Wapner, Jessica. *The Philadelphia Chromosome: A Genetic Mystery, a Lethal Cancer, and the Improbable Invention of a Lifesaving Treatment*. New York: Experiment, 2013.

Ward, Patricia Spain. "The American Reception of Salvarsan." *Journal of the History of Medicine and Allied Sciences* 36, no. 1 (January 1, 1981): 44–62. doi:10.1093/jhmas/XXXVI.1.44.

Weinberg, Robert A. *Racing to the Beginning of the Road: The Search for the Origin of Cancer*. New York: W. H. Freeman, 1998.

Weisz, George. *Chronic Disease in the Twentieth Century: A History*. Baltimore: Johns Hopkins University Press, 2014.

Weisz, George, and Jesse Olszynko-Gryn. "The Theory of Epidemiologic Transition: The Origins of a Citation Classic." *Journal of the History of Medicine and Allied Sciences* 65, no. 3 (July 1, 2010): 287–326. doi:10.1093/jhmas/jrp058.

Westfall, Catherine. "Rethinking Big Science: Modest, Mezzo, Grand Science and the Development of the Bevalac, 1971–1993." *Isis* 94, no. 1 (March 1, 2003): 30–56.

Wiebe, Robert H. *The Search for Order, 1877–1920*. New York: Hill and Wang, 1967.

Williams, Rosalind. "Cultural Origins and Environmental Implications of Large Technological Systems." *Science in Context* 6, no. 2 (September 1993): 377–403. doi:10.1017/S0269889700001459.

Wolff, Jacob. *The Science of Cancerous Disease from Earliest Times to the Present.* Translated by Barbara Ayoub. Canton, MA: Science History Publications USA, 1989.

Worboys, Michael. "Contagion." In *The Routledge History of Disease*, edited by Mark Jackson, 71–88. London: Routledge, 2017.

———. "From Heredity to Infection? Tuberculosis, 1870–1890." In Gaudillière and Löwy, *Heredity and Infection*, 81–100.

———. *Spreading Germs: Disease Theories and Medical Practice in Britain, 1865–1900.* Cambridge: Cambridge University Press, 2000.

———. "Was There a Bacteriological Revolution in Late Nineteenth-Century Medicine?" *Studies in History and Philosophy of Science Part C: Studies in History and Philosophy of Biological and Biomedical Sciences* 38, no. 1 (March 2007): 20–42. doi:10.1016/j.shpsc.2006.12.003.

Wright, Susan. *Molecular Politics: Developing American and British Regulatory Policy for Genetic Engineering, 1972–1982.* Chicago: University of Chicago Press, 1994.

Yates, JoAnne. *Control through Communication: The Rise of System in American Management.* Baltimore: Johns Hopkins University Press, 1989.

Yi, Doogab. "Cancer, Viruses, and Mass Migration: Paul Berg's Venture into Eukaryotic Biology and the Advent of Recombinant DNA Research and Technology, 1967–1980." *Journal of the History of Biology* 41, no. 4 (February 2008): 589–636. doi:10.1007/s10739-008-9149-9.

———. "Governing, Financing, and Planning Cancer Virus Research: The Emergence of Organized Science at the U.S. National Cancer Institute in the 1950s and 1960s." *Korean Journal for the History of Science* 38, no. 2 (2016): 321–49.

———. *The Recombinant University: Genetic Engineering and the Emergence of Stanford Biotechnology.* Chicago: University of Chicago Press, 2015.

———. "Who Owns What? Private Ownership and the Public Interest in Recombinant DNA Technology in the 1970s." *Isis* 102, no. 3 (2011): 446–74.

Zelizer, Julian E. "Reflections: Rethinking the History of American Conservatism." *Reviews in American History* 38, no. 2 (2010): 367–92. doi:10.1353/rah.0.0217.

———. *Taxing America: Wilbur D. Mills, Congress, and the State, 1945–1975.* Cambridge: Cambridge University Press, 1998.

Zelizer, Viviana A. Rotman. *Pricing the Priceless Child: The Changing Social Value of Children.* New York: Basic Books, 1985.

Zubrod, C. Gordon, S. Scheportz, J. Leiter, K. M. Endicott, L. M. Carrese, and C. G. Baker. "The Chemotherapy Program of the National Cancer Institute:

History, Analysis, and Plans." *Cancer Chemotherapy Reports* 50, no. 7 (October 1966): 346–539.

Zunz, Olivier. *Philanthropy in America: A History*. Princeton, NJ: Princeton University Press, 2012.

Zur Hausen, H. "The Search for Infectious Causes of Human Cancers: Where and Why." *Virology* 392, no. 1 (September 15, 2009): 1–10. doi:10.1016/j .virol.2009.06.001.

———. "Viruses in Human Tumors—Reminiscences and Perspectives." *Advances in Cancer Research* 68 (1996): 1–22.

Index

Duke University School of Medicine, 149
Dulbecco, Renato, 155, 188, 228, 301n10.
See also virology
Duran-Renynal, Francesco, 267n69. See also
virology

Eddy, Bernice, 56. See also virology
Edgerton, David, 283n55. See also history
education: campaign of, 64; of the public,
6, 30–33, 33f, 35–38, 51, 64–65, 85. See
also cancer; government; public; public
health
electron microscopy, 11. See also technology
electrophoresis, 191–92, 198; gel, 216,
307n28. See also technology
Ellerman, Vilhelm, 257n1. See also biomedical research
Endicott, Kenneth, 95, 103–4, 110, 116–18,
120–22, 125, 128, 151, 210, 279n65,
287n112. See also Cancer Chemotherapy
National Service Center; National
Cancer Institute
environmentalism: as an approach to
understanding cancer as a "disease of
civilization," 15, 18–19, 206–10, 230,
251n13; as a political movement, 206–10,
313n19; Proposition 65 (California) in
the, 207, 208f; and theories of carcinogenesis, 4, 6–7, 111, 317n53. See also
cancer; carcinogenesis; industrialization;
politics; society; urbanization
Environmental Protection Agency (EPA),
159, 207. See also government
enzyme: chemical, 48; restriction, 214–15;
reverse transcriptase as an, 140–44. See
also parasite; retrovirus; reverse transcriptase; viruses
epidemic, 55. See also disease; epidemiology; public health
epidemiology: data about cancer in, 24,
251n16; federal attention to, 205, 231;
international projects in, 144, 236; view
of leukemia in, 91, 111; yellow fever,
289n19. See also disease; epidemic;
public health
epistemology, 126. See also philosophy of
science
Epstein-Barr Virus (EBV), 131, 133–34,
179. See also lymphoma; viruses
Epstein, Michael, 131; The Politics of Cancer (1978) of, 207, 209

Erickson, Ray, 223. See also virology
eugenics, 50; stigma of, 139. See also
genetics
Eugenics Record Office, 168. See also
eugenics
eukaryotic cells, 11, 215, 319n77. See also
biology
Ewing, James, 33, 49, 51, 67–68. See also
oncology
experimental systems, 125–45, 184–85, 190,
199, 203; in biology, 288n6; and quantitative analysis, 147. See also biomedical
research; molecular biology; virology

Farber, Sidney, 89, 91–94, 154; "Jimmy
Fund" for leukemia chemotherapy
research of, 90f. See also Children's
Cancer Research Foundation; leukemia;
pathology
Federation of American Societies for Ex-
perimental Biology, 171–72
Fibiger, Johannes, 35. See also cancer;
parasite
First World War, 6, 8, 61, 63, 85
Flexner, Simon, 48–49. See also biomedical
research
Flow Laboratories (Maryland), 143, 171.
See also laboratories
Fogarty, Representative John, 151. See also
politics
Food and Drug Administration, 204,
252n28, 297n24. See also government
Fountain, Representative Lawrence, 102–3,
149. See also politics
France, 6, 23, 27, 55, 198, 232, 253n31
Frederick Cancer Research Facility, 171, 217,
225–27, 232. See also biomedical research
Friend, Charlotte, 56. See also Sloan-
Kettering Institute; virology
Fujimura, Joan, 312n10, 323n113
fundraising: for cancer, 63, 74–77, 83, 89; for
leukemia, 91. See also philanthropy
Furth, Jacob, 267n73. See also virology

Gallo, Robert, 143, 175, 225, 231–33. See
also virology
Garb, Solomon, 152–53, 160: Cure for Cancer: A National Goal (1968) of, 152
Gardner, Murray, 143. See also virology
genetic engineering: and gene cloning, 12;
research in, 11–12, 228. See also

DATE DUE

			PRINTED IN U.S.A.

Lightning Source UK Ltd.
Milton Keynes UK
UKHW010654110619
344187UK00001B/1/P